Office Skills

A PRACTICAL APPROACH

2nd Revised Edition
Now with GST

Margaret Horsfall • Shirley Cairns

The McGraw-Hill Companies, Inc.

Beijing Bogotà Boston Burr Ridge IL Caracas
Dubuque IA Lisbon London Madison WI
Madrid Mexico City Milan Montreal New Delhi
New York San Francisco Santiago Seoul
Singapore St Louis Sydney Taipei Toronto

Irwin/McGraw-Hill

A Division of The ***McGraw-Hill*** *Companies*

Reprinted 2002

National Library of Australia Cataloguing-in-Publication Data
Horsfall, Margeret
Office skills : a practical approach.

2nd revised ed.
Includes index
ISBN 0 074 70996 8.

1. Office management.2.Office practice. 3. Office management-Problems,excercises, etc. 4. Office practice - Problems, excercises, etc.1. Cairns, Shirley.11. Title.

651.3

Published in Australia by
McGraw-Hill Australia Pty Limited
4 Barcoo Street, Roseville, NSW 2069, Australia
Acquisitions Editor: Michael Tully
Production Manager: Jo Munnelly
Editor: Kate Ormston-Jeffrey
Designer: Tania Edwards
Typesetter: Midland Typesetters
Illustrator: Alan Laver and Tech View Studio
Proofreader: Julian Dahl
Indexer: Puddingburn Publishing Services
Printed by McPhersons Printing Group

CONTENTS

PREFACE

This second edition of *Office Skills: A Practical Approach* was written in response to the industry-driven introduction of competency-based training and the implementation of the National Competency Standards Curriculum within TAFE. The text incorporates 19 of the 'office procedures' cluster of competencies in the national framework. These have been reorganised to meet the packaging requirements of the National Administrative Competency Standards.

The text and workbook have been designed as an integrated learning package, providing students with a practical approach to applying the skills required to work in an office. The real-life approach of the book is enhanced by the realistic documents provided in the workbook. These documents were designed to simulate the office environment, and are an essential resource for the practical activities throughout the main text.

Increased emphasis has been placed on information technology in this second edition, as the use of electronic computer/Internet communication technology is growing dramatically in offices, and will become integral to all future areas of office work. Training in information technology skills will be of paramount importance to anyone wanting to apply for a position in an office.

More and more people, and businesses, are using the World Wide Web (WWW or Web) on the Internet, and this is having the most dramatic influence on all aspects of life. It offers users access to a wealth of information from all over the world. There are many positive aspects to using the Internet as an information resource, but it does have some serious flaws:

- The Internet contains a huge amount of useless and inaccurate information.
- There are no safeguards surrounding the information published on the Internet.
- Finding information can be time-consuming and fruitless due to the lack of a central, organised indexing system.
- Not all information is free, and some of it can be very expensive.
- At certain times of the day the Internet becomes congested and the flow of data slows down considerably.
- There are serious concerns about the security of information on the Internet: privacy of personal details; inappropriate material; harassment and even molestation are serious matters requiring immediate attention.

As the Internet is an electronic medium, it can provide current information that will be available long before it can be found in printed reference texts. This applies also to the other electronic tools we use in an office to improve the quality and speed of our work. For example, employees in nearly all large organisations send and receive messages using e-mail, which is described in some detail in Chapter 1. Computer-controlled telephone systems, facsimile machines and photocopiers are used widely in offices. Mobile phones linked by modem to a computer/laptop are becoming more common, and world area networks using the Internet will change the structure of offices and office work in the future.

If you are enrolled in a current office administration course you will be trained in the use of the computer and the Internet. This does not form part of the contents of this book. But to help you understand how information technology (IT) is now an integral part of all office operations and procedures, a description of how IT machines operate, and useful terminology, is incorporated into each chapter.

As more businesses and individuals use the Internet, some of the present problems will be eliminated, and most of the IT innovations mentioned in Chapter 4 will become commonly used.

The structure of the text and workbook

The text and workbook provide the tutor/supervisor with the flexibility of conducting:

- practical structured classes with learners working in small groups as in a real workplace; or
- self-paced sessions where learners work through activities at their own pace with the individual assistance of peers or the tutor/supervisor; or
- a combination of both.

Each chapter begins with a set of learning objectives to reflect the individual learning outcome requirements of the competencies. The text has been structured to develop the student's skills progressively. The structural components of the text are as follows:

- **Objectives** are provided on the first page of each chapter, incorporating the requirements of the learning outcomes of the related competencies.
- Each chapter begins with an **introduction** to the topic, providing insight into how the topic relates to situations in a real work environment.
- The essential **theory** to complete the assessments is presented concisely within each chapter but is not intended as reference book material where legislation and industrial relations are concerned, due to constant changes in these areas.
- **Information technology** has been incorporated in more detail in each chapter, as the use of electronic computer/Internet communication is increasing dramatically and will become integral to any office in the future.
- Practical **activities** are provided throughout each chapter, involving both group work and individual practice. An activity can be found at the end of each theory section for the practical application of the theory learned. Some activities provide the opportunity for each student to be observed working in a simulated office context, and the tutor/supervisor can use these for assessing the student's competency.
- A **case study** presenting an office scenario appears in each chapter, providing students with the opportunity to display problem-solving and decision-making skills.
- A **summary** of each topic appears at the end of each chapter for revision purposes.
- After students have read the text and practised their skills by performing the tasks in the activities, they can use the **review questions** at the end of each chapter for theory revision or group discussion. These can also be used for formal testing.
- A **competency self-assessment and/or tutor/supervisor assessment** form appears at the end of each chapter, providing students with the opportunity to rate their own competency in each skill or topic. The tutor/supervisor will then assess the student's competency.

Methods of assessment

As most tasks in a real workplace environment are performed as a part of a group or a team, the activities that will be used for official assessment of the competency of the student should ideally be performed as part of a group. But the structure of the book has been designed to cover all methods of assessment, as follows:

- **Group activity assessment.** This is primarily designed for students with little or no experience in the workplace.
- **Self-assessment.** The assessment of students with previous experience may take the form of self-assessment (where written activities can be modified for their own individual assessment) or, if in work, students may be given recognition of prior learning for individual office skills.
- **Peer assessment.** Experienced students may assess less experienced students (with approval from the tutor/supervisor), as group activity assessment is very time-consuming for the supervisor.
- **Co-assessment.** In some practical activities, more than one student may assess the skills and performance of other students (with tutor/supervisor approval).
- **Formal assessment.** Review questions at the end of each chapter may be used for formal assessment, or theory revision.

In the workbook, copies of letters and other documents are supplied for use in practical assessment activities.

Instructor's resource manual

An instructor's resource manual accompanies the text and the workbook, which contains:

- a wide range of further activities (which may be preferred by the supervisor, or used for extra assessment);
- solutions to activities;
- suggestions on how to set up equipment for practical assessment activities;
- notes to be used for transparencies for overhead projectors in lectures.

Note to students from the author

In my long career, I have worked for many years in different types of offices and in many different roles. I originally started work as a receptionist, moved up to personal assistant to the managing director in a textile company with branches all over the UK, and then changed to the public service. I moved overseas for a few years, and worked as a reporter in the law courts in Kenya, Africa (where I also started my teaching career). Since then I have worked on a tobacco plantation and a coffee plantation in the highlands of Papua New Guinea. On moving to Australia I worked as a temp in Melbourne and then in a permanent position as an administrative coordinator for a number of years with an insurance company. I have enjoyed almost every day of my working career and had some exciting times overseas. I completed further studies in Melbourne and commenced my teaching career 10 years ago. I have tried to convey to you in the chapters of this book the experiences and lessons I have learned in a wide and varied career in office work.

Giving you a little information about my career is to show you that office work can be rewarding and sometimes exciting. Once you have your office administration qualifications, you can enter many types of business (e.g. accounting, law, travel) that otherwise would be closed to you. Once you have entered the business of your choice, you may have the opportunity of promotion within the organisation. But like me you will have to train continuously to upgrade your skills in order to keep up with technological change, and especially if you want to obtain better positions.

I wish you every success in your career, and hope you enjoy it as much as I have done.

Acknowledgements

The author would like to thank the following people for their help and support, without which the book would not have been written.

- Dorothy Jackson, Head of Department, Administrative Studies Department, Holmesglen Institute of TAFE, Victoria, for providing the opportunity and resources for a simulated work environment, in which students performed activities similar to the ones outlined in each chapter of this book, and which provided the stimulus for writing the book.
- The reviewers of the manuscript, who provided encouragement and invaluable suggestions:
  —Bev Waldon, First Choice Training Group, QLD
  —Joseph O'Brien, West Coast College of TAFE, WA
  —Dianne Bessel, Launceston College of TAFE, TAS
  —Jan Bailey, Northern Melbourne Institute of TAFE, VIC
  —Elizabeth McEwan, Blacktown College of TAFE, NSW
  —Kate Neal, Adelaide Institute of TAFE, SA.
- The many students and teachers who have provided valuable feedback on the practical activities over the past few years.
- Renn Barker, Media Officer of the Equal Opportunities Commission, Victoria, for his help in verifying material, OzEmail for granting permission to use Figures 1.1 and 8.6, Australia Post for giving kind permission to reproduce Figures 1.4, 1.5, 1.6, 1.7, 1.8, 1.9 and 1.10, the Victorian WorkCover Authority for permission to use Figure 6.15, and Hertz Australia Pty Ltd for permission to use Figure 14.1. Screen shots reproduced by permission from Microsoft Corporation.
- A special acknowledgement is made to the following people for guidance and up-to-date information on occupational health and safety issues in Chapter 6:
  —Liz Bunker, Prevention Adviser, Comcare, Queensland
  —Jim Carmichael, Manager, OHS, Comcare, Queensland
  —Richard Sutherland, Manager of Prevention Unit, Comcare, Canberra.

Further thanks go to the editorial and production team at McGraw-Hill, particularly Michael Tully, Jo Munnelly and Kate Ormston-Jeffery—many thanks for your management skill, expertise, and friendly patience.

Margaret Horsfall
B.Bus. (Office Admin.), Monash University, Melbourne

section one

INFORMATION AND TECHNOLOGY

chapter 1

Mail and e-mail

chapter 2

Records management

chapter 3

Office equipment, introduction to the Internet

CHAPTER

1

MAIL AND E-MAIL

TEC301A, NOS244, 7365CP, INF101A, 201A, NOS107, 7365E

objectives

In this chapter you will learn how to process incoming and outgoing mail in a workplace environment, using 'hands-on' methods of practice and assessment, in order to complete the following learning outcomes:

Electronic mail (e-mail)—sending and receiving

1. Receive, process and distribute incoming mail

1.1 collect, open, check and register incoming mail to ensure accuracy of records
1.2 identify and distribute urgent and confidential mail to the nominated person/location promptly
1.3 sort mail and despatch to the nominated person/location promptly
1.4 identify, record and report any damaged, suspicious or missing items promptly

2. Receive, process, register and despatch outgoing mail

2.1 collect outgoing mail from each section
2.2 check and sort mail to ensure that all items are correctly prepared for despatch
2.3 fold, address and package mail
2.4 classify, weigh and stamp/frank mail according to carrier's requirements
2.5 record despatch details accurately
2.6 despatch mail to meet designated deadlines

3. Sort and despatch documents for bulk mailing

3.1 sort bulk mail into categories according to Australia Post specifications
3.2 batch mail according to Australia Post specifications
3.3 lodge batched items for delivery in time to meet designated deadlines

INTRODUCTION

Business organisations, small or large, cannot operate without information. Information is constantly flowing backwards and forwards in many different ways. Recent technological development means that now many business documents can be transferred electronically direct from one office to another. Facsimile machines and electronic messaging using computers, such as e-mail, are widely used in offices and private homes. Many service accounts such as telephone and electricity bills can be paid by telephone using electronic cards; payment does not need to be sent through the post. Despite such developments, we are a long way from experiencing a paperless office, and Australia Post still provides a very important service to business as well as private individuals. In fact, Australia Post has kept up with technological development and improved its service to individuals and business in many ways. Documents can now be sent electronically through Australia Post by people without a fax or a computer. But official letters, original documents, parcels, packages, advertising material and other items still need to be sent through Australia Post or courier services. These services are outlined in this chapter.

ELECTRONIC MAIL (E-MAIL)—SENDING AND RECEIVING

Electronic mail (**e-mail**) is the term used to describe the tool that allows users to exchange messages or computer files over a **computer network.** This can be a local area network (**LAN**), where all the users are in the same building and using the same computer system, or a network of unrelated computer systems that span the world (**WAN**).

When you send or receive an e-mail message it usually takes only seconds to find its way into the mailbox of the recipient, whether this is across the room or across the world. Anyone who has a personal computer and an attached **modem** can use e-mail. Telstra provides the service free of charge, for the cost of a modem and a telephone call. E-mail has unlimited uses, for example:

- Branches of organisations use e-mail notes, letters, memos, reports, financial details for inclusion in reports, with the immediate response of telephone conversations.
- All employees, including managers, send notes and letters to one another, and to people in other organisations, without having to dictate to a secretary. This saves time and energy.
- In the near future it will be common to correspond with clients via e-mail—answering product enquiries, sending quotes and generally keeping in touch. This again is as quick as a telephone conversation, but documents can also be printed out and hard copies stored in customer files for future reference and for legal requirements.
- It will become common in the near future to send e-mail containing audio and video as well as text and graphics.
- International backpackers and tourists now use e-mail to contact relatives in other

countries, instead of telephoning or sending a card or letter, as e-mail is being made readily available through hotels, motels, libraries and Internet cafes.

At present there are serious concerns about the use of e-mail that need to be addressed very promptly:

- There is no control over the useless, inaccurate and inappropriate material being sent around the world.
- There is a lack of privacy of information on the Internet.
- There is no control over harassment and even molestation of young people, which can happen when using e-mail and the Internet.
- There is still easy access for hackers and computer viruses.

Hackers

Hackers are computer operators (often experts or 'buffs') who carry out computer crime, such as obtaining huge amounts of money from organisations through data programs, or even from your bank account. Computer hackers are also hired to help prevent and detect other computer hackers. Many computer hackers are people who do it for fun or mischief.

Computer viruses

A virus is a small program that can be transferred into a computer system and affects the operation of the computer. It passes itself on from computer to computer in a similar way to biological viruses in animals. Viruses (or logic bombs or time bombs) are passed on contaminated removable disks or on e-mail and networks. The number of known computer viruses has grown from six in 1988 to many thousands in 1999.

There are also worms, which are passed over the Internet. The difference between **viruses** and **worms** is that worms do not contain deliberate destructive instructions but can cause havoc on your e-mail or the Internet. Similar to viruses are **trojans**. Trojans infiltrate a computer system and deliberately cause damage, but cannot propagate themselves.

Viruses are constantly changing. Antivirus software generally appears within weeks, but in the interim time and effort has been wasted and possibly documents lost, which can cause problems for you and your organisation. (Precautionary methods for preventing virus contamination of your computer and e-mail are listed in Chapter 3 under *Computer equipment*.)

Using e-mail

To use e-mail, you need software called a '**mail client**'. Popular e-mail clients for PCs are **Qualcomm Eudora** and **Pegasus Mail**, and for Macintosh, **Qualcomm Eudora** and **Claris Emailer.** You can also use the mail client that comes with your Web browser.

E-mail addresses

You will need an address to send mail, and you will need an address to receive mail. There is no official published listing of e-mail addresses as there is with telephone numbers. Each separate network connected to the Internet uses its own e-mail addresses issued by the Internet Service Provider (**ISP**) you choose. Your e-mail program uses **Post Office Protocol (POP)** or **Internet Mail Access Protocol (IMAP)** to transfer the message from the mail server to your computer. Your POP account is usually the same as your e-mail address.

It is therefore essential that you know your own e-mail address and the e-mail address of the person, or business, to whom you are sending a letter or a message. Your e-mail address is your **userid**, plus the **domain name** of the computer to which you connect to gain access to the Internet, separated by '@'.

Store frequently used e-mail addresses in a computer file and/or an address book. Most e-mail programs now offer an 'address book', and the most efficient method is to add addresses to this as soon as a message is received.

Address book

Your e-mail address book is used to store the addresses of people you regularly write to. You can enter other details there, such as the phone number and business address.

To delete mail, highlight the document and press 'delete' on the keyboard. This is then transferred to the **trash mailbox.**

Userid

When you want to use e-mail, you will need access to an ISP such as 'Big Pond' (Telstra). You will choose a userid as part of the registration process.

The computer network will request your account details and a password. The account name is your userid.

Domain names

A domain name is the Internet address of the computer you use to gain access to the Internet. All computers connected to the Internet have an Internet address, which enables other computers to locate them in order to pass on information.

Your e-mail address consists of your userid, the '@' symbol, and the domain name of the computer network you are using. For example, if I am using Telstra's 'Big Pond', my account details will be:

hors@bigpond.com.au

Below are some examples of domain names:

com commercial organisations
edu educational institutions (universities, colleges, schools etc.)
gov non-military government organisations
org non-profit organisations
au at the end of the address is the two-letter (ISO) country code, for example, **.au**—Australia, **.fr**—France, **.uk**—United Kingdom.

Nicknames

Nicknames allow you to save e-mail addresses with easy-to-remember tags. You can enter someone in your nickname file in your address book.

Sending e-mail

Your e-mail program will require you to provide the details listed below:

- recipient's name and address—make sure you supply both the userid and domain name;
- subject—subject headers are recommended;
- time, date and sender (including sender's e-mail address).

Mailing lists

Internet e-mail lists (mailing lists) allow discussion between many users on mutually interesting topics. You type your thoughts, send the letter to the mailing list administration program (**listserver**), which copies it and sends it to all the other people on the list. Your contribution is then read by everyone.

Mailboxes

Standard mailboxes in all e-mail systems are:

- **inbox** stores all incoming mail;
- **outbox** stores e-mail that is waiting to be sent;
- **sent mail** stores a copy of any e-mail that has been sent;
- **deleted items** stores items that have previously been deleted from other mailboxes;
- **drafts** stores letters that are not yet ready to be sent.

'Bounced' e-mail

If you make a typing error in an e-mail address, or some other mistake, your e-mail message will be returned to you (this is known as 'bouncing'). For example, if after sending a letter an error message appears on the screen, such as:

Network is unreachable or *Could not deliver because . . .*

this indicates that there are problems at the recipient's end. You should check the address carefully and try sending it again.

How to use e-mail

Figure 1.1 shows a computer screen with a typical e-mail message, and indicates how to use e-mail.

Security measures

- Do not reveal your password to anyone.
- Do not choose anything too obvious, such as your name, date of birth or driver's licence number, for your password.
- Never write down your password.
- Do not supply your password details via e-mail.
- Refer to methods for avoiding virus contamination in Chapter 4.
- You should be very cautious when you use your credit card to buy goods on the

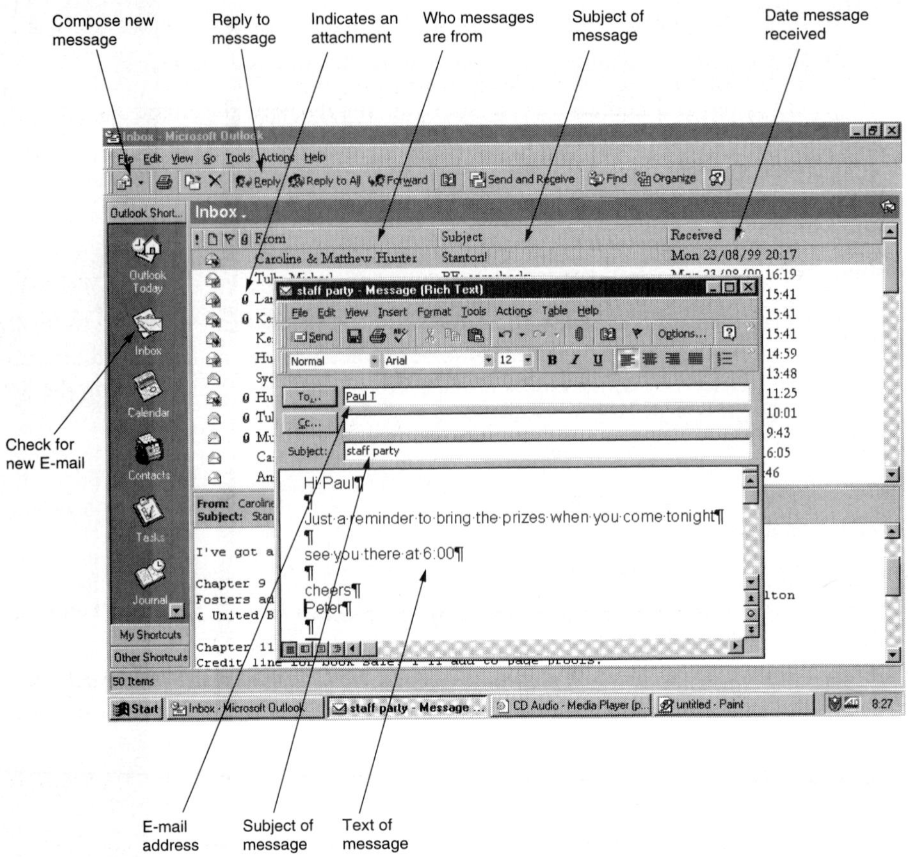

Fig. 1.1 How to use e-mail

Internet. There is a possibility that hackers (or unscrupulous dealers) will access your credit card number and build up a huge deficit, which you will have to pay.

Privacy

This is an important issue and concerns us all. In addition to using the above measures, there are many software packages that have been produced to monitor unsuitable material on the Internet. Not all are successful. One that has been produced in America for use on Windows and Macintosh is called **Pretty Good Privacy (PGP)**. This can be obtained in Australia from PGP International (www.pgi.com.au). You can download a free copy for personal use directly from the PGPi website, or you can buy it as a shrink-wrapped product.

E-mail etiquette

- Check for e-mail messages every day. Because e-mail is an instantaneous communication tool, senders expect immediate replies or may think that the message has not been received and send another one.

- For security reasons, do not send any message by e-mail that you do not want anybody else to read.
- E-mail letters are just the same as letters sent through the postal system, and should be checked for typing, spelling and grammatical errors and corrected. Most computers indicate when there is a typing error, or a grammatical error, and provide spell-check lists.
- Always check that your e-mail is correctly addressed.
- Use only short, concise sentences and paragraphs.
- If you are quoting material written by someone else, be sure to source your material.
- Do not make rude or libellous remarks about another person, or you could end up in trouble.
- Keep quotes as brief as possible, but long enough to make your point.

Glossary of basic e-mail terminology

Address book: your e-mail address book is used to store the addresses of people you regularly write to.

'Bounced' e-mail: mail which has been returned if an error has been made.

Computer virus: a small program that can be transferred into a computer system and affect the operation of the computer. It can be passed on from computer to computer in the same way as are viruses in animals.

Domain name: the Internet address (provided by your ISP) of the computer you use to gain access to the Internet.

E-mail (electronic mail): the tool that allows users to exchange messages or computer files over a computer network.

E-mail address: all users of e-mail need an address, which consists of a userid plus a domain name separated by '@'.

E-mail etiquette: based on the rules and regulations you should follow when sending or receiving e-mail.

Hackers: computer operators (often experts or 'buffs') who carry out computer crime.

ISP (Internet Service Provider): before you can use e-mail, you need access to an ISP that controls a WAN of systems.

LAN (local area network): all computer users are in one building using the same computer system.

Listserver: the mailing list administration program which copies your topic and sends it to all other people on the mailing list.

Mailbox: stores all incoming mail, mail waiting to be sent, mail that has been sent, and deleted mail.

Mail client: software needed to use e-mail.

Mailing lists (e-mail lists): allow discussion between many users on mutually interesting topics.

Nicknames: easy-to-remember tags used to save e-mail addresses.

POP (Post Office Protocol): your e-mail program uses POP to transfer a message from the mail server to your computer.

Trash mailbox: when you delete a message it is transferred to the trash mailbox.

Trojans: similar to viruses and worms, but cannot propagate themselves.

Userid: your ISP will issue you with an address. This consists of a userid and a domain name.

WAN (wide area network): unrelated computer systems linked together across the globe.

Worms: similar to viruses, but do not contain deliberately destructive instructions. They can still cause havoc on your e-mail or the Internet.

Australia Post electronic mail

FaxPost (within Australia and international). If you do not have access to a personal facsimile machine, documents may be sent from Australia post offices and some newsagents and milk bars (with a sub post office). The message is transmitted to a personal fax machine or to the fax centre nearest the receiver within Australia, or even overseas.

FaxPost transmits and delivers facsimile documents up to B4 size, as follows.

Delivery options	Lodged	Delivery within Australia	Delivery overseas
Over counter By post	Over counter at an Australia Post office *or*	$4.00 for first page, $1 for following pages	Available to private machines in all countries with international direct dial links with Australia; $10.00 for first page, $4.00 for following pages
To a private fax machine	By a private fax machine within Australia or overseas, to an Australia Post office	As above	

EDIPost. This is a service for businesses sending large volumes of accounts, statements or direct mail. Clients transmit information to an Australia Post computer. Australia Post then distributes the documents to EDIPost service centre sites in major cities, where they are printed, enveloped and placed in the mail stream.

Lettergram ComputerLink. This is software for businesses that regularly send forms, letters and lettergrams. These can be delivered by post, or fax within Australia or overseas.

INCOMING MAIL

Methods of receiving mail

There are various ways of receiving mail. The method your business chooses depends on the size of the organisation and the requirements of the specific business. Large organisations have their own mailroom with personnel and equipment such as a franking machine, an addressing machine, a folding machine, inserter, parcel-tying machine and so on, to assist in the handling of large volumes of mail. This chapter describes the procedures that receptionists or clerks would use where handling mail is only one of their many duties. Some of the common methods of receiving mail in this case are listed in Table 1.1.

Method	Procedure	User	Advantages
Electronic mail	Through a computer connected to a modem or mobile phone (e-mail, Internet) Facsimile machine	Almost all businesses, small and large Some private households	Almost instantaneous reception Quick replies and decisions can be made
Courier service	Courier van picks up document/s from sender and delivers to receiver within hours or (if interstate) next day	All businesses and private households	Quick delivery of bulky items that cannot be sent electronically Reasonable cost
Street mail delivery	'Postie' delivers mail direct to business or private location	Small businesses and private households	No time or effort wasted by addressee in picking up regular mail delivery
Post office counter	Collected by addressee at post office	Anyone who has no permanent address for a short period of time	Convenient for short-term movement either intrastate or interstate
Private box, locked bag	Post office puts all incoming mail in locked box or bag located in post office or on the wall outside. Mail is sorted into locked boxes and bags several times a day	Small and large businesses. Private individuals. Insurance companies. Public utilities or where a large volume of routine mail is received, such as payment of bills	Visits can be made at any time during the day to pick up urgent mail or large volumes of mail. Box may be cleared at the weekend, or at any time convenient to the business or individual

Method	Procedure	User	Advantages
RSD (roadside delivery)	Post office van delivers to a box at the roadside near the property	Small and large farms, or businesses in country areas located outside a town or city	Regular deliveries; owner may not be able to visit post office every day
Bus service (e.g. 'Redline' bus in Tasmania)	As well as delivering passengers, the bus picks up small and large packets, and even household goods	Small rural areas located some distance from a town or city where people cannot regularly make a trip into town or city	Very convenient for people or businesses who have no regular access to town and city. The goods required are too large for post office delivery. The service is also cheaper

Table 1.1 Methods of receiving mail

Activity 1.1

Collecting incoming mail

(This activity may be completed individually.)

Below is a list of different types of business organisations. Form groups of three, if possible, and discuss with the members of your group which method of collecting incoming mail each business will use. Write the method for each business on Activity sheet 1.1 in your workbook.

- a medium-sized business in a town near where you live;
- a large insurance organisation in the city;
- a farm in the country;
- a solicitors' partnership in the centre of town;
- a dress boutique in the centre of town;
- a large hospital;
- a small business on the edge of town.

Collecting, opening and registering mail

When you have received the mail, place it on a cleared part of your desk and sort it into order of priority:

1. electronic mail
2. courier delivery
3. express post

4. messenger post
5. letters marked 'urgent'
6. certified mail
7. security post
8. cash-on-delivery (COD) post
9. letters marked 'private and confidential'
10. airmail letters
11. routine mail
12. parcels and packages
13. magazines and journals
14. newspapers
15. advertising material.

Opening mail

Letters marked 'private and confidential' should be placed unopened in pigeon holes or on desks. If opened by mistake the envelopes should be resealed and marked 'opened in error'.

If your organisation requires you to open envelopes, use a letter-opener that slits the envelope without damaging the contents. If documents are damaged when opening the envelope, mark on the document 'damaged when opening' and the date.

When opening mail, check the contents carefully. Small enclosures such as cheques are often left in envelopes and can be lost or destroyed. Take all documents out of the envelope. Stamp the main document with the date and time received. Attach any enclosures and the envelope to this document.

Damaged incoming mail

Some examples of damaged incoming mail are listed below. A note should be made on each article and appropriate action taken for:

- badly packed goods resulting in damage to, or spillage of, contents;
- items missing from contents;
- suspicious envelopes or parcels;
- contents of packages prohibited by Australia Post but sent through the postal system;
- goods that have been delivered and have not been requested by anyone in your organisation;
- incorrectly addressed articles that have been sent to you by mistake.

Activity 1.2

Damaged mail

(You may discuss this activity in a small group, or you may complete it individually.)

Discuss with other people what action you would take in each case of the damaged mail listed above. Then on a separate sheet of paper write against each item the action you would take.

Registering mail

Some organisations require you to keep a register of all incoming and outgoing mail. This applies particularly to law and accounting firms, where receipt of documents may have to be recorded for legal requirements. Documents often go astray and a search through the register will confirm if, and when, the document was received. Receipt of valuable and important documents, and the method of postage, should also be recorded for future reference. An incoming mail register is illustrated in Figure 1.2.

Incoming mail register				
Date	Description of article (list in order of priority)	Article size (standard/measure)	To whom addressed	Action taken

Fig. 1.2 Incoming mail register

Activity 1.3

Processing incoming mail

(Refer to your tutor/supervisor for information on the mail system that has been set up for your use.)

Work in a group of three people. Each person will take six envelopes from the 'incoming mail' box and process them as follows.

1. Place these on your desk and sort them into order of priority (refer to the list earlier in this chapter).
2. List them in order of priority on the incoming mail register sheet provided in your workbook (Activity sheet 1.3).
3. Specify the size of each envelope. Write this in the column provided on the incoming mail register sheet. (Alternatively, read the section on standard articles later in this chapter and tick the envelopes that are standard sizes.)

4. Distribute the incoming mail, in the correct order of priority, into pigeon holes (or using the method of distribution provided by your tutor/supervisor).

Methods of distribution

Each organisation has its own method of distributing incoming mail. This depends on the type and size of the business. Large organisations have a centralised mail centre where incoming and outgoing mail is processed. Other organisations have pigeon holes into which incoming mail is sorted in order of priority. This is placed in a convenient location for personnel to collect their mail individually. In small organisations the person responsible for mail will place incoming mail on each individual desk.

Internal mail

Information sent from one person to another, or one department to another, within the organisation is circulated in the following ways.

Internal mail envelopes. These envelopes have spaces for up to 40 names. When a person receives an internal envelope his or her name is crossed out, the envelope is reused and another name is written in the space below. These envelopes are easily recognised as internal, and because they are used again and again contribute to the conservation of the environment.

Circulation slips: A memo, a trade journal or an article that needs to be read by a number of people can be distributed by means of a circulation slip. The name of the person who is circulating the document, and the date, are on the top of the slip, and the names of the people who need to read the document are listed on the left (see Figure 1.3). The first person reads the article, signs and dates it, and passes it on to the next person to read.

PLEASE CIRCULATE

From: ....

Date: ....

Circulate:	Date passed on:
Ardis, R.	....
Barker, A.	....
Brearley, G.	....
Chan, B.	....
Gerrard, M.	....
Leenstra, S.	....
Watson, M.	....

Return to: ....

Fig. 1.3 Circulation slip

OUTGOING MAIL

Organising outgoing mail is an important task. If this is one of your responsibilities and you do not handle it correctly, valuable and irreplaceable original documents could go astray and be lost for ever.

Outgoing mail is also an aspect of the public relations and corporate image of your organisation. Mail that is sent out badly presented and incorrectly addressed reflects on the efficiency of your business, and customers will feel that all your business practices may be badly managed. Time and consideration should be given to the appearance of all documents before they leave the premises. Once correspondence has been checked and signed, an envelope should be used that matches the letter in style, colour, quality of paper and size.

Letters should be folded to fit the envelope with only two creases, so that the document is opened in the correct way. Fewer creases make the document neat, tidy and easy to read. Examples are given in Figure 1.4.

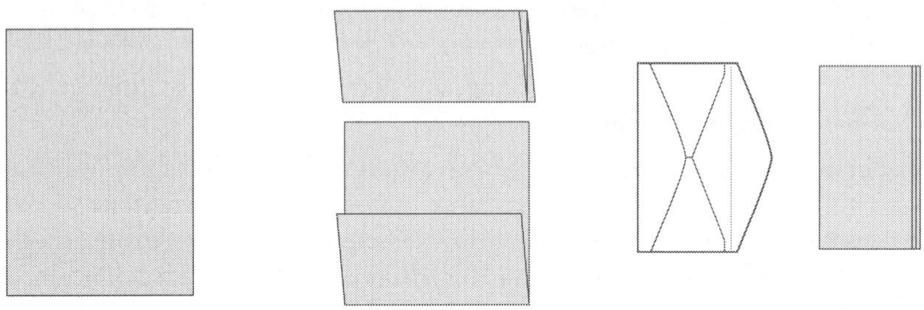

A4 letter folded to fit a DL (110 x 220 mm) envelope

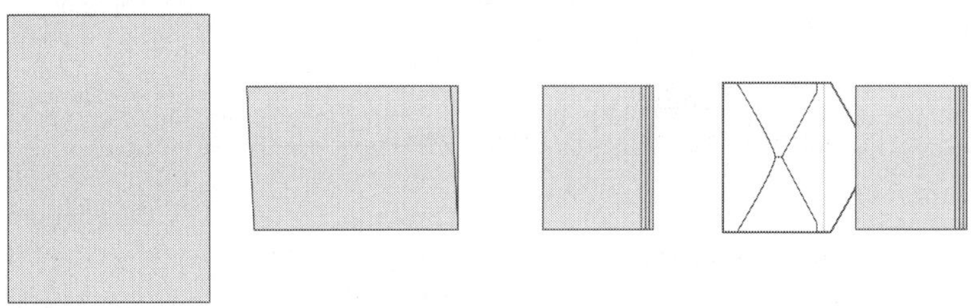

A4 letter folded to fit a C6 (114 x 162 mm) envelope

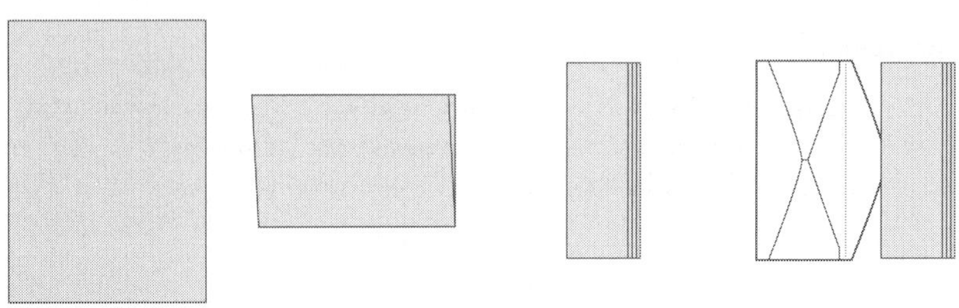

A4 letter folded to fit a 9 x 4 (100 x 220 mm) envelope

Fig. 1.4 Correct folding of letters

Correct addressing of outgoing mail

You should make sure that mail is addressed correctly and despatched efficiently in order for it to reach its destination in the shortest possible time and in good condition. The procedures for dealing with outgoing mail (which should be written down in your procedures manual) are outlined in the following pages.

Electronic scanning

Australia Post handles approximately four billion articles each year, and this is growing by about 200 million items a year. It has an excellent record of delivery, and most items that go astray have been incorrectly or insufficiently addressed. To make sure that letters are delivered as quickly as possible and cost-effectively, Australia Post has introduced advance letter-sorting technology. Optical character readers (OCRs) sort up to 35 000 letters an hour by electronically scanning the address. The machines rarely make a mistake if the envelopes are addressed correctly. Guidelines are given by Australia Post for either typing or writing addresses on envelopes.

Correct addressing practices

An address should contain the following elements:

- the full name of the addressee;
- the number and name of the street or, where applicable, the private box number;
- the name of the post office of delivery;
- the name of the state or territory;
- the postcode (very important when the machine is **optically scanning** the address).

Do not:

- use punctuation in the address;
- underline the bottom line of the address—use capitals.

Do:

- show flat numbers before the street number (e.g. Flat 6/306 Smith Street);
- include reference numbers or names (e.g. Attention Mr …) within the address zone;
- show an overseas destination as the last line in capitals;
- always include a return address.

Envelope zones

It is most important to observe the envelope 'zones', as indicated in Figure 1.5. The correct methods for machine addressing, for handwriting addressing and for addressing window envelopes are shown in Figures 1.6, 1.7 and 1.8.

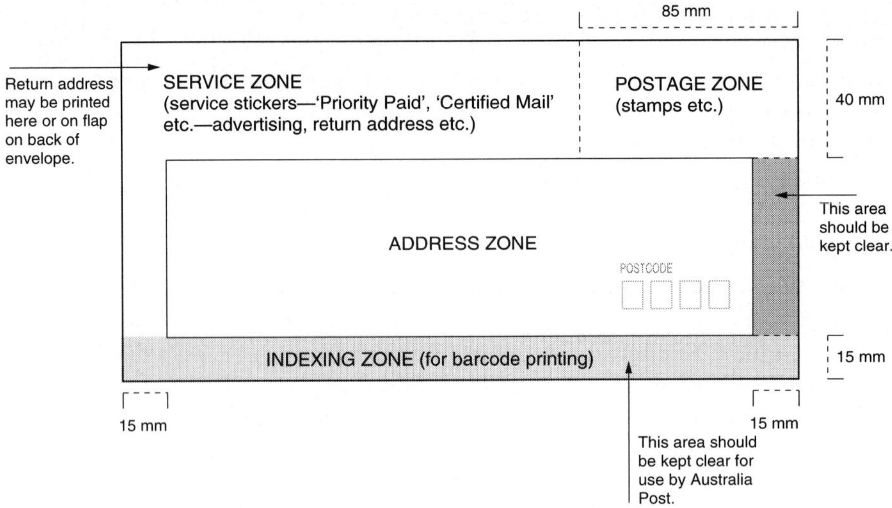

Fig. 1.5 Correct use of zones

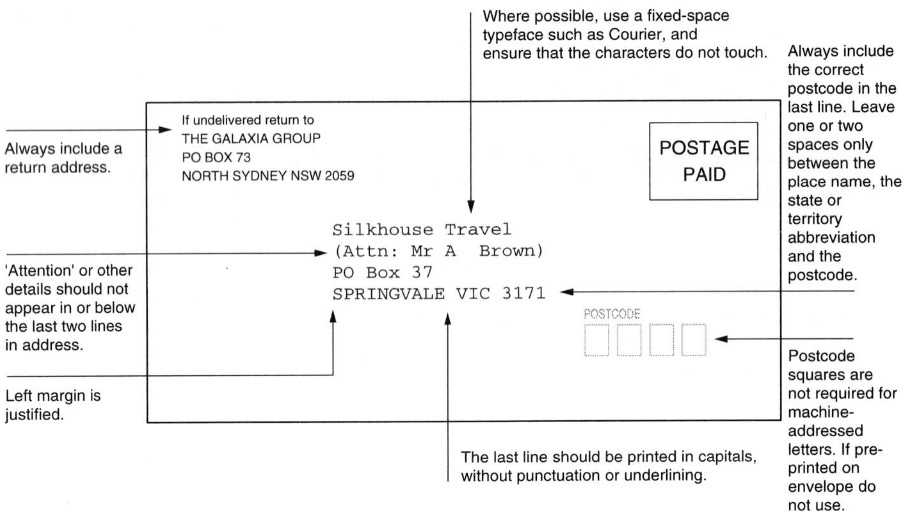

Fig. 1.6 Correct machine addressing

Fig. 1.7 Correct hand addressing

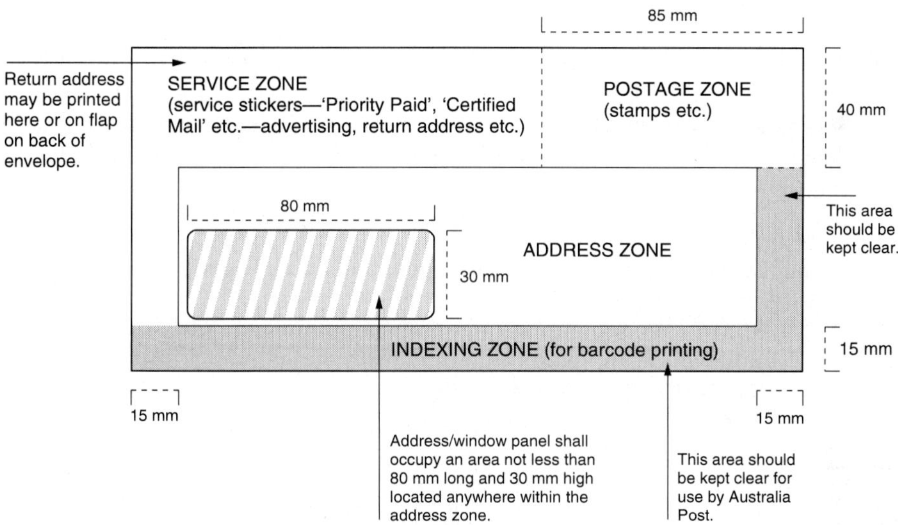

Fig. 1.8 Correct use of zones for window panel envelopes

Activity 1.4

Addressing outgoing mail

You will need for this activity:

- six standard-sized new envelopes for each student.

With the other members of your group, complete the following activities.

1. Choose six business names and addresses out of the Telstra *Yellow Pages*.
2. Type (or write very neatly) one of the business names and addresses from the *Yellow Pages* on each of the six envelopes. Check how to address envelopes correctly according to Australia Post.
3. Two envelope shapes, and two letters, are provided in your workbook (Activity sheet 1.4). Take the address from one of the letters and write it neatly in the correct way on one of the envelope shapes. Write the address from the other letter on the second envelope shape. Fold each envelope shape around the edges to make an envelope, and fold each letter in the correct way to fit the envelope. (How to fold letters correctly is illustrated in Figure 1.3.)

Place your six correctly addressed envelopes in the outgoing mail bin for use in Activity 1.5.

Methods for sending routine mail

The most convenient ways to send routine correspondence are those listed below.

Postage prepaid envelopes. These require no stamping, no weighing, no franking, no wrapping—simply pack and post. They may be lodged at any post office or street post-box. Postage prepaid envelopes are cheaper if bought in packs of 10.

Customised prepaid envelopes and business reply post. You can use personalised postage prepaid envelopes (printed with the business logo, company name and address). You will need to order a minimum of 500 envelopes and provide your post office with artwork for your logo, company name and address.

Stamp rolls. Simply peel the stamps off the roll, then stick and post. These are available in rolls of 100 and 200.

Franking machines. A franking machine allows firms to imprint the amount of postage on the envelope instead of using a stamp. The machine can be set with the date and amount of postage. This is very convenient for a business with large volumes of standard post. An amount is prepaid to Australia Post.

Express post. Use this for sending documents and small articles overnight to other capitals when next-day delivery is essential. You will need special prepaid envelopes and satchels, which can be ordered from your post office (see the post charges book for prices). Articles that may be sent by Express Post include:

- folded letters weighing up to 500 grams;

- documents (unfolded A4 pages) up to 500 grams;
- bulky documents and articles up to 500 grams;
- bulky consignments up to 3 kg;
- accounts and statements.

Envelopes and satchels may be lodged at any gold postbox before 6 pm or across the counter of your post office before the advertised closing time. Do not lodge in normal street postboxes.

Large letters. Large letters up to 500 grams automatically travel by air within Australia where this service applies. Ensure that the letters fit within the limits of Australia Post's 'standard article gauge'. For extra convenience use postage prepaid envelopes.

A standard article. Standard articles have a lower postage rate. Australia Post has produced a 'standard article gauge' to enable you to check the size of an article. To be a standard

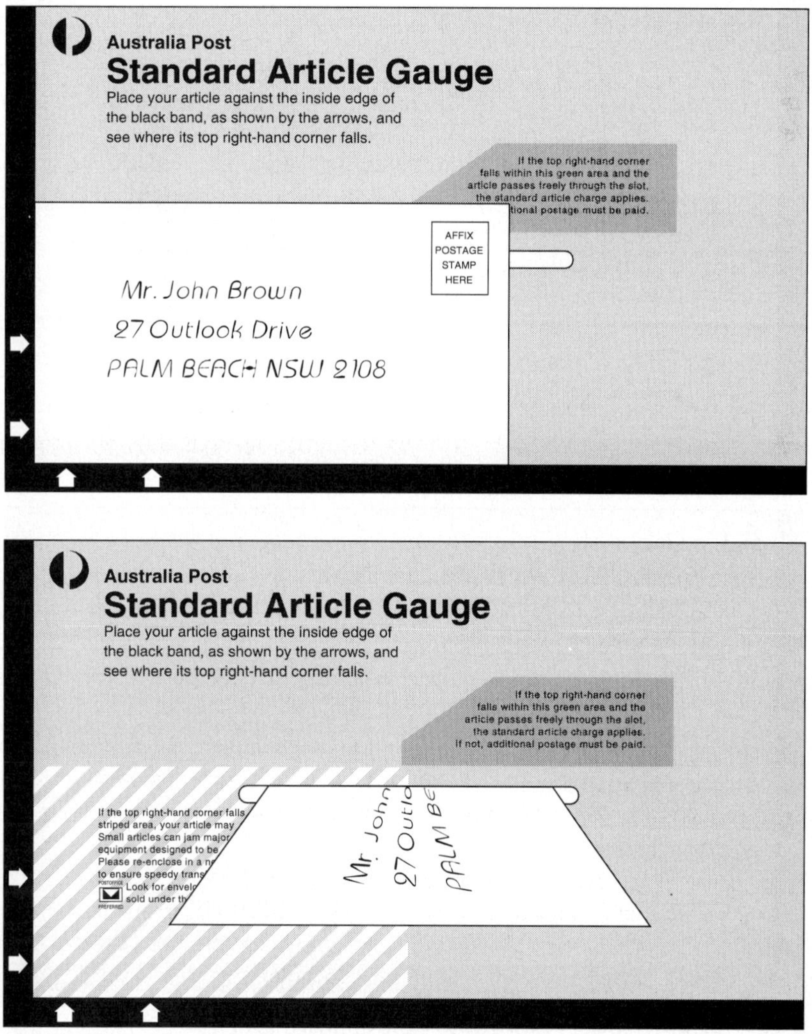

Fig. 1.9 Standard article gauge with standard article

article, the top right-hand corner of the envelope must fall within the green area, when the bottom left-hand corner is placed against the inside edge of the black band. The envelope must pass freely through the slot in the gauge and it must be oblong in shape, with its length at least 1.414 times its width. It must be no smaller than 88 mm × 138 mm; no larger than 122 mm × 237 mm; no thicker than 5 mm; and no heavier than 500 g.

A non-standard article. A non-standard article is any envelope weighing up to 500 grams which does not qualify as a standard article as described above. Large or thick envelopes, small packets and books are usually non-standard articles.

Non-standard articles may be sent through Australia Post in the usual way, but may also be sent in the following ways:

- **print post and advertising** — an economical way to send publications and periodicals anywhere in Australia (100 items minimum);

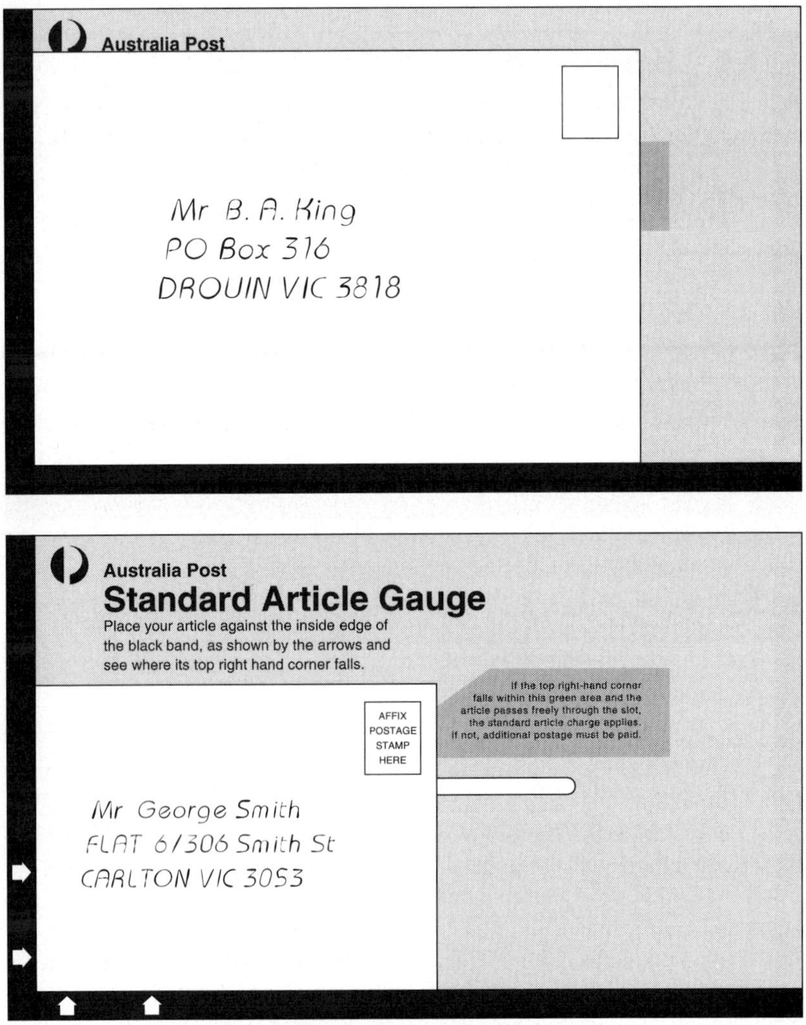

Fig. 1.10 Standard article gauge with non-standard articles

- **messenger post** — features radio-controlled motorcycles, cars and vans to send documents, packets and parcels anywhere in your metropolitan business district—fast. You can choose your delivery speed: direct; one hour; three hours; same day.

Activity 1.5

Processing outgoing mail (envelopes)

The materials you will need for this activity are:

- a post charges book from a post office (one to each group);
- a standard article gauge from a post office (one to each group).

With the other members of your group, complete the following activities.

1. Take 10 envelopes from the outgoing mail bin and enter them on the outgoing mail register sheet (Activity sheet 1.5) in your workbook (10 envelopes for each member of the group).
2. Enter on the outgoing mail register sheet the description of the article to be posted, to whom it is addressed, and the method of postage.
3. Check the envelopes with the standard article gauge.
4. Work out the cost of postage from the post charges book and write it in the last column of the outgoing mail register sheet.
5. Write today's date in the left-hand column of the outgoing mail register sheet.
6. Place each envelope in the correct mailing tray (provided by your tutor/supervisor) with other envelopes already in the tray, and batch them together with an elastic band.

This activity will be observed by your tutor/supervisor or by your peers.

Extra precautions

The options you can use for extra precautions when sending mail are as follows.

Security post. This is used for special secure handling, proof of posting and delivery. You would use security post, for example, when sending original documents such as legal documents, contracts, assignments, passports and birth certificates.

Certified mail. This provides a receipt as proof of posting and a signature on delivery. The service is more suitable for documents than for articles of value. In case of loss, up to $50 compensation could be provided if the fault is that of Australia Post. **Insurance** is needed for loss compensation when sending articles over the value of $50: for example, expensive presents, or articles that are being returned to the owner.

Packets and parcels

Use **Parcel Post** for parcels that weigh up to 20 kg (but special conditions apply to bulky parcels). Choose a delivery standard: surface parcels or air parcels. Weigh the parcel and attach the correct postage (use the post charges book available free from your post office).

Options for greater convenience. Packaging available from the post office includes **Postpak mailing boxes and tubes, bubble wrap, bonded bags and gusseted bags.** Postpak provides protection and eliminates wrapping, therefore improving efficiency. Prepaid Postpak provides extra convenience and simplified charging, or you can use COD and have Australia Post collect payment from your customers on delivery of the parcel. Other options are acknowledgement of delivery; electronic lodgement and delivery service (ELADS); and insurance.

Savings. Contact your local postal manager about discount prices and pick-ups for regular customers.

Local parcel delivery. Lodgement is made at the counter of the post office from which delivery will be carried out. For parcels weighing up to and including 500 g there must be a minimum of 10 parcels. For parcels over 500 g there is no minimum quantity requirement.

Packaging of packets and parcels. Special care should be taken when packaging items that are unsuitable for Postpaks. Breakable articles need special protection from the pressure of travel, and from each other. The most effective cushioning materials are plastic air-bubble sheets, shredded paper, woodwool, tissue paper, cottonwool, felt pads and corrugated cardboard.

Articles that cannot be sent through the post

Because they may cause damage to human life, aircraft, equipment or other people's parcels, some things cannot be sent through the post, such as material that is radioactive, flammable, explosive or corrosive; oxidising material; compressed gas such as aerosols; live animals or birds.

International post

There are several ways in which to send international post through Australia Post.

For **EMS international courier** use the special satchels (free from your post office) or your own envelopes with international express post stickers. Lodge the articles across the counter at your post office.

Express post international provides a worldwide delivery with one-rate prepaid envelopes. Discount prices apply for bulk purchases, and there is a guaranteed air despatch the next business day (though some exceptions apply). Post in the gold street postboxes by 6 pm or lodge at postal outlets before the advertised closing time.

For **Airmail** use airmail envelopes or a standard envelope with an airmail sticker (free from the post office). Gauge the weight, calculate the postage and attach this, then lodge the envelope at any street postbox or post office. For **economy air** use standard envelopes with an economy air sticker. Gauge the weight and calculate the correct postage.

International parcels

Use international parcel post for anything not prohibited for carriage by post (check with the post office if you are in doubt about what can be included in a parcel). Choose the delivery standard: airmail (attach the special sticker available from the post office); economy air (attach the special sticker available from the post office); or surface mail. Prepare the correct customs documentation to ensure prompt handling at the country of destination.

For **uninsured articles up to and including 500 g**, where the declared value is less than $500, attach a C1 form. Where the declared value is more than $500, attach a Customs Declaration—uninsured articles form C231. Write 'small packet' on the cover. For **uninsured articles over 500 g**, attach a Customs Declaration—uninsured articles form C231.

For **all insured articles**, attach a Customs Declaration—insured articles form C233, and write 'Insured value $A . . . ' in pen above the address. All the required forms are available at your local post office. Lodge them over the counter at the post office, or ring for a pick-up.

Activity 1.6

Processing outgoing mail (packets and parcels)

Work in a group of three. Take two packets or parcels out of the outgoing parcels box and weigh each article. On the outgoing mail register sheet (Activity sheet 1.6) in your workbook write a description of the article, to whom it is addressed, the method of postage to be used, and the cost of postage.

Bulk mail

Use bulk mail to save money on large mail lodgements (2500 letters). Choose regular or off-peak rates. For presorting and layout, obtain 'The guide to presort letters' from your local post office. Sort the letters using a letter gauge into small, medium and large sizes. Sort the consignment into postcodes or areas using the Australia Post national sorting plan. Bundle the consignment into lots of 50 or 100, label each bundle (labels are available free from the post office) and secure each with three rubber bands or stack in special Australia Post trays. Prepare a mailing statement using the standard Australia Post form (available from the post office). Lodge the consignment across the counter at a post office.

Activity 1.7

Checking the outgoing mail

With the other members of your group, check the outgoing mail trays.

1. Make sure that all the envelopes in the trays are the correct classification of postage.
2. Check that the envelopes are all facing the same way.
3. Batch all the envelopes together in each tray, and secure with a rubber band. This will be supervised by your tutor/supervisor.

Minimising wastage of paper

Some methods for minimising wastage of paper in mail handling are as follows:

- carefully opening large envelopes so that they may be reused by affixing a new label over the address;
- using internal envelopes up to 40 times;
- reusing Postpak envelopes and tubes by affixing a new label over the address;
- shredding unused paper to use for packaging;
- cutting unused paper into squares for use as scrap pads.

Activity 1.8

Minimising paper wastage

(This activity may be completed individually.)

There are suggestions above for minimising wastage of paper in handling mail. Discuss with your group any other ways in which you could minimise wastage of paper when handling mail. List your ideas on a separate sheet of paper.

Activity 1.9

Visiting the mailroom

With the other members of your group, visit the mailroom of your college, school or business. Observe how the staff perform the duties for incoming and outgoing mail discussed in this chapter.

Case study 1.1

Methods of postage

You are the receptionist for Business Stationery Supplies Pty Ltd. You have received the following articles for mailing today. State on Case study sheet 1.1 in your workbook which method of outgoing postage you would use for each item:

1. the details of a contract—this must be received in the next capital city today; it is not an original document;
2. an original legal document—this must be sent to the next capital city today to be signed and returned by the next day;

(continues)

3. an expensive pen which a client has left in your office—you will not be seeing this client again for some time (if at all) and feel you should return it;
4. an original document—this is being sent to a client for signature and return but is not urgent;
5. medical details being returned to a patient—this is not urgent but the documents are confidential.

Summary

- Many business documents can now be transferred electronically from one office to another, and from one country to another, through computers using e-mail and the Internet. But the bulk of mail will still be sent through Australia Post.
- There are various methods of receiving mail, including electronic mail, courier service, street mail delivery, picking it up from post office counters, private boxes and locked bags.
- When you have received the mail it should be sorted into order of priority—express, urgent, private and confidential, and then standard mail items.
- Mail should then be distributed either into pigeon holes or using the method specified by your organisation, with urgent letters placed on top.
- If letters are opened by mistake, they should be resealed and marked 'opened in error', and if contents are damaged when opened the envelope should be marked 'damaged when opening'.
- Each organisation has its own method of distributing incoming mail. This depends on the type and size of the business. Large organisations often have a centralised mail centre where incoming and outgoing mail is processed. In small organisations incoming and outgoing mail is often the responsibility of the receptionist.
- Internal mail is distributed in internal mail envelopes which can be reused up to 40 times, and articles and journals are distributed by means of a circulation slip.
- Processing outgoing mail is a very important task, and is an aspect of the public relations and corporate image of your organisation. Mail that is sent out badly presented and incorrectly addressed reflects on the efficiency of your business.
- The methods for sending routine mail include postage prepaid envelopes, business reply post, stamp rolls, franking machines and computer mail.
- A standard article is measured by means of an Australia Post standard article gauge.

- Extra precautions may be taken when sending articles through the mail by using security post, certified mail and insurance.
- Bulk mail can save costs through presorting and stacking into postcodes or areas, using the Australia Post national sorting plan.
- Methods for minimising wastage of paper include reusing internal envelopes, reusing large envelopes and bags by placing a new label over the address, and shredding unused paper to use for packing.

Review questions

(These can be used for group discussion or official testing.)

1. List the methods of receiving incoming mail.
2. How would you sort articles of mail into order of priority?
3. State what you would do with envelopes marked 'private and confidential'.
4. If you damaged a document when opening an envelope, what would you do?
5. How would you circulate internal mail?
6. Your manager has given you an article that should be read by all managers of all departments. How would you make sure that they all received this article?
7. List the various methods for sending routine mail.
8. What is a standard article? How would you check envelopes to determine whether or not they are standard size?
9. What is a non-standard article?
10. What is the rule when you are folding letters to mail out?
11. Describe briefly the options you can use for extra precautions when sending mail.
12. List articles that cannot be sent through the post.
13. Describe briefly the method for sending a large number of mail items regularly.

Competency self-assessment and/or tutor/supervisor assessment

When you have finished reading this chapter, and have completed the activities chosen by your tutor/supervisor (**practical in class, at work or theoretical**), fill in this self-assessment. Then hand it to your tutor/supervisor for official assessment.

Learning outcomes	**Self-assessment**		**Tutor/supervisor assessment**	
I have successfully completed the following learning outcomes	**Yes** (please tick)	**No**	**Yes** (please tick)	**No**
The basic terminology and methods of sending and receiving e-mail				
1. Receive, process and distribute incoming mail				
1.1 collect, open, check and register incoming mail to ensure accuracy of records	☐	☐	☐	☐
1.2 identify and distribute urgent and confidential mail to the nominated person/location promptly	☐	☐	☐	☐
1.3 sort mail and despatch to the nominated person/location	☐	☐	☐	☐
1.4 identify, record and report any damaged, suspicious or missing items promptly	☐	☐	☐	☐
2. Receive, process, register and despatch outgoing mail				
2.1 collect outgoing mail from each section	☐	☐	☐	☐
2.2 check and sort mail to ensure that all items are correctly prepared for despatch	☐	☐	☐	☐
2.3 fold, address and package mail	☐	☐	☐	☐
2.4 classify, weigh and stamp/frank mail according to carrier's requirements	☐	☐	☐	☐
2.5 record despatch details accurately	☐	☐	☐	☐
2.6 despatch mail to meet designated deadlines	☐	☐	☐	☐
3. Sort and despatch documents for bulk mailing				
3.1 sort bulk mail into categories according to Australia Post specifications	☐	☐	☐	☐
3.2 batch mail according to Australia Post specifications	☐	☐	☐	☐
3.3 lodge batched items for delivery in time to meet designated deadlines	☐	☐	☐	☐
Date:		Tutor/supervisor signature:		

CHAPTER

2

RECORDS MANAGEMENT

INF102A, NOS108, 7365K, INF202A, NOS208, NOS302, 7365KB, INF301A, NOS302, 7365KC

objectives

In this chapter you will learn how to operate an effective records management system, using 'hands-on' methods of practice and assessment and involving handling, processing and maintaining records, in order to complete the following learning outcomes:

1. **Records handling**
   File documents to maintain access to and security of documents while implementing safety procedures
   - 1.1 classify and sort documents to maintain integrity of each of the following systems: • alphabetical • numeric • alpha-numeric • geographic • subject • chronological
   - 1.2 refer classification uncertainties to an appropriate authority
   - 1.3 index and file documents in correct location, such as: • lever-arch files • filing cabinet

   Identify and retrieve files/documents in a paper-based system while implementing safety procedures
   - 1.4 locate and extract specified files/documents within designated timelines
   - 1.5 record file/document movement and destination in accordance with security procedures
   - 1.6 despatch file/document to nominated person ensuring that confidentiality procedures are followed
2. **Records processing**
   Update and modify existing organisational records to allow for easy access
   - 2.1 enter records in file register regularly and accurately to ensure integrity of filing system
   - 2.2 collect, sort and place individual items in filing system
   - 2.3 cross-reference files as directed for easy access

   Archive files as directed
   - 2.4 remove inactive files from working system, record on register and store/archive
   - 2.5 remove dead/obsolete files from system record on register and archive/destroy as directed
   - 2.6 suggest possible methods of recycling while ensuring confidentiality and security on the system
3. **Records maintenance**
   Assemble new files and integrate into filing system
   - 3.1 collect and collate information to be incorporated in a new file
   - 3.2 create a new file promptly and index accurately to maintain integrity of system
   - 3.3 record the new file in the system register
   - 3.4 establish multiple files from a source file to maintain integrity of system

   Identify and archive inactive and dead files

INTRODUCTION

All businesses use stored information to support the day-to-day operation of their primary function of manufacturing goods or providing services. Data are recorded in an electronic database system or in a paper-based system. This information is classified, indexed and stored efficiently and effectively so that it can be retrieved quickly when it is required—in one hour, one day, one week, or even years in the future. This is called records management.

A record is any documented information used in the following ways:

- letters to clients confirming an order of goods or replying to a request for information;
- internal memoranda from management to personnel communicating information on corporate policies and procedures;
- reports from departmental managers to top-level management on the progress of production and/or sale of goods or services;
- sales reports from management to department managers communicating information on future goals and deadlines;
- invoices and statements to customers;
- purchase orders for supplies;
- internal stationery requisitions;
- personnel information/data recorded in the human resources department for internal use;
- names and addresses of customers/clients stored in an internal telephone directory or a computer database;
- details of books and journals on microfiche or computer database in libraries;
- insurance policies and other legal documents necessary for the running of the business.

THE FUNCTIONS OF RECORDS MANAGEMENT

A records management system performs three functions for the business, as follows.

Primary function

The primary function of records management is to control the creation, care, storage and retrieval of essential information/data required by personnel at all levels of the organisation, in order to:

- make decisions;
- provide information on decisions already taken;
- record transactions for the business and those who deal with the business;
- provide a basis for management planning;
- provide a source of material for research into better methods and procedures;
- satisfy legal and taxation requirements.

Secondary function

The secondary function of records management is to provide essential information/data for:

- statistical analysis;
- input into administrative services;
- updating supplies, maintenance and purchasing;
- preparation of reports for management.

Tertiary function

Records management allows analysis for internal audit for evaluation of the records management system itself.

CENTRALISED AND DECENTRALISED FILING SYSTEMS

Whether your records management system is a paper-based or an electronic database one, who operates it and how it is controlled will depend on the size and type of the organisation you work for and whether the system is centralised or decentralised. In any kind of business both systems can be operating at the same time.

Centralised filing systems (computer/paper-based)

In a centralised filing system, all records are kept in one location and are controlled by one person (or a number of persons) using one procedural system.

Electronic records management

Computer database filing systems are often used in centralised filing systems. In a computerised system it is easy for anyone to access the system from different work stations anywhere in the building. For example, in large organisations such as insurance companies, Telstra or the Commonwealth Bank, huge computer data banks, containing customers' names, addresses and transactions, are easily accessed by hundreds of people using the same centralised computerised system.

Advantages of a computerised filing system. In a computer filing system customer records are immediately accessible, and are usually easier to find. Creating and filing documents in a computer system are illustrated in Figure 2.1.

Relational database systems. Developed to cut administration time in half, the database can be customised to handle all the paperwork of the business. You only need to enter data once and then point and click to format it in any way you choose. Each job has a unique number and can be viewed in several layouts ready to print—quotations, invoices, letterheads, faxes, WIP list and other documents which include the company logo. The system stores all these documents together and will find them when required by searching on different criteria—title, job number, subject etc.

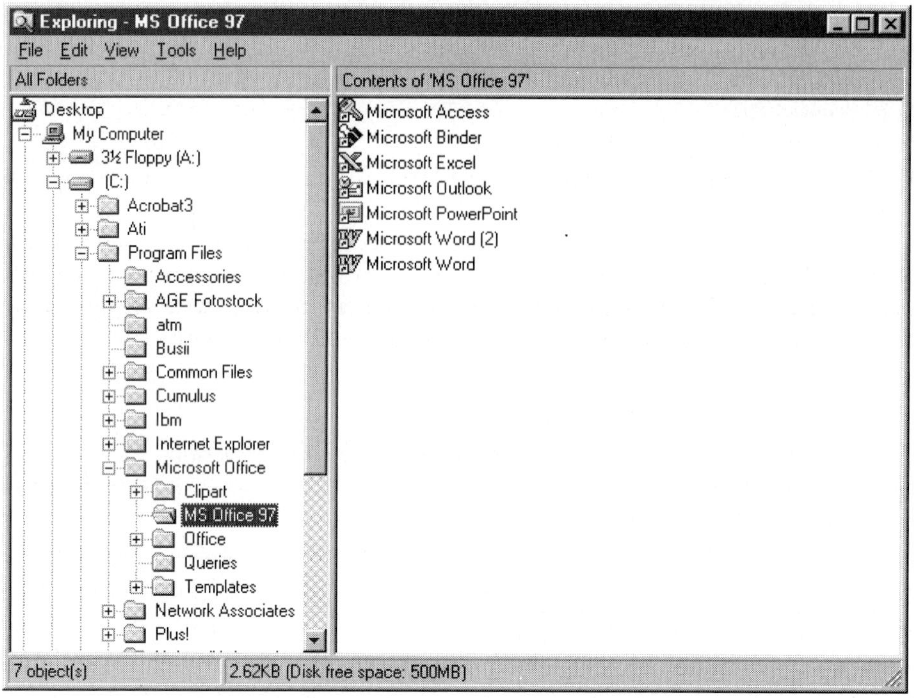

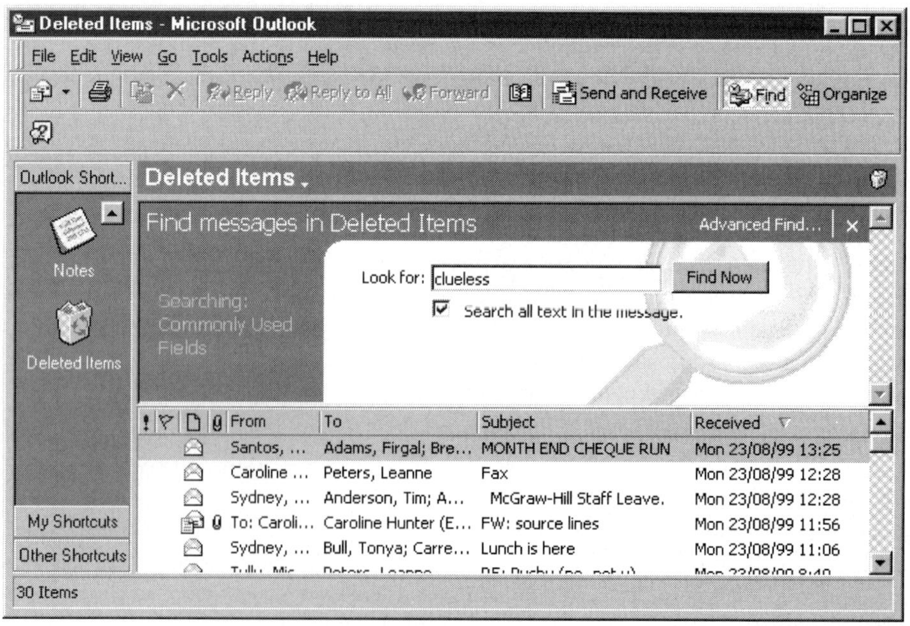

Fig. 2.1 A computer document creation, filing and retrieval system in 'Microsoft Office'

One relational database is '**FileMaker Pro**', which organisations can use to set up complex databases from scratch. Or they can use one of more than 40 templates included with the program.

Speech recognition. Documents can now be dictated directly onto computer, edited and stored in a computer file, and then used like any other file (see Chapter 3).

Finding missing files in a computer-based system. In an effective records management system, one of the basic procedural principles is that it must be simple and easy to use. Finding missing files in Windows 95 has been made easier by the introduction of a powerful 'Find' tool which will find missing files. Even if you can't remember in which folder you saved a file, or you know the contents but not what it was called, so long as you can remember at least one or two words in the file, 'Find' will locate it for you.

Some **disadvantages** of computer filing are:

- Documents could be lost in power failures and surges.
- Documents could be damaged by virus and worm contamination.
- Time may be wasted because of power failures and virus contamination, with consequent loss of business.

Precautionary methods for the prevention of the above are described in Chapter 3.

Paper-based records management

In a paper-based system, anyone requiring a document in a specified file must make an official request for release of that file, which is recorded manually, and all this takes time.

In most organisations, electronic and paper-based systems operate side by side at present. This could change to computer systems only in the future. Examples of a centralised computer-based system and a paper-based system operating together are in medical and dental centres, where patients' personal records are stored on computer for easy access at reception, and records of tests and doctors' notes are stored in a paper-based centralised system in a room at the back of the reception area.

Decentralised filing systems

In a decentralised system, each department, office or section controls and maintains its own filing system and filing procedures.

AN EFFECTIVE RECORDS MANAGEMENT SYSTEM

Each business chooses the best system and procedures to meet the demands of the individual requirements of the organisation. The records management system you choose for your business must incorporate four basic procedural principles. It must:

- be simple and easy to use;
- be accessible;
- have a high retrieval speed;
- have safety and security.

Simplicity and ease of use

Procedures should be established and entered in the office procedures manual so that each person in the office who uses the system is familiar with the procedures chosen for your particular type of business. For example, it may be more effective to index and file your customers/clients in alphabetical order, or numeric order, or a combination of both. This depends on whether your business has a huge client base of individual people divided into smaller groups controlled by one person, or whether it deals with large corporate clients controlled by a number of people.

Your system may be a centralised computer-based one, with access to all personnel. Or it may be a centralised paper-based system, where people have to request files. Whatever system you use, all the people using it should be trained in the correct procedures required to maintain a smooth flow of instant information. This is entered in the corporate procedures manual to be used by all staff.

Accessibility

The most important factor of any records management system is ease of retrieval. Documents that are filed incorrectly and cannot be retrieved or located when needed to communicate with clients, or to make a decision, could result in such problems as:

- damaged corporate image as a result of apparent inefficiency and/or indifference;
- reduced customer confidence, which may lose you a sale or a contract;
- heightened stress for yourself and any others involved in trying to locate the information.

The procedures you set in place must incorporate some method of locating documents when required, including those which are being used by people who have taken them out of the system. (This should be set out in your filing procedures manual.)

High retrieval speed

Retrieving information from a file as quickly as possible is the ultimate goal of an effective records management system. A quick, confident, accurate answer to a query from a customer/client, either on the telephone or at reception, will give you a positive image and improve the corporate image of the business.

Safety and security

One of the controversial issues today, with so many personal records being stored on computer systems, is the safety and security of that information. 'Computer hackers' are working continuously to obtain access to confidential data records.

To make your computer system as safe and secure as possible, you should make sure that all staff who use your computer database system observe the following rules, which should be entered in the procedures manual.

Security in a computer-based management system

- Staff should only have access to data by using their personal security password or number.

- Staff should strictly follow the security procedures so that unauthorised personnel do not have access to this information.
- Disks containing confidential data should be locked away when not in use.

Security in a paper-based records management system

- Confidential files should be colour-coded for easy identification when out of the system.
- A list should be compiled of staff with access to the confidential files, so that unauthorised personnel cannot retrieve these records.
- All filing cabinets containing confidential files should be locked when the person responsible for them is away from the area, and confidential files should not be left on desks unattended.

Record life cycle

Records management covers the complete life cycle of a record. It includes:

- external or internal creation
- utilisation
- filing procedures and systems
- filing supplies
- filing equipment
- storage of files
- control
- retrieval
- disposal.

Creation

A record is created in a number of ways.

External

- A letter is received from a customer/client.
- A letter in reply is dictated, transcribed and typed on a letterhead, and copies are made before the letter is mailed.
- Messages are taken over the telephone from customers/clients.
- Diaries or computer systems contain records of appointments.

Internal

- Memoranda are written or typed and distributed from one department to another within the organisation.
- Stock-control cards are entered to maintain a supplies system.
- Personnel details are stored on a computer database or cards in the human resources department.

Utilisation

- The information stored in these records is used to create more information, or is used as a basis for decision-making.

- A letter is sent in reply to a letter received from a customer and copies are filed for future action.
- Information in an internal memo received from the manager of the sales department by the manager of the production department is used to make decisions on manufacturing and production of goods. It is then filed away for future reference.
- A message you take from a client is handed to your manager for action on a query regarding delivery of goods.
- The appointments made by you for clients, and entered in your diary and your manager's diary, are used to gain new business or increase existing business.
- Stock-control cards, or computer database systems, are controlled and maintained so that supplies are received in plenty of time to ensure the smooth operation of the business.
- Stationery supplies such as letterheads, copy paper, staples, folders, pens and pencils must be ordered before stocks are exhausted. Records stored in the human resources department are used to calculate and pay out wages and salaries, or to select an employee for a position in the organisation.

Records handling

Your duties as a filing clerk using a paper-based system are to control and maintain documents in a clean and complete condition ready for instant retrieval and use by management and other personnel on demand. This involves:

- using the correct filing procedures;
- preparation of documents ready for filing;
- indexing documents;
- coding documents;
- placing documents in the correct filing system;
- keeping track of any files that have been taken out of the system;
- arranging for the disposal of non-active or 'dead' files (after this has had the approval of your manager or supervisor).

Filing procedures

Above it is mentioned that an effective records management system must incorporate factors such as ease of use, accessibility, high retrieval speed and safety and security. These factors must be included in the system most suited to your type of business. The system may have been worked out by you, or by someone previously in your position as filing clerk.

Procedures manual. Once the details have been organised and approved, these should be entered in the office procedures manual for reference by other people using the system. A procedures manual should include a section on records management, incorporating:

- the corporate style of correspondence;
- the systems of classification and types of records stored in each section;
- the procedures used for creating files;
- the procedures used for storage and retrieval of documents;

- the procedures for retention and disposal of documents;
- the procedures for release of files, including a list of names for the release of confidential files.

Preparation of documents for filing

Before filing documents, you must prepare them by inspecting, indexing and coding them.

Inspecting documents. First look for a release mark, which is the authorisation for the document to be filed. This could be the initials of the person who has written the letter, or a rubber 'file' stamp. Then you prepare documents by:

- removing all staples and paper clips from them;
- repairing any damaged sheets with sticky-tape;
- placing documents together that relate to one file.

Indexing documents (key word). Indexing is the process of deciding the order in which names and titles of organisations are to be filed. The order is determined by the most important part of the name or title or the **key word** or **key number**, as shown below:

J B Morrison and Associates Pty Ltd (Morrison is the key word)

Next, look at the parts of the name or title (**indexing units**) and make a decision on the order in which the units will be arranged.

Coding documents. Coding documents is deciding in which order you are going to index the units of the name or title. For example, when filing in alphabetical order of title, you should mark above the words each unit 1, 2, 3, 4 and so on, as shown below:

Unit	2	3	1		4	5	6
	J	B	Morrison	and	Associates	Pty	Ltd

When filing in numeric order or subject order, underline or highlight the key word or number. This will help when returning the documents to the filing system.

Colour-coding documents. One of the most helpful aids in any filing system is colour-coding both copy documents and sections of the system. In our example organisation, Business Stationery Supplies Pty Ltd, we colour-code our copy letters according to the different classifications of filing:

- alphabetical classification—white
- alpha-numeric classification—green
- subject classification—yellow
- numeric classification—pink
- geographic classification—orange
- chronological classification—pink

Copies of documents are made by word processors or clerical officers and placed in the relevant filing trays. The filing clerk can see at a glance the documents for each system, which makes sorting much easier.

Another use for colour-coding is to clarify different files:

- confidential—red
- administration—yellow
- customer—blue
- staff—orange

Again it is easy to classify the types of files when they are out of the system and sitting on desks. Colour-coding different sections of lateral files clarifies different sections and makes it easier to find specified files.

Filing classifications

Documents can be classified in six different ways:

- alphabetical
- subject
- geographic
- alpha-numeric
- numeric
- chronological

The alphabetical classification

Examples of listing items in alphabetical order are the classified sections of the newspapers and the telephone *White Pages*. In the classifieds, under each subject, such as 'positions vacant' and 'cars for sale', each item is listed in alphabetical order. In the *White Pages*, names and organisations are listed in alphabetical order in sections divided with primary guides from A to Z.

Primary guides (key guides). Primary guides divide files into sections within each classification. For example, in a card-index box a special card with a label indicating the letter (A, B etc.) is placed at the front of each section, or a tab is attached to the first suspension file in each section of the filing cabinet drawer. This clarifies each section and makes a specific file easier to find (see Figure 2.2).

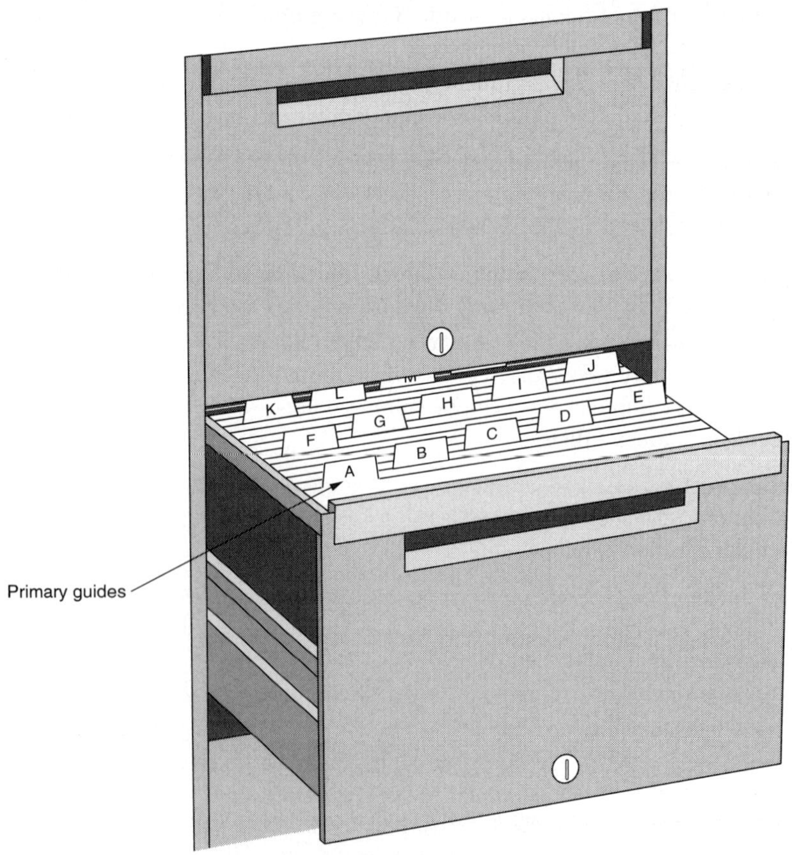

Fig. 2.2 Alphabetical classification *(continues)*

Fig 2.2 (cont.)

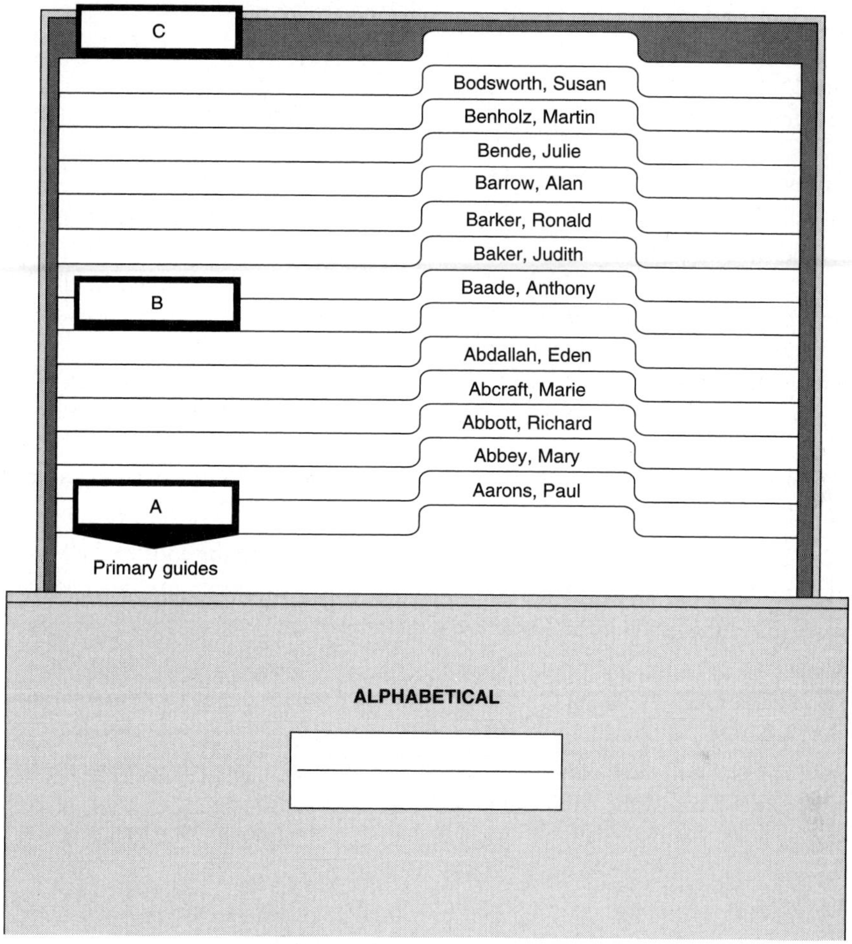

Fig. 2.2 Alphabetical classification

Activity 2.1

Finding names in the telephone directory

Refer to the Telstra *White Pages*, under the title 'How to find names in the *White Pages* directory', and read how names are sorted into alphabetical order. This will help you to understand the alphabetical classification. On Activity sheet 2.1 in your workbook list the rules for alphabetical filing as indicated in the *White Pages*.

Alphabetical classification rules. There are six basic rules for alphabetical filing.

Rule 1: Alphabetical order. Names and titles are arranged by units.

Name	Unit 1	Unit 2	Unit 3
Antell, J	Antell	J	
Crawford, Alan	Crawford	Alan	
Dayton Distributors	Dayton	Distributors	
Harper Furnishings	Harper	Furnishings	
Reynolds Dress Collection	Reynolds	Dress	Collection
Reynolds Dress Shop	Reynolds	Dress	Shop

Note: When names in the first unit are the same, the second and sometimes the third unit determines the order.

Rule 2: Nothing comes before something. A single letter, for example, comes before a full name or word.

Name	Unit 1	Unit 2	Unit 3
J	J		
J. A. Security	J.	A.	Security
Jim's Greek Tavern	Jim's	Greek	Tavern

(Consider all letters in a possessive, e.g. Jim's, but disregard the apostrophe.)

Rule 3: Single letters. Single letters that are not abbreviations should be considered as separate units, even if the letters are joined together.

Name	Unit 1	Unit 2	Unit 3
KB International	K	B	International
Q & L Music	Q	L	Music

Rule 4: Last name first. Names of persons are indexed by the last name as the first unit. Other parts of the name are indexed in natural order.

Name	Unit 1	Unit 2	Unit 3
Davis J	Davis	J	
John Davis	Davis	John	
Martin Davis	Davis	Martin	
James M. Davis	Davis	James	M.

Rule 5: First word first for business titles. Treat each word of a business title as a separate unit and index the name in the order in which it is written.

Business title	Unit 1	Unit 2	Unit 3
Freeman Dresses	Freeman	Dresses	
Freeman Travel Agency	Freeman	Travel	Agency
Joseph Freeman Wholesale	Joseph	Freeman	Wholesale
Sara Freeman Flowers	Sara	Freeman	Flowers

Rule 6: Numbers. If a business title contains a number, index the number as if it were a word and treat it as one unit.

Business title	Unit 1	Unit 2	Unit 3	Unit 4
A1 Industries	A	One	Industries	
24 Hour Printing	Twenty	Four	Hour	Printing
21 Club	Twenty	One	Club	

You may be unsure how to classify unusual names and titles. Below are variations to the above six rules.

Classification uncertainties

Prefixes. Prefixes such as 'De', 'Van', 'Mc', 'Mac' are considered as part of the name to which they are attached, and are sorted exactly as written.

Name	Unit 1	Unit 2	Unit 3
James Dee	Dee	James	
James Delaport	Delaport	James	
James De La Porte	De La Porte	James	
James D'Lando	D'Lando	James	

Mc and Mac: 'Mc' is treated as though spelt 'Mac'. Names such as 'Mace' and 'Mackay' are sorted with those names which begin with 'Mc' and 'Mac'.

Hyphenated names. Hyphenated names are treated as one unit. Ignore the hyphen.

Name	Unit 1	Unit 2	Unit 3
Gina Knapp	Knapp	Gina	
Rosa Knapp-Forbes	Knapp-Forbes	Rosa	
Robert J. Krups	Krups	Robert	J.

Activity 2.2

Alphabetical classification

You are the filing clerk for Business Stationery Supplies Pty Ltd. On Activity sheet 2.2 (1) of your workbook there are details of 20 letters which have been sent to your clients. These include the following:

- the date the letter was written;
- the name of the person to whom the letter was addressed;
- the title and address of the customer;
- the subject heading, including:
  (a) order number, (b) subject, (c) invoice number.

1. Index and sort the names of the people to whom each letter was addressed into alphabetical order and list them on sheet 2.2 (2) of your workbook.
2. Index and sort the titles of the organisations into alphabetical order and list them on sheet 2.2 (3) of your workbook.

The numeric classification

In the numeric system of classification, each file is allocated the next number in sequence (see Figure 2.3). Numeric classification and alphabetical classification are the most common ways of filing documents. Classification of customers and clients in most large organisations is made including both a number and the name or title and address of the customer/client. Examples of this are:

- telephone bills—account number and name and address of customer;
- RACV membership—member number and name and address of customer;
- Australian Taxation Office—tax file number, sequence number, name and address of customer;
- rate and evaluation notice—assessment number and name and address of customer;
- insurance policies—policy number and name and address of client;
- invoices and statements—number, name and address of customer;
- libraries—use the Dewey decimal classification for indexing books and other materials.

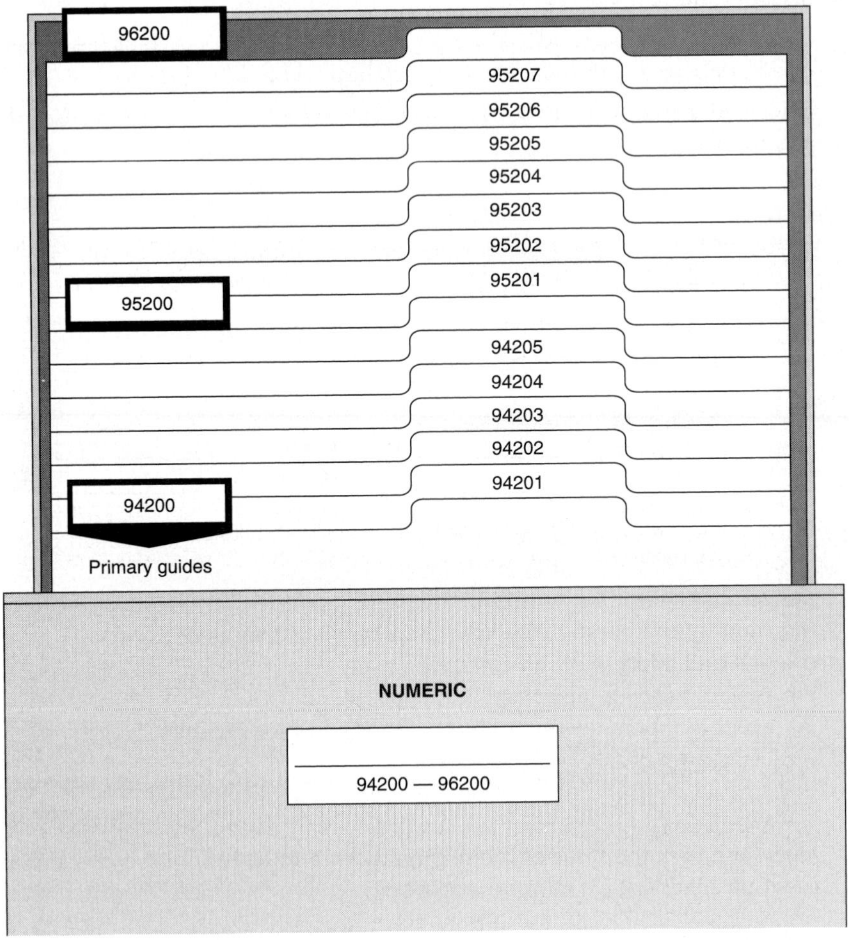

Fig. 2.3 Numeric classification

Activity 2.3

Numeric classification

Index and sort the letters on Activity sheet 2.2 (1) of your workbook into numeric order of invoice number. List them on Activity sheet 2.3.

The alpha-numeric classification

The alpha-numeric system of classification uses a combination of both letters and numbers (see Figure 2.4). For example, in a hardware store, a certain type of nut and bolt is given the stock control letter of NB and each different size is given a number (NB100, NB101, NB102 and so on).

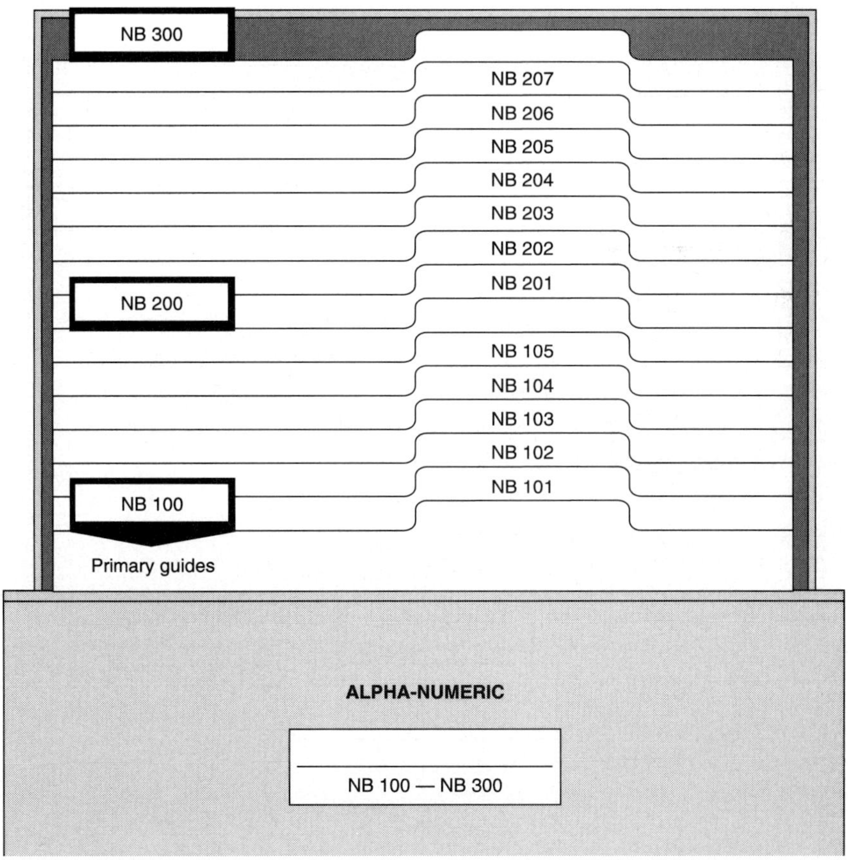

Fig. 2.4 Alpha-numeric classification

Activity 2.4

Alpha-numeric classification

In our example organisation, Business Stationery Supplies Pty Ltd, each product is given letters (e.g. CS—computer software) with the customer order number after it. This is shown in the customer letter subject heading in the letters for filing on Activity sheet 2.2 (1). Index these order numbers in the correct alpha-numeric order and list them on Activity sheet 2.4.

The geographic classification

The geographic system of classification is filed according to geographic location. Documents are indexed using cities, towns and suburbs in the different states of Australia. For

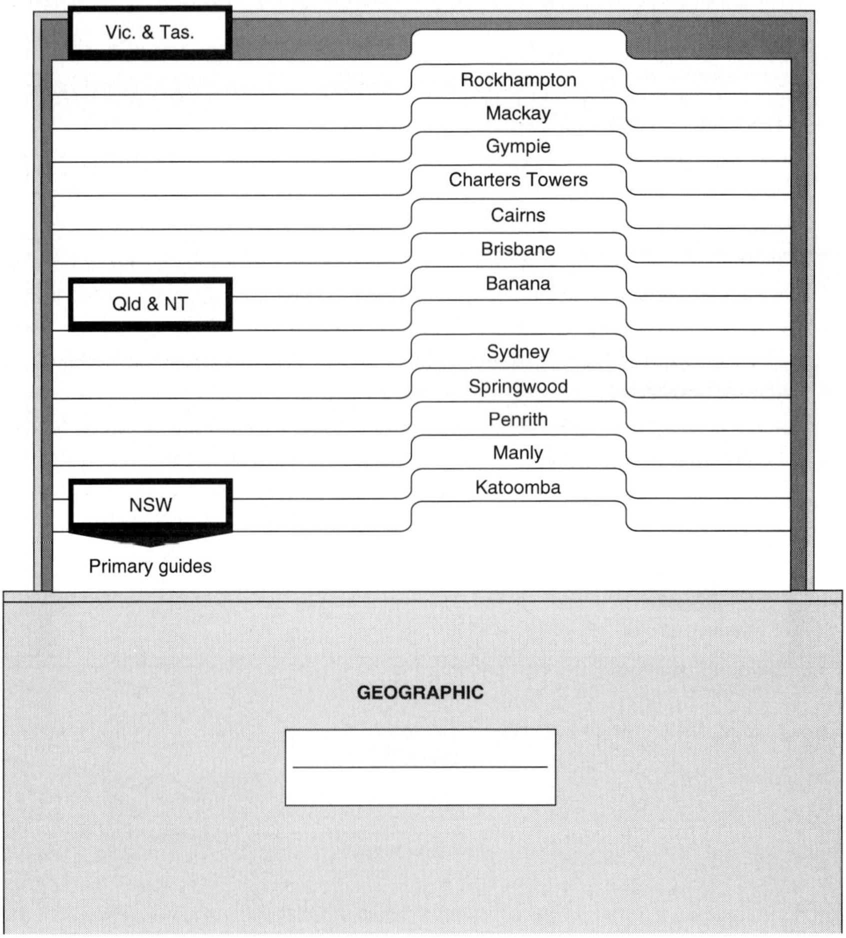

Fig. 2.5 Geographic classification

example, most large organisations divide their sales operations into regions within each state. Business Stationery Supplies Pty Ltd divides its operations into four regions: Queenland and Northern Territory; Victoria and Tasmania; New South Wales; and Western Australia and South Australia.

Documents such as sales reports are indexed and filed using the state as the primary guide, with documents filed in alphabetical order of suburb (see Figure 2.5), and then alphabetical order of customer, in each section.

Activity 2.5

Geographic classification

Index and sort the states and suburbs (or town, city) on the list of letters on Activity sheet 2.2 (1) into geographic classification of state (in alphabetical order), and alphabetical order of suburb within each state. List these on Activity sheet 2.5.

The subject classification

A simple example of subject classification is when you divide your filing system at home into sections with headings such as 'Taxation', 'Electricity', 'Rates', 'Gas', 'Holidays', 'Investments'. A clothing manufacturing business could classify the copies of letters to customers into categories of goods such as 'Silk blouses (yellow)', 'Long cotton skirts (print)', 'Leggings, wool (black)'. These would be used as primary guides with the customers' names and addresses indexed in alphabetical order in each subject section. This would be useful in ascertaining the popularity of a particular article and would help in maintaining manufacturing and stock records.

The goods supplied by Business Stationery Supplies Pty Ltd are, for example, 'Computer software', 'Office equipment', as shown in the subject headings of the list of letters on Activity sheet 2.2 (1). In Figure 2.6 we use the hardware store example, with 'Plumbing equipment', 'Building equipment', 'Electrical equipment' as primary guides.

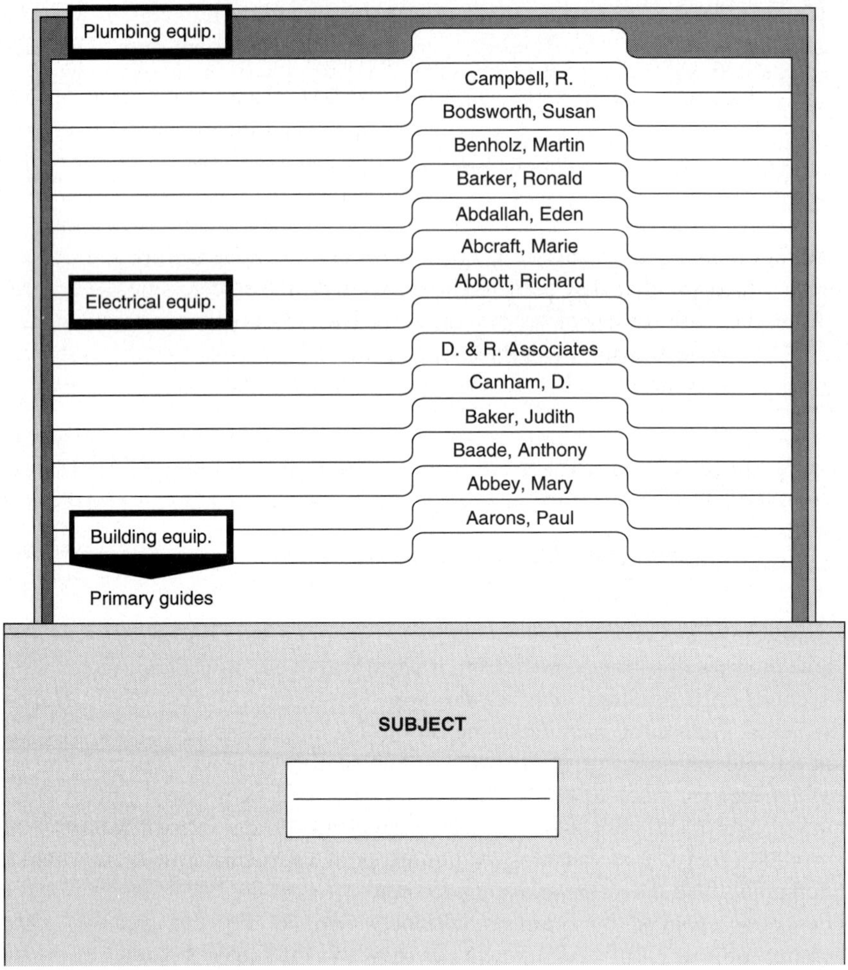

Fig. 2.6 Subject classification

Activity 2.6

Subject classification

Index and sort the letters on Activity sheet 2.2 (1) into alphabetical order of subject, with the titles of organisations filed in alphabetical order within each subject. List these on Activity sheet 2.6.

The chronological classification

Documents in the chronological classification are sorted in date order. The most useful application for this system is in a 'diary file'. This is sometimes called a 'tickler file' or a 'bring-forward file'. Documents that need attention in the future—one day, one week, one month, or even years ahead—are filed in the relevant date and retrieved on that date. This system is also used as a reminder and can be a useful tool for better personal effectiveness. It can be used in conjunction with a computer diary system, but the actual documents are filed in the diary file ready for dealing with when retrieved. One drawer in a filing cabinet is filled with suspension files labelled with primary guides for each month. Thirty-one suspension files are installed after the first month labelled with the dates 1 to 31. There are also suspension files at the back of these files labelled with the dates of the next three or four years. Documents are filed in the relevant day of the month, with enough time allowed for processing the documents. Every day when you arrive at work you retrieve the documents filed for that day, place them on your desk for processing, and place the suspension file (with the day of the month) at the back of the primary guide for the next month. The days of the month are then moved forward, which is why this is sometimes called a 'bring-forward file'. This system is illustrated in Figure 2.7.

Activity 2.7

Chronological order

Index and sort the letters on Activity sheet 2.2 (1) into chronological order of date. List these on Activity sheet 2.7.

Filing supplies and equipment

When you have classified, indexed and sorted your documents, these need to be labelled and stored in folders and placed in containers or cabinets in a convenient location for easy access. You may have inherited the equipment you use and have had no involvement in choosing the methods of your particular kind of records management. But even old and inconvenient equipment can be improved with more modern types of folders and modern labelling techniques. Some of these are described below.

Vertical files. The most common method of storing customer records is in metal storage cabinets (see Figure 2.8) with two, three or four drawers. Suspension files containing documents are placed one behind the other.

Lateral files. Although metal cabinets have been the traditional method of storing documents, today there are many types of lateral filing cabinets which save space and use different methods of colour-coding for ease of retrieval. In these, suspension files containing documents are placed next to one another. Some examples are given in Figure 2.9.

Visible index files. Index cards, or strip index cards, which contain the names, addresses and telephone numbers of customers/clients, may be stored in a rotary file (see Figure 2.10) or a flat file which sits on your desk for easy access when dealing with enquiries on the telephone or at the reception desk.

Disk storage. As computer disks are a form of magnetic media, care should be taken to protect them from heat and light or any other kind of damage (see Chapter 3). There are many different types of disk storage, from small boxes to multi-drawer.

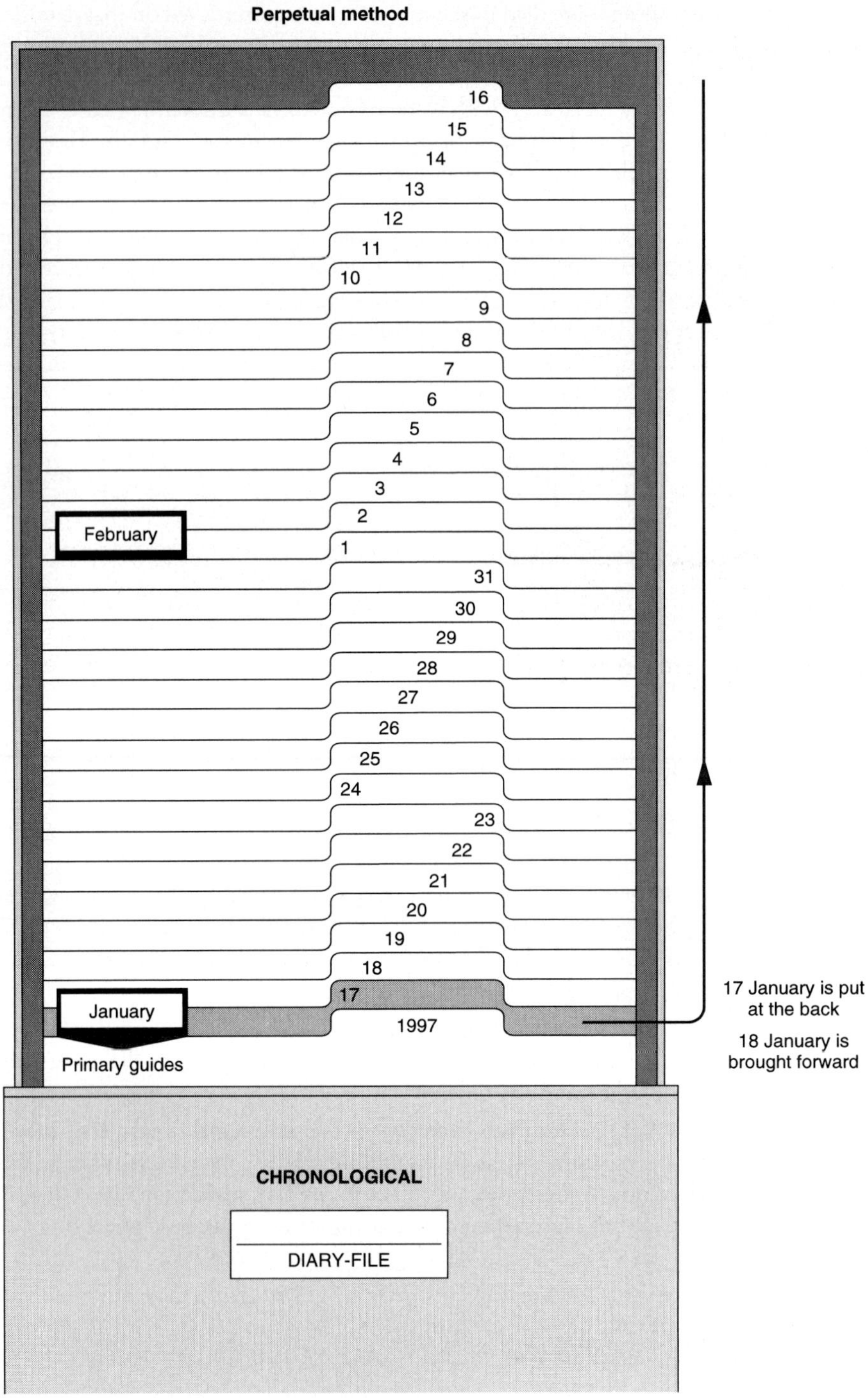

Fig. 2.7 Diary, tickler or bring-forward file

Fig. 2.8 Metal filing cabinets

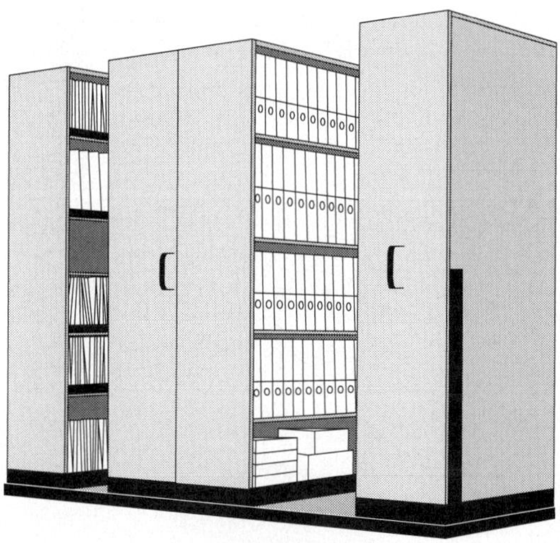

Compactus

Fig. 2.9 Lateral filing equipment *(continues)*

Fig 2.9 (cont.)

Colour-coded lateral filing

Trolley file

Fig. 2.9 Lateral filing equipment

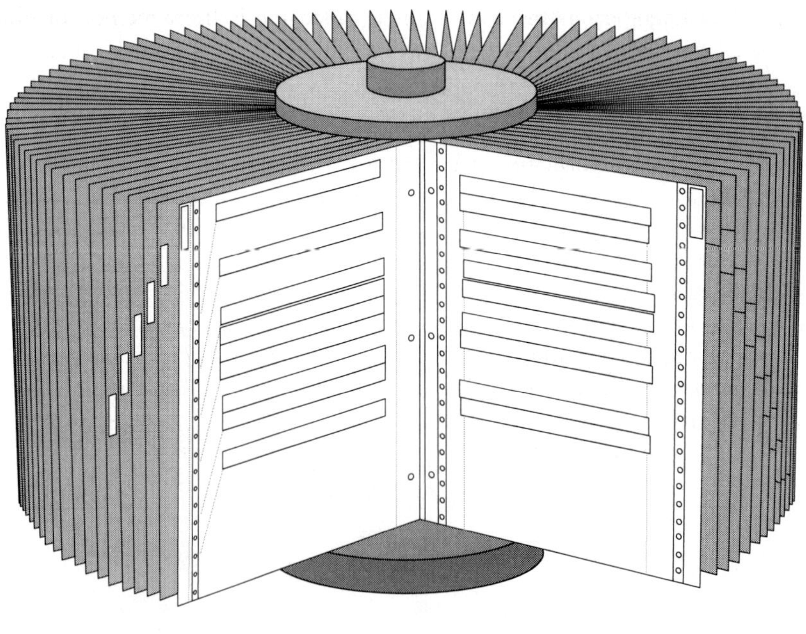

Fig. 2.10 Rotary file

Activity 2.8

File storage equipment

In groups, visit a large stationery supplies organisation and take note of the different types of storage equipment and supplies available today. List as many different types as you can find on Activity sheet 2.8 in your workbook.

RECORDS PROCESSING AND MAINTENANCE

You have established the methods and procedures for your filing system. You now have to control the system by updating and modifying records to allow for entering new customers' details, taking out and returning files, locating misplaced or lost files, and cross-referencing files.

Direct and indirect filing

The direct method of filing is when a document is accessed immediately from storage without reference to another source of information. An example is when a letter is filed and retrieved using only a name or company title. If reference is made first to a number in a card index and then to the name, this is called the indirect method of filing. A method of cross-referencing is used when classifying in the indirect method of filing.

Cross-references

Sometimes a record of information may need to be stored in more than one place, and this could cause problems when trying to retrieve the information. For example, stationery goods that are ordered by a large organisation may be delivered to different locations with different company titles. A centralised accounting system operates from head office because of economies of scale, which means that, due to the large quantities ordered, the goods are bought at greatly discounted prices. One invoice is sent for all the goods, so you need to have some method of linking the goods sent to the different locations with the one invoice number. The usual method is to attach a cross-reference sheet or card to the main head office company title, indicating the other branch titles and the invoice number, as shown in Figure 2.11.

Cross-reference

Name/title:	Southern Cross Transport
See:	Banbury Cross Transport
See:	Scorcher Ignition Products
Refer:	Invoice Number 27683
Date:	3 January 199X

Fig. 2.11 Cross-reference sheet

Activity 2.9

Cross-references

On Activity sheet 2.9 in your workbook, cross-reference Banbury Mobile Mechanics with A Pro Auto Electrics and Invoice Number 24736 from the list of letters on Activity sheet 2.2 (1). Use today's date.

Movement of files

To ensure that files that are borrowed from the system are returned to the correct location, there are several practices you need to put into operation. These are detailed below.

Updating and modifying records

Identifying, locating and retrieving files and documents

Your filing system is now organised into the procedures and systems most suited to your particular business. All the classifications are colour-coded and labelled and everything is set up efficiently. You now need to make sure that it stays that way. There are filing practices you need to put into operation when taking files out of the system and returning them to the system to make sure that everything continues to work efficiently and effectively.

Good filing practices

- Work out regular times to do your filing. This need not be every day, but documents should be grouped together in classifications ready for filing so that items can be located if required.

- At the end of each filing classification, set up a 'miscellaneous' folder for documents that do not relate to a particular client or number. File the documents in the miscellaneous file in alphabetical order for ease of retrieval.
- Do not overcrowd files or cabinets. This makes it hard to find anything. Always leave working space. Once you have filled a drawer or lever-arch file, make up another one. Set up a new folder when a current one becomes too thick.

Procedures for placing documents in a folder

- Place correspondence in the folder with the top of the letter on the left.
- Place the most recent letter on top (chronological—most recent date; numeric—the highest number). This means that all documents are facing the same way and easy to read.
- Before placing documents in the folder, fold the paper loosely in half from top to bottom and make a small indent with your thumb and first finger in the middle of the page down the left-hand side. This is a guide when punching two holes in the left side of the paper and ensures that all the documents in the file are placed together neatly.
- Make new labels and replace folders when they become torn. Folders can be used many times.

File/document movement

Files in some systems are recorded in another location such as a file index box or a computerised system as a double-check in case one file is lost. There are some simple aids to help prevent the loss of files (see Figure 2.12).

File movement register. A file movement register can easily be compiled by using a ring-binder with loose sheets which record details of files that have been taken out of the system, by whom, the date they were taken, and when they were returned.

File movement marker. At the same time as a file is recorded in the file movement register, a file movement marker is placed in the location from which the file has been removed so that anyone searching for that particular file knows who has taken it.

Locating missing files

If you have checked the file movement register and still cannot locate the file, there are several things you can do.

- Check the files just before and after the file you are trying to find; it may have been filed incorrectly.
- Check the bottom of the filing cabinet drawer; the file may have slipped under the other files.
- Check the other drawers; the document may have been placed in the wrong classification system.
- Check your desk and your boss's desk. The file may have become mixed up with other documents.

If you still can't find the file, you may need to investigate your file movement register to

FILE MOVEMENT MARKER				
TITLE	DATE REMOVED	DATE RETURNED	NAME	SIGNATURE

FILE MOVEMENT REGISTER					
Title of document	Location field (cabinet, drawer, lever arch)	Date filed	Documents taken	Date	Date returned
Alphabetical (mauve)					
Numeric (blue)					
Alpha-numeric (yellow)					
Geographic (green)					
Subject (pink)					
Chronological (white)					

Fig. 2.12 File movement marker and file movement register

see who has previously borrowed the file. They may have borrowed it again without authorisation.

Setting up new files

As your business expands, it will become necessary to set up new files for new customers/clients, or to restructure existing files. This may entail setting up an entirely new system if you have acquired a number of new customers because of a sales drive or an expansion into a new product or products. Or you may just make new files when you obtain new customers who will be placed in your existing system. The title and address of a new customer will be placed in your file register, and a new folder labelled with the title and containing the new documents will be placed in the correct location in the alphabetical classification and cross-referenced in all other locations.

Activity 2.10

Placing new files

Obtain three business names and addresses from the *Yellow Pages*. Indicate where you would place these in the titles you have indexed and sorted on Activity sheet 2.2 (3) of your workbook.

Records retention and disposal

There are five stages of records retention:

- **Active records** are files which are used regularly. They could be documents that have been recently created, or they could have been in the filing system for a long time. Active records should remain in the current system.
- **Semi-active records** are files which are used only occasionally and can be placed in a less convenient location.
- **Inactive records** are files which are not in current use and are being retained only for reference purposes. These can be placed in archive boxes and stored in a less convenient location.
- **Long-term records** are documents which have to be kept for a period of time for legal requirements. These are also placed in archive boxes and stored in a less convenient location.
- **Archive records** are records which are retained only because of their historical interest. These can be placed in a storeroom or off-site location.

Records disposal

All records pass through the life cycle from 'creation' to 'disposal'. As part of records management, you regularly check the stages of all files, and any that have reached the inactive, long-term or archive stages should be removed from your current filing system. The reasons for this are that:

- more space becomes available for active and new files;
- retrieval time becomes quicker, as it is easier to locate your current files if old files are not taking up space in filing drawers;
- control of records is more efficient and effective when active and inactive files are separated;
- cost savings can be made to the business if less storage equipment is used.

A decision is made by management, or your office supervisor, when the life of a record has ended and these documents have to be disposed of. Once records are outside the retention period, and are no longer active or semi-active, the documents are destroyed by shredding (particularly if they contain confidential information) or recycling, as authorised by management or your supervisor. (Refer to Chapter 3 for methods of recycling documents to help conserve the environment.) A note of records that have been destroyed should be made in a records disposal register in case this information is required for legal purposes. A records disposal register sheet is shown in Figure 2.13.

RECORDS DISPOSAL REGISTER					
No.	Title of document	Description of documents	Decision made by:	Disposal action	Date

Fig. 2.13 Records disposal register

Activity 2.11

Records processing and maintenance

In this activity you will be creating a new customer file containing the following source documents, which will be used for the practical filing activities in Activity 2.12. The folders and filing trays you will need will be provided by your tutor/supervisor.

Using the blank documents for Activity 2.11 in your workbook, photocopy each one onto the correct colour-code sheet of A4 copy paper. Create new documents as follows:

1 copy letter (alphabetical; white)
(sheet 2.11 (1))

1 invoice (numeric; pink)
(sheet 2.11 (2))

1 purchase order (alpha-numeric; green)
(sheet 2.11 (3))

4 sales reports (geographic; orange)
(sheet 2.11 (4–7))

5 copy letters (subject; yellow)
(sheet 2.11 (8–12))

1 customer stores requisition (chronological; blue)
(sheet 2.11 (13))

1. Type (or write very neatly) in the correct address format (as discussed in Chapter 1) *one* business title and address onto each of the documents (use the same business for each document).
2. On the invoice, in addition to the title and address, type (or write neatly) in the correct location an invoice number taken from the invoice number folder provided.
3. On the purchase order, in addition to the title and address, type (or write neatly) in the correct location an order number taken from the order number folder provided.
4. Now fold and 'nick' the documents as described in this chapter, punch two holes in the left side of each document and, with an 'Arnos clip', file them in a manila folder.
5. Label the folder on the front and down the right-hand side with the title of the new business.
6. Using the cross-reference sheets (2.11(14)) in your workbook, cross-reference one document with your invoice and one document with your purchase order.
7. Show your completed new customer folder to your tutor/supervisor for approval and assessment.
8. Enter the title of the file in the new business register provided in alphabetical order.
9. In preparation for Activity 2.12, take the documents from your folder and place each in the correct classification labelled tray, ready for filing.
10. Place the manila folder and Arnos clip in the location indicated for recycling at a later date.
11. Take one set of documents out of the 'inactive/archive' file and archive as directed by your tutor/supervisor.
12. Take one set of documents out of the 'dead/inactive' file and destroy these documents as instructed by your tutor/supervisor (shred, or place in a recycle bin).
13. Enter the documents to be archived and destroyed on the records disposal register sheet 2.11 (15) in your workbook.

Activity 2.12

Records handling

(Filing equipment will be set up by the tutor/supervisor for the use of students in this activity.)

In this activity you will be filing and retrieving documents in the same way that you would in a real office, with real equipment, and using simulated documents. Documents that have already been filed will be sitting in the filing cabinets, and you will file your documents with these as you would in a real office filing system.

Form into a group of two or three people. More than three people will not be able to access the filing cabinets at the same time (or any other filing equipment made available for your use), as it will be too crowded. Process documents, file and retrieve in the following way.

1. Take the following documents from each filing tray labelled with the relevant classification; then index, sort and file the documents as instructed below.
   (a) Alphabetical classification (colour-code: white)
   Take six copy letters from the filing tray labelled 'Alphabetical filing'.
   Sort and index into alphabetical order of business title with other copy letters in the file.
   (b) Numeric classification (colour-code: pink)
   Take six invoices from the filing tray labelled 'Numeric filing'.
   Sort into numeric order (lowest number at the bottom to highest number at the top) with other invoices in the file.
   (c) Alpha-numeric classification (colour-code: green)
   Take six purchase orders from the filing tray labelled 'Alpha-numeric filing'.
   Sort into alpha-numeric order (alphabetical first and numeric second) with other purchase orders in the file.
   (d) Geographic classification (colour-code: orange)
   Take four copy sales reports from the filing tray labelled 'Geographic filing'.
   Sort into numeric sales region, and then alphabetical order of suburb within each region, with other sales reports in the file.
   (e) Subject classification (colour-code: yellow)
   There is one tray for each subject. Take one copy letter from each tray labelled 'Subject filing'.
   Sort into subject heading and then alphabetical order of company title within each subject, with other letters in the file.
   (f) Chronological classification (colour-code: blue)
   Take one internal stores requisition from the filing tray labelled 'Chronological filing'.
   (This is filed in the 'tickler/diary file' for the date on which you expect to retrieve all the above documents.)

If you are unsure how to index or classify any of these documents, ask another member of your group or your tutor/supervisor for clarification.

2. Index and sort the documents you have taken out of the filing trays provided for you, labelled with the relevant classification. (The documents in each classification will be colour-coded as indicated in each section above.)
3. Enter these documents in the correct classification in the left-hand column of your file movement register (Activity sheet 2.12).
4. File the documents (in the correct order with other documents already in the files) in the correct storage systems and locations provided for you. (This will be explained in detail by your tutor/supervisor.) Enter each location in the second column of your file movement register sheet.

CHAPTER

3

OFFICE EQUIPMENT, INTRODUCTION TO THE INTERNET

TEC101A, NOS109, 7365H, COM301A, NOS250, 7365CS, TEC201A, NOS210, 7365HB, TEC301A, TEC204A, NOS323, 7365HC, NOS132

objectives

In this chapter you will learn how to operate a range of office equipment in accordance with manufacturers' instructions using 'hands-on' methods of practice. You will also get basic information on how to use the Internet, and learn the basic terminology and security measures for using the World Wide Web and the Internet, in order to complete the following learning outcomes:

1. Office equipment—routine tasks
Use relevant occupational health and safety practices

- 1.1 demonstrate correct power access procedure
- 1.2 follow safety procedures recommended in the equipment manuals

2. Operate a range of office equipment to complete routine tasks

- 2.1 select appropriate equipment for the task
- 2.2 locate and check for availability
- 2.3 use in accordance with manufacturers' instructions
- 2.4 identify any equipment faults and report to the nominated person
- 2.5 follow shutdown procedures
- 2.6 introduction to the Internet
  - joining the Net
  - choosing an ISP
  - understanding the Web
  - web browsers
  - search engines
  - glossary of basic computer and Internet terminology
  - security measures on the Web and the Internet

3. Office equipment—non-routine tasks
Operate a range of office equipment to complete non-routine tasks

- 3.1 use various functions of the equipment simultaneously

4. Identify and rectify or report minor faults

- 4.1 undertake routine maintenance according to manufacturers' operating instructions
- 4.2 identify equipment faults accurately and describe to nominated person/section
- 4.3 rectify minor faults following manufacturers' operating instructions

5. The implications of copyright legislation as applied to an office

- 5.1 describe the purpose of the Copyright Act
- 5.2 explain the implications of the legislation for office copying

6. Apply recycling techniques to minimise paper wastage

7. Bind documents

8. Office supplies—stock control

INTRODUCTION

In today's technological age you will be using an increasing number of items of office equipment and machines to enable you to complete your work. You will be provided with furniture which enables you to use this equipment. How you operate the equipment and how you sit at the desk can be a hazard to your health. A hazard is anything that could cause harm, and when you are using equipment in an office this includes such factors as:

- incorrect use of electrical machines, causing accidents;
- a badly designed workplace, causing accidents;
- badly designed chairs and desks, affecting posture;
- incorrect use of computer screens and keyboards;
- chemical solvents, including toners and cleaning materials;
- too much noise and poor lighting.

While the chance of a fatal injury occurring in an office is slight, the injury rates to back, neck, shoulder, arm, wrist and eye are high. These cause work-related illness, resulting in pain, stress, visits to the doctor and hospital, absence from work and sometimes loss of position.

OCCUPATIONAL HEALTH AND SAFETY PRACTICES

According to the *Occupational Health and Safety Act* covering work practices in each state, an employee shall not:

(a) wilfully or recklessly interfere with or misuse anything provided in the interest of health, safety or welfare;
(b) wilfully place at risk the health or safety of any person at the workplace.

(Refer to Chapter 6 for further details.)

Safety awareness includes using equipment and machines according to manufacturers' instructions, which are outlined for each item of equipment illustrated in this chapter. It is your responsibility as an employee to make sure you operate equipment in a responsible and safe way. There are some general rules common to the use of all electrical equipment:

- Make sure the equipment is positioned with enough space around it for people to use it.
- Check that all power leads are connected correctly to power points and telephone wall sockets.
- Make sure that all leads are stored at the back of the machines so that people do not trip over them.
- Check that all equipment is placed on level surfaces and that there is no danger of equipment being pushed or knocked off desks or tables.

- In order to protect your neck and back, position equipment so that you do not have to stretch or twist to use it.

Shutdown procedures

- Always turn off the power at the wall switch before taking out or putting in a power plug.
- Protect equipment when you know a thunderstorm is approaching by turning off power switches and unplugging power cords from all outlets.
- At the end of the day, you should get into the habit of switching off all electrical equipment where possible. You may have to leave the facsimile machine, the answering machine and the computer connected so that messages can be received after business hours. A power-surge protector should be connected permanently to the power points used for this equipment.

Activity 3.1

Equipment safety procedures

(This activity may be completed by one person.)

Form into groups of three people (or you could do this activity by yourself if you are at work). Visit the main office in your area (or the office you work in) and list the equipment used in that office. Write next to each item answers to the following:

(a) Is the item of equipment in an easily accessible position with enough space for people to use it?
(b) Are the power leads connected correctly and are they out of the way at the back of the item of equipment?
(c) Is it clean and well maintained?
(d) Is it positioned on a level surface and not in danger of being pushed or knocked?

The kind of equipment you will use, and the understanding you will need in order to operate the equipment, will depend on:

- the specific nature of your job;
- the equipment required for that job;
- the amount of centralisation of services in your organisation;
- the responsibility you have for using equipment;
- the level of technology associated with the equipment;
- the range of equipment available in the particular organisation you work for.

OBTAINING THE RIGHT EQUIPMENT

One of your duties as an administrator may be to obtain the equipment for your office or department. There are many things to consider before ordering new equipment to make sure you are obtaining the correct machine for the job. In order to avoid costly errors, you should analyse the reasons for buying the equipment. Are you buying it:

- to save costs?
- to improve efficiency?
- to improve client service?
- to keep up with the competition?

'Green' technology

When you are buying new equipment you should consider the newer 'green' technology. These machines use less power (which lowers electricity bills), and because of this are considered to be more reliable.

If you really don't need a particular piece of equipment, don't get it. If you really do need it, you should undertake a feasibility study to discover which particular model, type and size will be the best for your requirements.

Specifications

- What size is the most suitable for the space available?
- Is it fast enough for your requirements?
- Where is it going to be located?
- Which members of staff are going to use it, and how many?
- What are the electrical and environmental considerations?

Cost analysis

- What options does your budget allow you? Is it better to buy or lease?
- Will this equipment contribute to profit?
- What are the ongoing costs, such as electricity, service and consumables?

Renting/leasing equipment

In a lease agreement the equipment is leased for a specific period of time, and at the end of the period there is usually an opportunity for the lessee to buy the equipment. With a rental agreement the equipment is returned at the end of the rental period.

Advantages of renting/leasing. There are several advantages to renting or leasing equipment:

- Technological equipment soon becomes obsolete, and with a lease agreement equipment may be updated regularly.
- On-site repairs or replacement of parts is the responsibility of the supplier.
- After-installation support is provided by the supplier.
- There is no capital outlay for purchase of equipment.
- Often tax deductions may be claimed.

Service maintenance agreements

You can arrange with the supplier for a regular preventive maintenance, which includes periodic cleaning, lubrication and replacement of worn-out or defective parts, for a fee. Factors to consider when choosing a service arrangement are:

- the costs for a maintenance contract compared with an 'on-call' service arrangement;
- the effectiveness of regular maintenance as a preventive measure;
- how often the equipment is used and the volume of production;
- the speed with which service contractors respond to a call.

Supplier analysis

It is worthwhile comparing packages offered by different suppliers, as this can save you a lot of money. The up-front cost may not be the lowest when you compare what is included in the deal. You should take into account the following factors:

- *Reliability.* Reliability of the equipment is one of the most important considerations. A machine being out of production for long periods of time can cause serious problems. Carry out an analysis of equipment used in other departments and make comparisons on the reliability of the different types of equipment.
- *Training.* This is very costly, and initial training by the supplier can save costs later.
- *Service maintenance.* Service maintenance can be costly if your equipment has a high production rate. It is often better to lease the equipment, making the costs in the long run lower.
- *Supplies.* What supplies are required for the equipment you are considering? Is it a popular model, or will it require specialised supplies (which are very costly)?

Activity 3.2

Choosing the right equipment

(This activity may be completed by one person.)

This can be used as an extension to Activity 3.1 and performed with the same group. Each member of the group should choose one item of equipment used in the office you visited in Activity 3.1. On Activity sheet 3.2 in your workbook, state the following information (you will need to ask the office supervisor's permission to obtain this):

- the type and model;
- whether it is leased or purchased;
- whether it has a service maintenance agreement;
- the volume of production—high, medium or low;
- whether the operators find it reliable;
- the type of training that was provided.

This activity could be developed (with Activity 3.1) into a group project for Chapter 12.

CONTROL OF OFFICE EQUIPMENT

Equipment in a large organisation is usually centrally controlled in order to keep a record of the hours of usage, cost of operation, and times for replacement. An inventory should be kept of all equipment for administration and accounting purposes, including the following details:

- date of purchase or lease;
- cost at time of purchase or lease agreement;
- current book value for depreciation purposes;
- maintenance records of costs incurred for servicing and repair of all equipment;
- record of manufacturers' serial numbers, so that the location of each item of equipment is known at all times to prevent loss;
- for photocopiers etc. a record of how many copies are made by each department, for costing purposes;
- a record of the dates and times of service maintenance calls, and of any parts replaced.

Activity 3.3

Control of office equipment

(This activity may be completed by one person.)

Working in the same group as in the previous two activities, complete the equipment control card in your workbook (Activity sheet 3.3) for one piece of equipment in your area (classroom, nearest office or the office you work in). Place your completed card in a box in alphabetical order with the other students' cards. Label the box 'Equipment control cards'. These cards should be kept to record the future maintenance of equipment in your area.

OPERATING OFFICE EQUIPMENT

Operators of all equipment should receive training and instruction in the use of individual machines according to the manufacturer's user manual. Occupational health and safety procedures should be observed as instructed by the manufacturer. Training should be provided:

- initially by the supplier;
- through training sessions by the organisation;
- by a skilled operator who has been trained to use the equipment;

- by using an up-to-date manufacturer's manual;
- in health and safety procedures;
- in preventive maintenance procedures.

Below are instructions on the operation and maintenance of the more common types of office equipment, including the following:

- computers
- photocopiers
- answering machines
- overhead projectors
- shredders
- facsimile machines
- printers
- laminators
- telephone systems
- dictation machines
- binders
- electronic typewriters

COMPUTER EQUIPMENT AND THE INTERNET

(Refer to Chapter 6 for the relevant information on health and safety practices when using a computer and computer furniture.)

Computers, the Internet, and e-mail (like the telephone) are becoming integral to all offices of the future, and many offices are using them today. Therefore, training in the operation of the computer and the Internet will be prerequisites for an office position. Training in the use of the computer and the Internet does not form part of this book (this will be undertaken in another section of your course). But this chapter gives you tips on what to look for when choosing a computer and an explanation of how the Internet works with a glossary of simple terminology. (E-mail is described in detail in Chapter 1.)

Choosing a computer

When you use a computer in an office, this has been chosen by management of the organisation you work for, to meet their requirements. But you as an operator need to know a little about the machine you are using. It is somewhat like driving a car: you don't need to know how the engine works, or how the electrical parts are wired up, but you do need to know the size of the engine and the layout of the interior of the car, to see if it is big enough and powerful enough to meet your needs.

The same applies when you are choosing a computer for your personal needs. You see computers advertised for sale on television and you see them for sale in computer shops, but do you know what a Pentium processor is, or MB of RAM, or what 2GB means?

If you are buying a computer for the first time you should shop around and ask a lot of questions. You should first ask yourself the questions listed under *Obtaining the right equipment* above. Should you rent or lease and, most importantly of all, what are the service maintenance agreements and the training provided? How are you going to get help if you come across a problem? This can be very expensive.

Before you buy you should use the following checklist:

- Be wary of special offers that demand up-front payment—this is often a sign that a

dealer has a cashflow problem.
- Check with the state consumer affairs department to ensure that no warnings have been issued against the dealer.
- For greater protection, run a computer search on the dealer.
- Negotiate to obtain the kind of warranty cover you think you will need (including a replacement machine should yours fail). Buy somewhere else if the dealer refuses to meet your warranty conditions.
- Learn as much as you can to ensure that the computer you buy suits your requirements.

After you buy:

- Check before you leave the shop that the box contains all the components.
- Leave the machine running for a few days after purchase in order to 'stress-test' it.
- If you are not happy about the quality of repairs or service you have received, never be afraid of expressing dissatisfaction.
- Don't accept excuses if you think you are being given the 'run around'.
- Don't be afraid to escalate a complaint in order to get satisfaction.
- Don't tolerate a machine that constantly breaks down.

Many computer companies make you special offers, and some of them are good. But you should know what you are buying. The computer may seem to be a bargain, but what software is provided in the total package price? Software can be very expensive if bought individually, and can cost more than the computer hardware.

Computer technology is changing all the time, and the computer you bought this year could be outdated next year. Do you want a computer just for personal use at home, for writing letters and documents? Or do you want to use e-mail and surf the World Wide Web? A lot depends on how much money you have available for a computer. Do you want to work from home, or start your own business? Should you buy a laptop or a palmtop with built-in modem if you want to use remote access?

If you are going to use the Internet, the speed at which your computer performs the tasks required is crucial if you want to avoid slow operation and high cost. You need to spend a lot of time and effort getting answers to these questions. The market is saturated with different computers with different capacities of operation. A computer's performance is largely measured by the speed at which it performs a task—particularly when you are using the Internet, as time costs you money.

A minimum guide for PCs (personal computers) using the Net: they should have at least a Pentium 133, with 16 MB (megabytes) of RAM (random-access memory), an 8× CD-ROM (read-only memory) and should be running Windows 95 as the operating system. Below are some of the features of more up-to-date and more expensive computer packages (see also Figure 3.1).

Hardware includes
- AMD K6-2 350MHz CPU
- 512k Cache
- 64MB SDRAM
- 8MB 3D AGP Graphics
- 15" Colour monitor
- 32× CD-ROM

- Infinity speakers
- 104KB & mouse
- 56K Fax/Modem

Software includes

- Windows 98 and Lotus Smartsuite Millenium version
- IBM Via Voice 98—Office edition
- Netscape Navigator
- Microsoft Internet Explorer
- Norton Antivirus (OEM version)

Fig. 3.1 Up-to-date computer package

Figure 3.2 illustrates the capabilites of a reasonably priced up-to-date notebook.

- Pentium Processor 300MMX
- 32MB SDRAM (MAX 256MB)
- 4GB Drive
- 24× CD-ROM
- 2MB Neomagic Graphics
- 12.1" TFT Screen
- Integrated 56K Fax/Modem
- SoundCard
- Altec Lansing Audio
- Windows 98

Fig. 3.2 Up-to-date computer notebook

Speech recognition

Not long ago, the quality of computer speech technology was poor, had limited vocabulary and required specialised hardware. But recent developments, such as faster processors, improvements in speech algorithms, advances in signal processing and features such as USB (universal serial bus), fast RAM and faster and bigger hard-disk drives, have led to dramatic improvements in the quality and number of speech-enabled applications for personal and business use.

The Pentium III processor was developed, in part, for speech technology applications, and software has quickly followed. Commercial speech applications are now relatively cheap and offer extremely high accuracy with improved vocabulary. Speech recognition (or speech to text) captures human speech and translates it to a digital-text format. This allows voice commands to control the computer (command and control applications). These commands can be used to trigger actions such as opening files or dialling a telephone. Dictation applications can capture speech data and transcribe and store them as text file, to be edited and used like any other file.

In continuous-speech recognition, the voice sounds natural as there is no need to pause between words. In discrete-speech recognition there are pauses after each word or phrase. Using only your voice, you can:

- create, edit and format documents, letters, e-mail, presentations etc.;
- navigate through your desktop and control applications;
- talk to discussion groups on-line.

One such product on the market is Philips' **FreeSpeech**, which is not expensive. FreeSpeech 98 adapts to your voice, dialect and natural dictation style, becoming progressively faster and more accurate in its recognition of your speech. A playback feature audibly plays back words and sentences while simultaneously highlighting the text on screen, allowing you to overwrite and insert dialogue. An example of such software is shown in Figure 3.3.

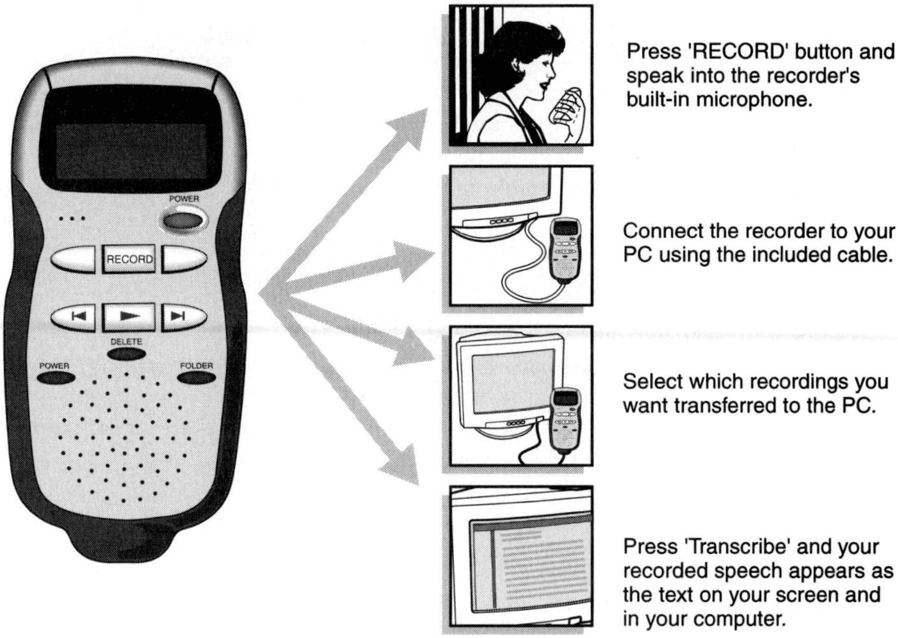

Fig. 3.3 'Dragon' continuous-speech recognition dictating hardware and procedures

Introduction to the Internet

The Internet is a network composed of many individual computer networks linked all over the world.

Network. A network is a collection of computers linked directly, usually by cable, in a room or a building. This is referred to as a local area network (**LAN**). Each computer connected to a LAN shares information and communicates with every other computer, as well as sharing printers and programs. When two or more LANs are connected you create a wide area network (**WAN**). WANs are connected by dedicated telephone lines, or even satellites. The Internet is a massive WAN, connecting many hundreds of computer networks around the world.

Joining the Net

At home, to get on-line you will need a computer and modem. Nowadays many computers and notebooks have in-built modems. You will also need an account with an Internet service provider (**ISP**). This will already be in place if you are using the Internet at college or at work. Most Internet accounts today are **PPP** (**point-to-point protocol**) accounts; these are user-friendly and allow you to do most of the things you want to do on the Net.

A modem. This is a device used by a computer to send and receive data over a telephone line. Your ISP has a computer that is permanently connected to the Internet. You use a modem to connect your computer to the ISP's computer via the telephone network.

Choosing an ISP

You should choose an ISP that is in your local area, otherwise you could be paying STD charges. Most ISPs charge a set fee for a month, with a number of hours included in this fee and an hourly rate from then onwards. This can range from 60 cents to $10. Some ISPs will charge only a monthly fee with no time charge, so you have to choose very carefully. There are also 'on' and 'off' peak rates and discount rates. Some ISPs charge for technical support. Below are some of the questions you should ask different ISPs before you make a choice:

- How much does your service cost?
- What is your modem-to-user ratio?
- Do you connect direct to Telstra or do you go through another server?
- Do you have an overseas link of your own?
- Is the account a PPP (point-to-point protocol) account, with a direct connection to the Net from your computer via modem and phone line?
- Do you provide technical support, and what are your hours?

Understanding the Web

The World Wide Web (the **Web**) uses browsers to search the Internet for a variety of uses. You can book your holidays or buy any number of other consumer goods from the vast number advertised on-line via the Web. You can browse through libraries of updated information, you can gamble, read newspapers. You can buy or do anything via the Web. What makes this possible is **Hypertext**, which incorporates links in the text and enables users to move to one part of a document and return to where they left off. Web documents can contain links that automatically connect the user to other documents anywhere on the global Internet. These Hypertext links are viewed using web browsers.

Web browsers

Internet packages usually contain web browsers such as Netscape Navigator or Internet Explorer. You can obtain information about web browsers from **FAQ** files or the Web-specific newsgroups. Every document or service available on the Web has a **URL** (universal resource locator). URLs are often used in conjunction with one of several commands, such as the http://www.name.com address.

Search engines

Searching through the huge volume of information available on the Internet using browsers can be a time-consuming occupation, as information is not classified. Search engines do this for you. Just click on 'Search', enter your topic, and your search engine will find it. You can choose from a number of these:

'Ask Jeeves'	www.askjeeves.com.au
Yahoo	www.yahoo.com.au
Excite	www.excite.com

and there are many more.

Usenet. This is a collection of electronic noticeboards (called newsgroups) used for discussion among people interested in the same topic, and covers a wide range of subjects. You can read articles through a 'newsreader'.

Internet relay chat (IRC). This allows many users on different systems at different locations to converge on one 'room' and have a discussion, similar to a conference call or part line. IRC is used both for entertainment and serious discussion purposes.

Downloading. To download files, you need software called an FTP client. Some computers connected to the Net have large archives of files. These computers are called FTP sites. Files at many FTP sites are freely available to the public. A virus checker should be used to check downloaded files to make sure they do not contain viruses.

Figure 3.4 shows you how to use the World Wide Web on the Internet.

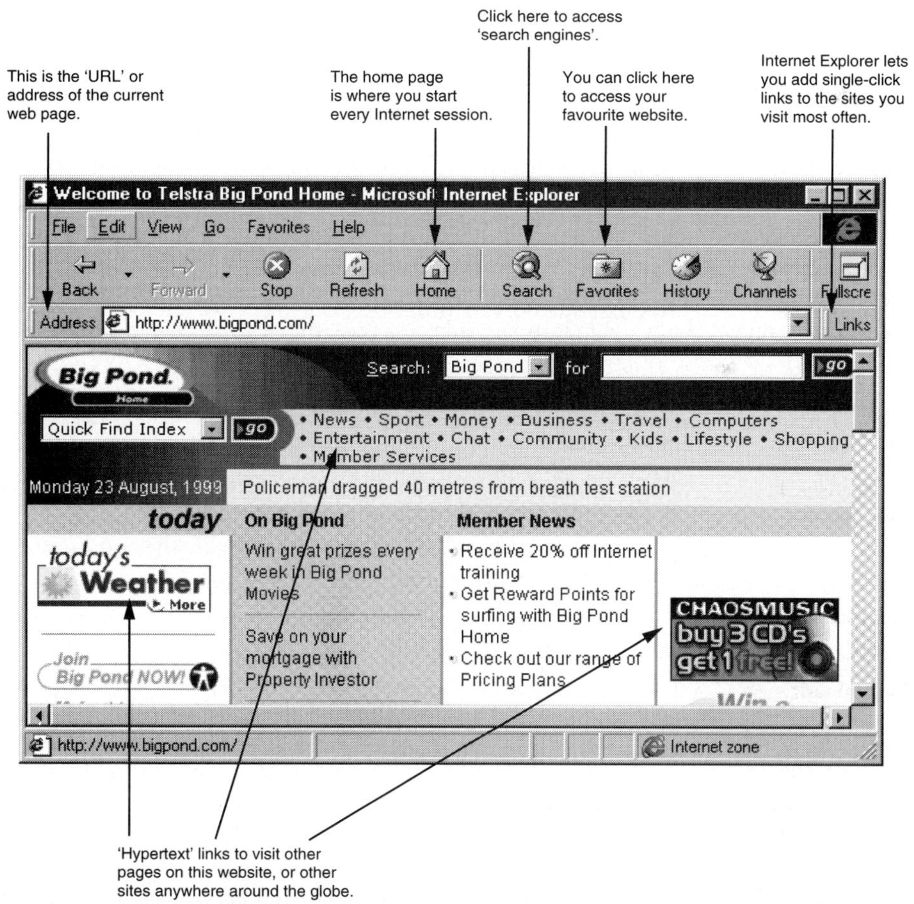

Fig. 3.4 Understanding the World Wide Web (Source: Adapted from Mark Neely, *The Internet guide for teachers, students and parents, 4th edn*, MaxiBooks, Kiama, NSW)

Glossary of basic computer and Internet terminology

Account: your ISP will provide you with an account to access the Internet

Anonymous FTP: the process of connecting to other computers on the Internet that allow public access

Archie: a service available to all Internet users which is used to search for files or directories on other computers

Article: the name used to refer to messages posted on the Usenet news system

Ascii: the American Standard Code for Information Interchange, the main printable character set used by today's computers

Backup: a copy of a file, directory or disk made for safe-keeping

Baud rate: the number of bits per second transmitted through a modem

Binary file: a computer file which contains characters other than pure (ascii) text

Bit: 'binary digit', the smallest unit of measurement of computer data

Bps: bits per second, the speed by which modems are rated: the amount of data a modem can send and receive each second

Bus: a pathway between CPU and the devices connected to it

Byte: a byte is usually made up of 8 bits. The size of a computer file or disk is referred to in bytes. A kilobyte is a thousand bytes, a megabyte is a thousand thousand bytes (e.g. 32MB SDRAM)

Cache/Cache memory: very fast memory is fitted between the CPU and the bus, and information is read into this fast memory while the CPU is processing data—**caching (e.g. 512k cache)**

Chat: the act of 'talking' over the Internet; many chats use **IRC (Internet relay chat)**

CPU (central processing unit): this is the piece of hardware that houses your CD-ROM and your floppy disk unit, and the control buttons; interprets and executes program instructions and communicates with the input/output units and secondary storage

Domain name server: the computer that remembers the domain names of other machines and their IP address

Download: loading information from another computer into your own; the opposite is upload

Encryption: method of coding data so that it can be read only by the people it is intended for (in theory)

Http: Hypertext transfer protocol, which regulates how information is transferred over the World Wide Web

Hyperlink: a touch-sensitive spot on a Web page linking it to another page or site

Hypertext: documents that contain links to other documents; forms the basis of the World Wide Web

IP (Internet protocol): one of the many standards that regulate the way in which information is passed between computers on the Internet

ISP (Internet service provider): a company that provides Internet access

Modem: a device used to connect two computers via a telephone line

Packet: the term given to a unit of data sent over a network

POP: point of presence—gives local access to a network service. Also called **PSTN**

PPP: point-to-point protocol—a direct connection to the Net from your computer, via modem and phone line

RAM (random access memory): the primary storage unit for data and instructions

ROM (read-only memory): a user may read or access a program direct but may not write on ROM as it is dedicated to a particular application

Router: a device that transfers data between one or more networks

Server: the computer running the server software, or software that is used to provide access to the Internet (a web server)

URL (uniform resource locator): the address system used on the Web

World Wide Web (WWW): a hypertext-based system linking information and files on different computers around the Internet

Security measures on the Internet

- You should be very cautious when you use your credit card to buy goods on the Internet. There is a possibility that unscrupulous dealers or hackers will access your card number and build up a huge deficit, which you have to pay.
- Always back up files on the computer in case these are lost through power failures or other computer crashes.
- Install a power-surge protector in case of power failures and 'brown-outs'.
- There are 'disaster recovery' software packages which also recover and restore services and have integrated antivirus facilities, and inform you of impending problems before a server crashes.
- Protect against viruses by installing a virus scan which detects 100 per cent of viruses on floppy disks, Internet downloads, e-mail attachments, Intranet shared files, CD-ROM and other on-line services.
- You should be very wary of pirated software and floppy disks obtained from an unreliable source, as these could contain viruses.

Computer maintenance

Computers combine electronic components with some electromechanical parts. The motherboard, video and network cards carry only an electrical current, while the hard and floppy disk drives use moving parts, which can wear out. To make sure that your computer has a longer life, there are a few simple maintenance procedures you should follow.

- **Power surges** (which are the result of too much power) and brown-outs (too little power) can damage your computer. A power-surge protector should be fitted to the power point to protect computer equipment against these surges.
- **Static electricity** caused by synthetic carpets can be discharged through the first metal you touch. This is often the computer, so a rubber mat should be placed under it to prevent this problem.
- Computers can cause **excessive heat** if your room is not well ventilated. Do not place computers in a sunny location and make sure you leave at least 30 cm space around computer equipment.
- **Moisture**—including high humidity, condensation and spilt liquids—is a common cause of machine damage. Make sure there is adequate space around the equipment. The use of a special keyboard seal can prevent spilt coffee or other liquids from damaging the keyboard.
- **Dust** can often lead to electrical components 'shorting out', which can cause a fire. Dust can also lead to failure of floppy disk drives and backup tape drives. Use dust covers where appropriate and a soft brush and cloth to keep the computer and printers as free of dust as possible.
- Use the correct **start-up and shutdown procedures** as instructed in the user guide to avoid corrupting the disk, and limit the number of times you switch the computer off and on.

Storage and maintenance of floppy disks

Floppy disks are not so easily damaged today as they were a few years ago, when the softer 5.25-inch disks were more commonly used. These were more easily scratched and bent. Care should still be taken with the 3.5-inch disks in use today. To prolong use and protect your disks from corruption, the following steps should be taken when handling them:

- Do not expose disks to liquids, or to excessive heat and sunlight.
- Do not place disks on the VDU while you are operating the computer, as magnetic radiation can corrupt them.
- Use the special adhesive labels and a soft felt-tip pen for identification.
- Do not open the metal section of a disk, as this will expose the electromagnetic surface.
- Use a special disk regularly for the identification of viruses.
- Store all disks in a specially designed box and lock away any disks that hold confidential data.

Printers

The printer is the most common output device from a computer. Printers are capable of producing high-quality letters, business reports and other documents at very high speed. There are two categories of printers: impact and non-impact.

Impact character printers

Impact printing devices print one character at a time at speeds of 10–500 characters per second. The fastest of these is the wire- or dot-matrix printer, which prints characters made up of a pattern of dots formed by the ends of small wires arranged in a rectangular pattern. Certain wires are extended to make the pattern of a number, letter or special character. This character is then pressed against an inked or carbon ribbon. The number of wires (up to 100) determines the quality of the printing. Dot-matrix is not letter-quality printing and is used when the appearance of the work is not so important, such as for sheetfed printing. Letter-quality impact printing is produced by daisy wheels or moving chains or drums pressed against an ink or carbon ribbon. See Figure 3.5.

Fig. 3.5 Examples of dot-matrix print (left) and letter-quality print (right)

Included in the routine maintenance required for impact printers in electronic typewriters is changing daisy wheels if a different font is required, and replacing them when they are damaged. If the print on a dot-matrix printer becomes faint or irregular, this means that the dot-matrix head has become damaged and needs replacing. This requires the services of a maintenance mechanic.

Non-impact printers

Non-impact printers use laser beams, inkjets, and electrostatic or thermal methods of printing.

Inkjet printer. The inkjet printer discharges tiny droplets of electronically charged ink onto the paper. The size of the droplet used affects both printing quality and speed.

Maintenance involves replacing the inkjet cartridge when printing becomes patchy or faint. Care should be taken not to touch certain areas of the cartridge or it could be damaged. Instructions are given in the cartridge package. When replacing copy paper in sheetfed printers, always fan the paper to prevent sheets sticking together and jamming the printer.

Thermal printer. A thermal printer uses a series of heating elements, which form character patterns on special sensitive paper.

Laser printer. A laser printer uses a laser beam to form characters by projecting a pattern onto a drum surface. Toner is attracted to the area exposed by the laser beam, brushed across the drum's photoconductive surface and transferred to the paper. Laser printers produce high-quality copies, and using a desktop publishing package in conjunction with a colour laser printer and an electronic scanner can produce work comparable to professional print jobs. Large organisations often produce their own publicity material and other publications in this way.

Routine maintenance involves replacing copy paper and toner and keeping the equipment clean. Non-routine maintenance may involve identifying faults involving the replacement by a mechanic of more sophisticated parts, such as the drum. This usually entails putting in a call to the supplier's service department and filling in an equipment faults report sheet, an example of which is given in Figure 3.6.

<table>
<tr><th colspan="9">Equipment faults report sheet</th></tr>
<tr><td colspan="2">Date</td><td colspan="2">Meter</td><td></td><td></td><td colspan="3">Job No.</td></tr>
<tr><td colspan="2">Eng. No.</td><td>PM</td><td>1</td><td>2</td><td>3</td><td>4</td><td>Call</td><td>18</td></tr>
<tr><td colspan="9">Reason for call.</td></tr>
<tr><td colspan="9">Cause.</td></tr>
<tr><td colspan="9">Action.</td></tr>
<tr><td colspan="9"></td></tr>
<tr><td colspan="9">Parts & consumables used</td></tr>
<tr><td>Drum</td><td>C.B</td><td>UFR</td><td>LFR</td><td>OSR</td><td>WEB</td><td>STARTER</td><td>Blk</td><td>Col</td></tr>
<tr><td colspan="3">Part Description</td><td>Qty</td><td colspan="4">Part Description</td><td>Qty</td></tr>
<tr><td colspan="3"></td><td></td><td colspan="4"></td><td></td></tr>
<tr><td colspan="3"></td><td></td><td colspan="4"></td><td></td></tr>
<tr><td colspan="3"></td><td></td><td colspan="4"></td><td></td></tr>
<tr><td colspan="3"></td><td></td><td colspan="4"></td><td></td></tr>
<tr><td colspan="3"></td><td></td><td colspan="4"></td><td></td></tr>
<tr><td colspan="3"></td><td></td><td colspan="4"></td><td></td></tr>
</table>

Fig. 3.6 Equipment faults report sheet

Photocopiers

Health and safety issues

Photocopying machines have many computerised parts, and in order to protect the machine and yourself there are several operating precautions you should observe.

- Install the copier on a stable and level surface.
- Do not place heavy objects on the machine or subject it to any physical shock.
- Do not place a coffee cup or any other liquid-filled container on the photocopier. Spillage can damage the electrical parts and insulation.
- Do not drop paper clips, staples or any other metal object into the photocopier.
- Do not watch the lamp during the photocopying process with the cover open. This can damage your eyes.
- Do not turn off the power switch during the copying process.
- Always disconnect the power plug before opening the document cover to clean, remove jammed paper, replace toner bottles and so on.
- Always check that the power switch is in the 'off' position before reinserting the plug.

- When operating the copier for an extended time in a small room, make sure the room is well ventilated. Photocopiers produce small quantities of ozone gas. These are normally harmless, but in an enclosed space they could have adverse effects.
- Do not try to open any fixed cover or alter the copier in any way.

See Figure 3.7 for an example of a photocopier.

Fig 3.7 Photocopier with bin sorter shown on left

Routine maintenance

If you want to make clean copies of documents, you must regularly clean the document cover and the platen glass (see Figure 3.8). Clean with a soft, slightly moistened cloth. Do not use cleaners with benzene or thinners, as these will damage the copier.

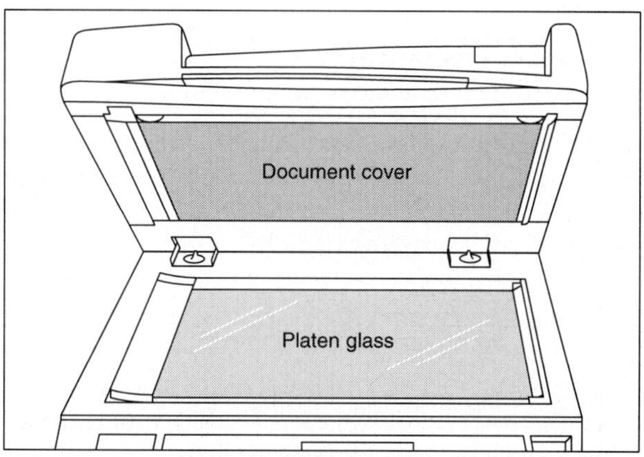

Fig. 3.8 Document cover and platen glass

Replacing toner cartridge

Toner is the fine black powder that forms the characters and images of copies. It has to be replaced regularly. The toner cartridge is replaced in similar ways in all modern photocopiers (see Figure 3.9). Refer to instructions given in the user manual. Replacing the drum and cleaning cartridges should be performed as instructed in the user manual.

1. A message or flashing light appears on the instruction screen.

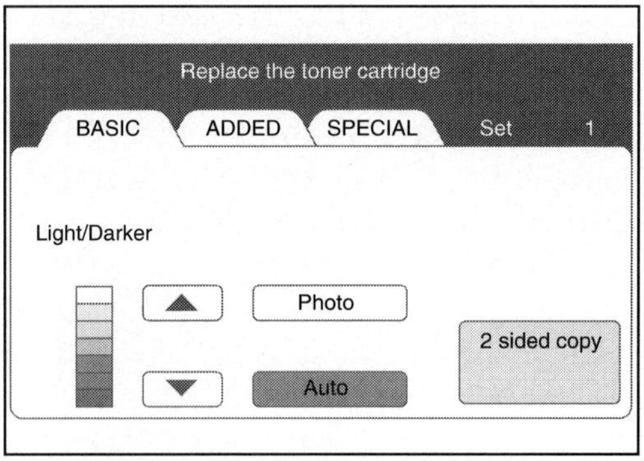

2. After switching off the power, open the front cover.

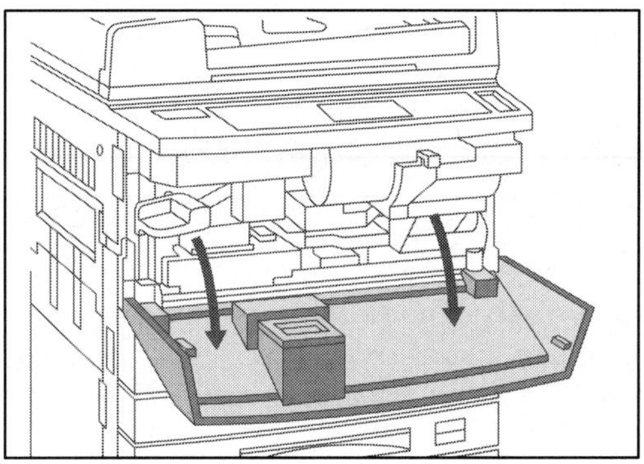

3. Swing the toner box out, take out the old cartridge and replace with a new one.

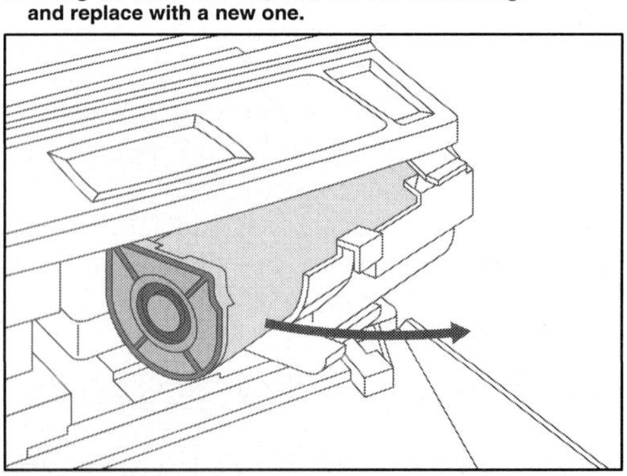

Fig 3.9 Replacing toner cartridge

Adding paper

The size and type of your photocopier will depend on the sheet capacity. Trays can hold 250, 550 or 1500 sheets of paper and up to 2000 sheets in a high-capacity feeder. When replacing paper in the paper trays (or replacing paper in printers), follow this procedure:

- Open the package at the seam and always stack in the same way as when removed from the package, so that the upper side is used first, or your printer may be affected.
- Keep the rest of the paper wrapped until required and store in a cool, dry place.
- Fan the sheets to prevent sticking.
- Make sure the paper is straight and stacks neatly, with edges aligned.
- Slide the track back in until it stops.
- Do not load paper over the level line in the tray (see Figure 3.10).
- Do not mix different-sized paper in the same tray; this can cause paper jams.
- Use the bypass tray when making transparencies, or when copying onto oversized paper or card.
- Always store leftover paper in the wrapping. Paper that becomes moist from humidity can jam the machine and cause poor copying quality.

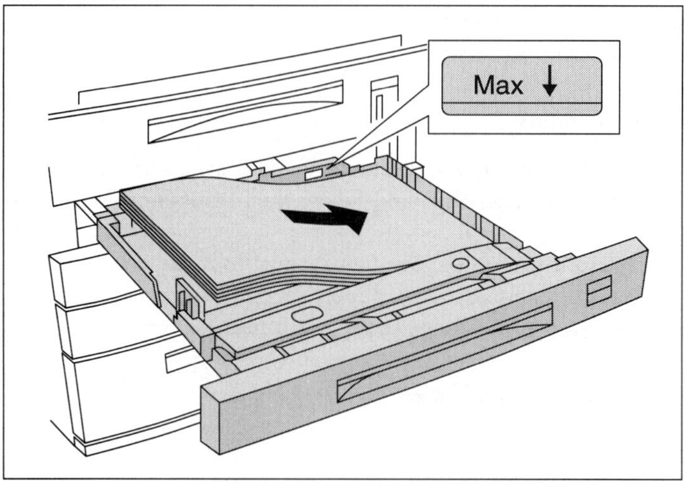

Fig. 3.10 Photocopier paper tray

Clearing paper jams

To clear paper jams:

- Switch off the machine.
- Open the front cover and lift the document cover.
- Be careful when putting your hand inside the copier, as certain parts are very hot when in use.
- Remove all the jammed sheets of paper. If you tear them, you must remove all the small pieces.
- If the copier seems excessively hot, turn off the power switch and disconnect the power plug without opening the front cover. Then contact the service centre.

Using the bin sorter

Sorting refers to making several copies at one time, which are delivered to respective bins (e.g. when making 10 copies of a 25-page report).

Stacking refers to making multiple copies of the first page of a document and stacking them in the first bin, then making multiple copies of the second page and stacking them in the second bin, and so on (e.g. when making 10 different 10-page handouts for a 10-person business meeting).

Changing the staple cartridge

A bin staple sorter automatically staples the sorted documents. When it is time to change the staple cartridge a sign will appear on the operation panel.

Making copies

Modern photocopiers have sophisticated facilities for making copies of documents. These copies can be so good that it is difficult to identify which is the original and which the copy. All modern photocopiers have a control panel which gives you instructions on copying features. Refer to the manufacturer's manual of your particular copier model to find out how to produce excellent copying results.

Making same-size copies

Before making copies, check that the relevant paper tray is sufficiently full. If you have to refill the tray, fan the paper first, and follow the instructions given above on adding paper.

Refer to the relevant section in your user manual to make same-size copies on same-size paper using auto (automatic paper supply section) function and to make same-size copies on different-sized paper (using paper tray button).

Making reduced or enlarged copies

To use the 'auto %' (automatic reduce/enlarge) function, or the preset or variable-percentage function, refer to the relevant section in your user manual.

Activity 3.4

Operation of the photocopier

Form groups of three. Refer to the manufacturer's user manual for the specific photocopier provided for your use. With the members of your group, perform the photocopying tasks outlined below, while being observed by your tutor/supervisor. You are required to copy, reduce and enlarge the document provided for your use.

1. Check that the photocopier is connected correctly according to occupational health and safety practices.
2. Refer to the routine maintenance section of the manual and, using the cloth provided, make sure the copier is clean to ensure best copy quality.
3. Perform these photocopying tasks, following manufacturer's instructions:
   - Using Activity sheet 3.4, make a photocopy of the facsimile form in the same size. (Place the original in the correct position against the 'length scale guide' to ensure a clean copy.)
   - Reduce the document 50 per cent.
   - Enlarge the document 50 per cent.
   - Make one copy darker and one copy lighter.
   - Refill the paper tray according to instructions.
   - Check how to replace the toner according to instructions, and replace the toner if necessary.
   - Check how to remove jammed paper according to instructions, and remove any jammed paper if this is necessary.
   - Photocopy any picture onto a transparency (provided by your tutor) according to instructions, for use on the overhead projector.
4. Report any machine faults on the facsimile form. Retain the form for use in Activity 3.6.
5. When all projects have been completed, use the correct 'shutdown' procedures.

Copyright legislation

The *Copyright Act 1968* (Cwlth), with amendments to the Act up to 1989, is designed to protect and reward creativity and intellectual effort. This means that a work cannot be copied without the owner's or the publisher's permission. In general, copyright covers a work from when it was created to 50 years after the death of the author. Once the copyright expires the work goes into the 'public domain', and can be copied freely. Copyright covers books, articles, poems, instruction manuals, plays, film scripts, cartoons, paintings, musical works, films, photographs, sound recordings and broadcasts, and computer software, among others.

There is no formal registration required for a work to be copyright. The only requirements for protection under the Act are:

- that the work is original;
- that the author is an Australian citizen or permanent resident, or citizen or resident of a country to which the Act extends.

Many countries have similar copyright laws, where publications are protected in the same way. The copyright symbol '©' identifies a work that is protected, but the symbol is not a requirement for protection. It is generally used with the name of the copyright owner and the year of first publication or production.

Up to a 'reasonable portion' of a work may be photocopied. This is defined in relation to a literary work published in an edition of 10 or more pages to include:

- up to 10 per cent of the number of pages in that edition;
- up to one chapter, where the work is divided into chapters (even if more than 10 per cent);
- more than a 'reasonable portion' if the work cannot be obtained within a reasonable time.

Infringements of the Copyright Act could result in fines, and organisations are required to display a warning notice next to the photocopying machine.

It is illegal to make copies of some documents, including:

- banknotes and cheques;
- bank and government bonds and securities;
- passports and identification cards;
- postage stamps.

Penalties of fines or imprisonment may be imposed on those found guilty of such copying.

Activity 3.5

Copyright

With other members of your group, discuss the copyright rules applying to the following publications:

- two textbooks;
- two reference books from the library;
- two equipment user guide manuals.

On Activity sheet 3.5 in your workbook, provide the following information for each publication:

- the title and country of publication;
- the copyright symbol, the owner and the year of publication;
- how many copies you are allowed to make of each publication.

Other office equipment

Telephone systems

The operation of a telephone system, details of occupational health and safety issues and maintenance issues, and the operation of answering machines, are dealt with in Chapter 8.

Facsimile machines

Facsimile machines have revolutionised communication between organisations within Australia and internationally, and are becoming more commonly used in homes. Electronic mail permits the transmission of letters, memos and other documents from one terminal or computer to another in a different office, city, state or country. Using a facsimile machine, pictures and drawings as well as text may be transmitted almost instantaneously. There are basically two types of facsimile machines:

- those which use an impact printer and a paper roll (see Figure 3.11); and
- much more sophisticated ones using a laser printer and individual sheet feed, with many facilities using computer memory.

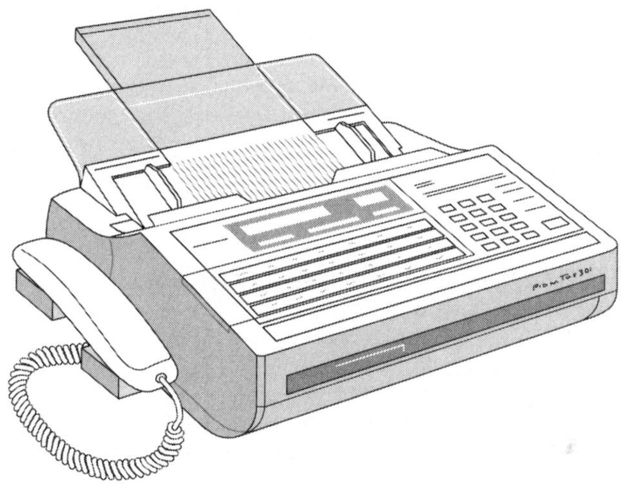

Fig. 3.11 A small facsimile machine

Health and safety practices

- Do not install the machine near heating or air conditioning units.
- Avoid exposure to direct sunlight.
- Install the machine on a flat surface and leave at least 10 cm space between the machine front and other objects, and between either side of the machine and other objects; 20 cm of space is required between the rear panel and other objects.
- Do not block ventilation openings.
- Do not place heavy objects on the machine or spill liquids on it.

General precautions apply, as mentioned at the beginning of this chapter.

Bound documents

In the previous pages we have talked about producing documents on the computer or an electronic typewriter, and printing these documents. Of the different methods of printing, if quality printing is required inkjet or laser printers should be used. Even better-quality documents can be produced using desktop publishing, although large quantities of printing may be produced more cheaply by an outside printer because of economies of scale.

Desktop publishing (DTP)

Using a computer software package and a laser printer, you can produce excellent-quality documents. Professional-looking documents can be produced incorporating different layouts, fonts and graphics. Many organisations now produce their own internal magazines, newsletters and external advertising material at a much lower cost than they could using outside printers.

There are one or two DTP software packages incorporating templates and many other features. One example is 'Microsoft Publisher', which uses such features as:

- a design gallery, which provides drag-and-drop access to an enormous range of significant design elements and makes the block-by-block construction of complex documents simple;
- almost automatic support for colour separations and bureau output;
- rotating all text and graphic blocks;
- easier-to-edit wizards;
- high-quality colour proof or output to bromide or film;
- excellent printed design guide.

E-mail and Internet documents

There are also computer software packages such as 'CorelDraw' and 'PageMaker', which have new features using advanced technology to boost productivity levels, such as intuitive web tools for greater impact on the Internet and interactive special effects for striking visual images.

Scanners

When creating documents you can add the colour and vitality of images, which will give a professional touch to your documents. A flatbed scanner plugs into your computer, and using the appropriate software enables you to:

- image edit through Windows software;
- use OCR (text recognition);
- scan to print/fax a large library of backgrounds and stock photography;
- automatically blend your pictures with text to create calendars, greetings cards, newsletters and business documents, incorporating logos, graphics and photographs.

Using an SCSI card with your scanner and computer enables you to 'daisy chain' other external devices, such as SCSI zip drive, tape drive, CD writer and slide scanner.

Relational database systems

As mentioned in Chapter 2, software packages such as 'FileMaker Pro' will produce any kind of document using complex databases and more then 40 templates.

Binders

Once you have produced your document, you will need to place this in a folder to complement the professional appearance of your specific kind of document. As mentioned in Chapter 9, you should consider the reader of your document.

- Are you sending out a legal document to a client?
- Are you binding a report to be given to the board of directors of your organisation, or another organisation?
- Are you sending out a report on your products to a prospective customer?
- Is your document being handed out at a large conference interstate?
- Do you need to make your document stand out from a number of other documents?
- Does your document need to have a very businesslike appearance?
- Does your document need to reflect your product (e.g. children, a particular sports group, teenagers, older people, a particular kind of food or article of clothing)?

All these points need to be taken into account when you produce and bind your specific document. This does not mean you have to buy an expensive binder. Using some of the computer software features mentioned previously, or even hand-drawing borders, graphics, and using different kinds of fonts and some imagination, you can produce professional-looking documents. I have seen students produce attractive bound documents using the above features and a cheap folder.

How you present and package a document can have as much impact on the reader as the contents of the documents. For example, how you organise, present and package your assignments, your long report, your travel folder or your business documents can have an effect on the results of your assignments.

Folders

We are lucky that nowadays there is a wide range of reasonably priced folders which you can buy in stores, supermarkets or newsagents. Large stationery stores stock a rainbow of different-coloured and different types of folders.

Binding machine

There are various methods of binding documents. Reports and other material requiring a professional presentation are greatly enhanced when bound, and are less likely to be lost. Two types of binding machines are:

- **spiral binders**, in which holes are punched (see Figure 3.12) and then a plastic backing and documents are threaded through the holes; and
- **thermal binders**, in which papers are bonded together with a heated adhesive seal.

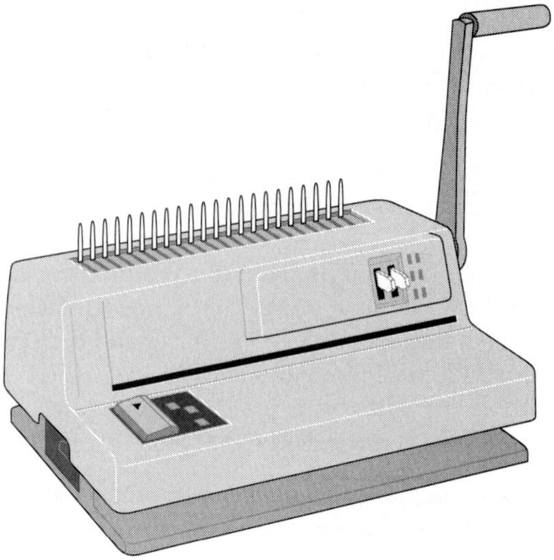

Fig. 3.12 A spiral binder

Activity 3.6

Bound documents

(You may discuss this activity in your group, but complete the questions on your own.)

With the other members of your group (or on your own) visit a large stationery store, a supermarket or a newsagent, and look at the different types of folders displayed on the shelves. List, on a separate sheet of paper, how you would present, organise, and which kind of folder or type of binder you would use for each of the documents listed below:

- a travel folder containing all the documents your boss requires for a two-week trip overseas;
- a long report of 20 pages on health and safety issues in your office to be handed to your managing director;
- an advertising campaign of 10 pages, including statistics, graphs and illustrations. This is to persuade large distributors or large retail stores (not consumers) to stock your product. How you decide on the presentation and the binding of your documents depends on the type of product you choose;
- a five-page legal document to be sent to a client;
- a folder to contain your curriculum vitae (résumé). Again, this will depend on the type of job you are applying for and the kind of skills the position is demanding.

Overhead projector

An overhead projector (see Figure 3.13) is useful for sales meetings and lectures, where lecture notes and images are projected onto a screen. Maintenance includes replacing the lamp and keeping the lens and glass clean.

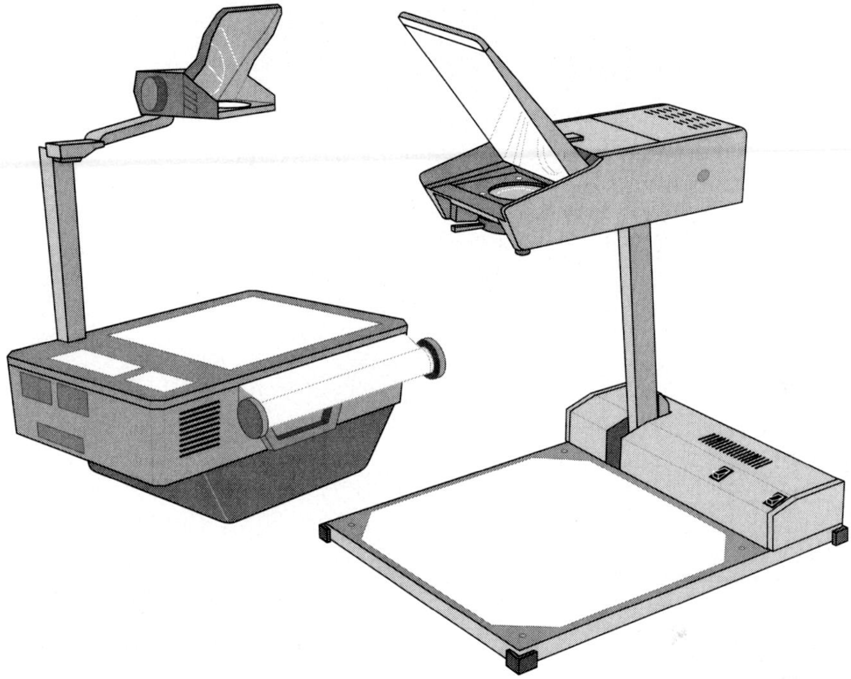

Fig. 3.13 Overhead projectors

Copy board

A copy board is an electronic whiteboard with a built-in printer, which copies notes using thermal paper.

Shredding machine

There are various sizes and makes of shredding machines (see Figure 3.14). These are used to shred paper from confidential dead files into strips that are almost impossible to read. The shredded paper can be recycled.

Calculator

There are many sizes and types of calculators, ranging from pocket calculators to desk calculators (see Figure 3.15).

Fig. 3.14 A shredding machine

Fig. 3.15 Two examples of calculators

Electronic typewriter

Although computers are widely used, electronic typewriters are still in use for simple tasks.

Laminating machine

Laminating is the process of coating paper with transparent plastic. This is useful where a document receives considerable handling, such as for identity cards, driver's licences, price lists, photographs and posters. See Figure 3.16.

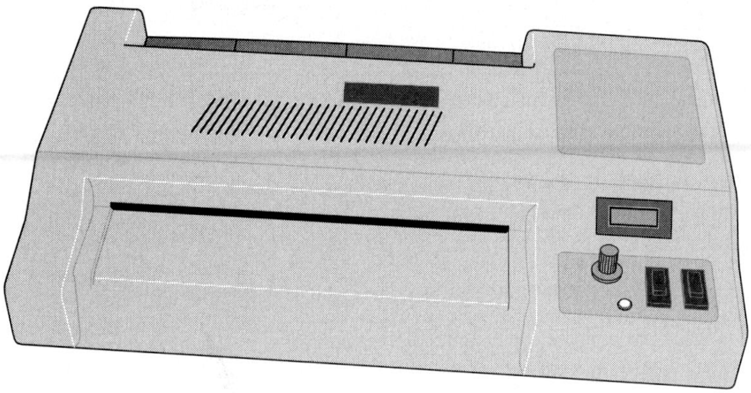

Fig. 3.16 A laminator

Lettering machine

Lettering machines produce words and symbols on a plastic strip, which is then fixed to shelves or folders. This gives a professional appearance.

Dictating and transcription machine

Electronic dictating machines are now very small and can be hand-held when dictating letters and other documents, for use in any location. The cassette can be posted back to the secretary for transcription if necessary.

Refer to the manufacturers' manuals for operation and maintenance of all the above equipment.

Activity 3.7

Office equipment routine tasks

With other members of your group, carry out the practical tasks outlined below. These will be observed by your tutor/supervisor, and an assessment/observation sheet is provided to record the completion of these tasks.

Binding machine
Check that the binding machine is connected correctly according to occupational health and safety practices. Refer to the manufacturer's instructions and bind together the documents you produced in Activity 3.4. Report any faults on the facsimile form (sheet 3.4 in your workbook).

Overhead projector
Check that the overhead projector is connected correctly according to occupational health and safety practices. Refer to the manufacturer's instructions and perform the following operations.

1. Clean and set up the projector.
2. Before you switch on the projector, find out how to replace the lamp.
3. Project the transparency onto the screen and adjust the focus.

Report any faults on the facsimile form. When you have finished, use the correct shutdown procedures.

Facsimile machine
Check that the facsimile machine is connected correctly according to the manufacturer's instructions and complete the following operations.

1. Before the machine is switched on, check that there is sufficient paper (roll or sheets) in the tray.
2. Complete the details on the facsimile form and send it to the number supplied by your tutor/supervisor.

Report any faults to your tutor/supervisor, and when finished use correct 'shutdown' procedures.

Shredder
Shred any unwanted or spoiled paper.

Case study 3.1

Obtaining the right equipment

You are the administrative coordinator for Business Stationery Supplies Pty Ltd. The photocopier used in the general office has a high volume of production and has been in use for a number of years. It has developed many faults requiring the attention of the maintenance mechanic, which wastes time and costs money. You feel that a new one is required urgently.

You know that the chief accountant is very careful when spending money and will have to be persuaded to buy a new photocopier. You are going to speak to her tomorrow and need to be well prepared, so you write down the important items you will discuss with her.

(continues)

1. List the reasons you will give to persuade her to buy a new photocopier.
2. To obtain her permission you will tell her that you intend to carry out a feasibility study to discover which particular model, type and size will be the best one for your requirements, and how you could save costs. What would you include in your feasibility study?
3. Give her the reasons for your choice of buying, renting or leasing the equipment.
4. You will also tell her that you will carry out a supplier analysis to obtain a photocopier on the most favourable cost conditions. What factors would you take into consideration when carrying out a supplier analysis?

OFFICE SUPPLIES

In order to operate office equipment it is necessary to keep a stock of stationery supplies. In a large organisation there may be a centralised system of storing office supplies, but you will still need a small stock in your office for immediate use. Listed below are the more common items of stationery supplies:

Supplies for use in office equipment
photocopy paper
photocopier toner
facsimile letterheads or rolls
letterheads
printer ink cartridges
floppy disks and CDs
imaging film rolls
typewriter cassettes

Supplies for use in the office
memorandum forms
writing pads
staplers and boxes of staples
telephone message pads
correction supplies
scissors
rulers, pens, pencils
notepads
paper clips
compliments slips
suspension files and other filing supplies
manila folders

Office supplies for a range of specific tasks

The range of supplies you store will depend on the type and size of your business. In addition to the common items of stationery supplies mentioned above, there will be special items that relate to the specific type of business.

- An **insurance company**, for example, will require quite a lot of space to store brochures describing insurance products; application forms for each product; claim forms; and other relevant documentation.
- An **accounts department** may store journal and accounts ledgers; payroll summary sheets; and wage advice slips.

- A **law firm** will store a special weight and type of paper for reproducing legal documents; a supply of standard legal documents; special folders and envelopes; and legal tape for storing and sending legal documents.
- A **travel agent** stores travel brochures on stands, and the brochures will require ordering and replacing as customers take them. A basic stock of special forms used for confirming booking of hotel rooms, car hire and travel insurance will also need to be stored.

Activity 3.8

Special office supplies

Form a group of three, then each member should select three different business organisations from the *Yellow Pages*. In the space provided on Activity sheet 3.8 in your workbook, make a list of the special office supplies that would be required for each organisation.

Different types of paper

The weight of paper is given in grams per square metre (gsm). The average weight of office paper is 73 gsm. A pack of paper is 1000 sheets (two reams). Heavier-weight paper is better quality. The sizes and types of paper required for different tasks are listed below.

Bank paper. 44 gsm bank paper is usually used for file copies of original letters, or for other office tasks that require paper of medium quality.

Bond paper. Bond paper is heavier than bank paper, and is used for letterheads, reports, forms, photocopying and other documents.

Onion-skin paper. This is used for second sheets because of its strength. It stands up to file retrieval better than bond or bank paper, and is available in two weights.

Parchment and Cambridge paper. These are heavy-quality and expensive papers. They are used only for personal letterheads of top executives or for special jobs.

Laser paper. This is specially designed for high-speed laser printers, desktop laser printers, and high-speed photocopiers.

Recycled paper. This is made out of paper that has already been used. Recycling bins are provided in large organisations for used paper, which is taken away and reprocessed, usually without bleaching, into reusable paper. Because it is unbleached it is grey, and cannot be used for top-quality purposes, but it can be used in many other ways. The use of this paper helps to conserve the environment.

Paper sizes. The most common size of paper is A4, which is used for letters and many other documents. Some paper sizes are illustrated in Figure 3.17.

Envelopes

There are many different sizes and types of envelopes for use in different tasks. Some are listed here:

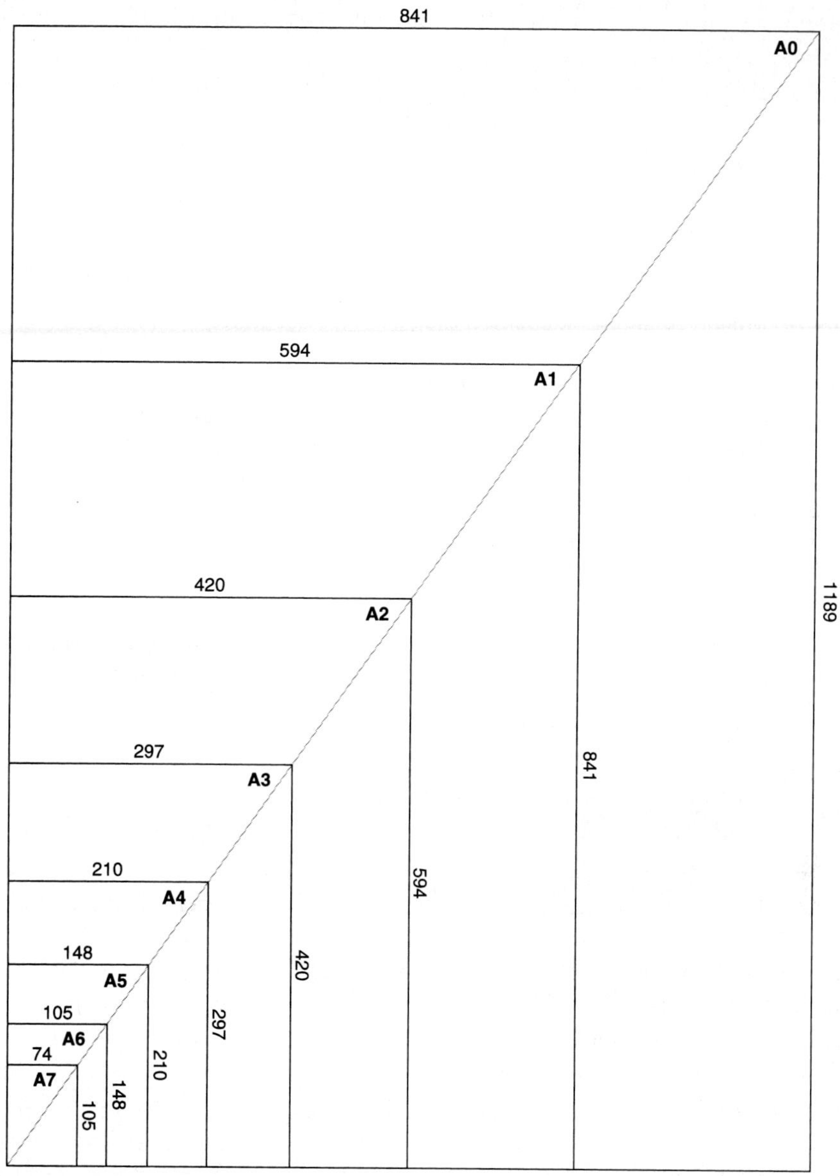

Fig. 3.17 Paper sizes (in millimetres)

- white envelopes: (lick and stick, self-seal, cartridge) for sending high-quality documents;
- manila envelopes: for lesser-quality purposes;
- window-face envelopes: for sending invoices and statements and bulk documents;
- airmail envelopes: usually made from lighter paper to keep costs down when sending letters overseas;

- carton envelopes: self-adhesive, for protecting documents (usually an invoice) attached to a parcel;
- internal envelopes: can be used up to 40 times for distributing mail within the organisation; help to conserve the environment;
- jiffy bags: padded envelopes for use when sending items that require protection.

Centralised supplies

Large organisations will have a centralised store of office supplies, with individual departments or offices obtaining supplies through a stock requisition. The *advantage* of centralised stock control is that stock may be ordered in volume (using economies of scale), enabling the purchaser to receive discount prices.

Disadvantages. The disadvantages are that you have little control over:

- the quality of items you order;
- any specialist items you may require;
- time delays for items urgently required;
- flexibility and adjustment to particular needs;
- negotiating agreements to meet individual needs.

If you are working for a small organisation you may have the responsibility for ordering stock direct from suppliers. It may also be your responsibility to store and control these supplies.

Activity 3.9

Office supplies

You are starting your own small business, and will need a small store of basic stationery items as listed here:

- A4 copy paper—white
- A4 copy paper—different colours
- suspension files
- crystal tabs
- pens, pencils, rulers etc.
- fax rolls
- paper clips
- computer disks.

Form into a small group and visit three of the following stationery suppliers: your college stationery shop, a large stationery supplier, a small stationery supplier, a supermarket, a large store.

When you visit each supplier, make a note, against each of the items above, of the following:

- the quality (poor, good, excellent);
- the minimum quantity you may buy (and the quantity in a pack or box);
- the cost for one item and the cost for a pack or box.

After each group has completed the above activity, bring this information to your next class. Analyse and discuss the quality, quantity and costs with other groups.
Write down which supplier you would prefer to buy your supplies from, and why.

Control and maintenance of stationery supplies

The expense of storing and maintaining stationery supplies is high, and some method of controlling stock is required. There are three ways of doing this.

The open office supplies room

In an open supply room, staff may enter and take items as required. A member of staff is given the responsibility of stocking, maintaining and reordering supplies.

Advantages

- There is no delay for individual staff requiring supplies.
- There is less administration time and paperwork needed.

Disadvantages

- Items of stock are used more quickly through wastage and pilferage.
- The person responsible for stock control must check supplies more often.
- On-the-spot demands for large quantities may cause stock shortages.

The closed office supplies room

In a closed supply system the room is locked, and staff obtain supplies by requisition to the person responsible for stock control. Requisitions are attended to at specified times.

Advantages

There is:

- more efficient record-keeping;
- reduced wastage and pilferage;
- more efficient reordering.

Disadvantages

There is:

- a delay for staff in obtaining supplies;
- more administration and paperwork.

The controlled office supplies room

In the controlled supply room, members of staff enter the room and fill in a requisition for stationery supplies under supervision of the supply officer. They are allowed to take the supplies immediately.

Advantages
There is:
- no delay in staff obtaining stationery supplies;
- efficient stock control through on-the-spot checks of stock taken;
- easy recording of costs for each department;
- reduction in wastage and pilferage;
- monitoring of the system by the supply officer.

Disadvantages. There are few disadvantages to this system, provided stock levels are maintained. It combines the advantages of the first two systems without the disadvantages. An exception is when the supply room is open only at set times, in which case there is some delay in staff obtaining supplies.

Stock control

Control and maintenance of stationery supplies should be carried out at least once a week. This is done using an inventory control.

Perpetual inventory

Computer-controlled perpetual inventory systems are used by department stores. This is operated at the checkout when articles are bought. Barcodes are scanned with a laser reader, which automatically deducts the article from the total balance of the inventory. The computer keeps a record of minimum and maximum stock levels and, when these fall below the basic stock level, makes a note for reordering.

Using stock-control cards for office stationery is a simple but effective method of perpetual inventory for small quantities of stationery items. Each time a person takes an item of stationery out of the store, the date and amount taken is recorded and deducted from the balance in stock. The balance shown should be the amount left on your shelves.

Activity 3.10

Stationery supplies store

In a group, arrange to visit the stationery supplies store of your college, school or work. Discover the following details of how the store is controlled and maintained, and make notes about them:

- the kind of special supplies the store needs to stock for this particular business;
- the size of the supplies room;
- whether it is centralised or decentralised;
- the method of storing stock (open, closed etc.);
- the method of controlling stock.

This could be developed into a group project for Chapter 12.

Terminology for stock control

Lead time level. This is the amount of stock used during the period between placing an order and receiving the goods (the minimum level of items to be kept in stock).

Safety stock level. This is an amount of stock kept in case there are unexpected delays in the delivery of goods from the suppliers.

Basic stock level. This is the amount of stock required for work already commenced in the office.

Desired inventory level. The equation for this is:

lead time level + safety stock level + basic stock level = desired inventory

Example:	lead time level	=	10 reams (minimum stock level)
	safety stock level	+	5 reams
	basic stock level	+	10 reams
		=	25 reams (maximum level) desired inventory

Desired order quantity. This is the quantity of stock to order, which minimises the quantity of stock that should be held to prevent overstocking, but is sufficient to allow for lead times and unexpected delays.

The equation for desired order quantity is:

desired inventory – current stock = order quantity

Example: desired inventory = 25 reams – 5 reams = 20 reams (order quantity)

Just-in-time ordering (JIT). This method is used to order items just before they are required, which minimises the levels of stock stored. Car manufacturers use this system, as parts for cars are expensive and huge amounts of capital could be tied up in stock. One disadvantage of this method is that it does not allow for delays in delivery of goods.

Storage of stock

It is important to store stock in the correct way so that items are easily accessed, are not damaged, and do not cause health and safety hazards. Some examples of storage methods are:

- a small room with fitted shelves (if there is a large volume of supplies);
- a compactus (if there is a medium volume of stationery supplies);
- a steel cupboard (for small quantities).

The same rules for storage apply to all methods.

- Do not store anything on the floor of a storage room, as certain articles could be damaged by damp or flooding.
- Use a ladder to reach high shelves. Do not stand on boxes or swivel chairs.
- Do not leave floppy disks and other computer equipment where they could be damaged by heat or light.
- Store heavy items on the bottom shelves so that they are easier to lift out.
- Small items should be placed in boxes so that they do not become scattered over shelves or knocked onto the floor.
- When new supplies are received, place these at the back of existing stock so that old stock is used first.

- Make sure there is space between items so that they are not damaged.
- Keep everything visible and clearly labelled.
- Keep similar items together, or all items together that are used by a specific department or section (e.g. financial documents or sales documents).

Reordering of stock

Once certain items reach their minimum level it is time to bring them back to the maximum by reordering these items. There are many ways of reminding yourself when stock needs to be reordered. If you use the perpetual inventory system, either the computer will tell you when to reorder or the balance on your stock-control card will indicate when to reorder. If you do not use either of these methods, you can place a coloured card at the minimum stock level to remind you when to reorder. There are three ways in which you would reorder stock, as follows.

Purchase order. Stock for a centralised store will be bought from a supplier through a purchase order (see Figure 3.18). Savings can be made if large quantities of supplies are ordered in advance, but storage can be costly to the business through use of space, wastage and tying up capital.

Purchase requisition. Stationery supplies for a department or office can be ordered from the central store through an internal stores requisition form (see Figure 3.18).

Direct purchase. If special items are needed that are not kept in the stationery store, these may be ordered direct from a supplier. Catalogues are provided by the suppliers illustrating a complete range of the stationery and equipment available. A copy of one or two of these catalogues or 'buyers' guides' should be kept in the stationery supplies store.

Business Stationery Supplies Pty Ltd

163 Collins Street Melbourne Vic 3100	404 Rutherglen Road Sydney NSW 2100	234 Waverly Road Brisbane QLD 4100	604 Gordon Street Perth WA 6100

PURCHASE ORDER

Date: .... ORDER No: ....

Deliver to:

Please supply the undermentioned goods:	Quantity	Price

Please quote order number on delivery note

INTERNAL STORES REQUISITION

DESCRIPTION —Order alphabetically as per catalogue	QUANITITY	STORE USE ONLY	
		ISSUED	VALUE

	DEPT. No.	ORDER. No.	ACC. No.	AMOUNT

Department .... Date ....

Authorised .... Received ....

Fig. 3.18 A purchase order form and an internal stores requisition

Activity 3.11

Stock control

One of your tasks is to control and maintain stationery supplies for the offices of Business Stationery Supplies Pty Ltd.

In your workbook there is a stock control card for A4 white copy paper (Activity sheet 3.11 (1)) and a purchase order form (Activity sheet 3.11 (2)). The maximum stock level for this item is 250 reams and the minimum stock level is 50 reams. Enter the following quantities taken out on the dates indicated. Work out the perpetual inventory and indicate the final balance of stock. Indicate also when you would reorder stock. Enter the amount and other details on the purchase order form.

1.	3 January	30 reams allocated to administration department
2.	6 January	10 reams taken by managing director's secretary
3.	9 January	45 reams allocated to sales department
4.	10 January	25 reams taken by purchasing department
5.	2 February	20 reams taken by administration department
6.	7 February	40 reams allocated to computer section
7.	14 February	6 reams taken by sales secretary
8.	20 February	4 reams taken by sales representative

Case study 3.2

Stock control of office supplies

You are the administrative coordinator for Business Stationery Supplies Pty Ltd. As your organisation is a wholesale distributor of stationery supplies to retail business, supplies of stationery items required internally by each department are obtained from the warehouse individually, as and when required, through an internal requisition form. If special items are required (which are not supplied by the warehouse) they are ordered direct by individuals from an outside supplier.

You have received an internal memorandum from the chief accountant today which states that this method of controlling office supplies is inefficient and far too costly. She has asked you to set up a centralised office supplies system to cut costs of stationery supplies, and to let her have a plan of this as soon as possible.

There are five departments in your business: sales, purchasing, accounts, human resources (personnel), and administration. The personnel in each department include: managers, executive assistants (sales representatives in the sales department), personal assistants, word-processor operators, receptionists and clerks.

Write your plan and include the following:

(continues)

1. the size and type of storage area you would use for this business;
2. a diagram of the shelves showing the correct method of stacking items of stationery;
3. the method you would use to control and maintain supplies ('open', 'closed' or 'controlled'); and, depending on your choice, the number of staff required to operate the supplies room;
4. the method of stock control you would use;
5. the method you would use to reorder stock from the warehouse and for special supplies direct from an outside supplier.

Summary — Office equipment

- A hazard is anything that could cause harm when using equipment in an office.
- According to the *Occupational Health and Safety Act* covering work practices in each state, an employee shall not wilfully or recklessly misuse equipment or place at risk any person in the workplace.
- For all electrical equipment, always use shutdown procedures at the end of the working day (as outlined in the manufacturer's manual) to prevent damage to equipment and the risk of fire.
- Considerations that should be taken into account when obtaining office equipment include: specifications; cost analysis; whether to buy, rent or lease; 'green' technology; reliability of the equipment; and service maintenance.
- You should use a checklist of items before you buy a computer and after you have bought it to help you obtain the correct computer for your particular needs.
- The Internet is a network of computer networks composed of many individual networks linked all over the world.
- In order to join the Net you will need to choose an ISP (Internet service provider).
- Web browsers and search engines help you to find information quickly and easily.
- Before using any item of equipment, an operator should be trained to use the equipment correctly and to use individual machines according to the manufacturer's user guide manual.
- Computer maintenance includes providing protection from power surges and static electricity, and from excessive heat, moisture and dust.
- Floppy disks should be stored in special boxes to prevent corruption through exposure to liquids or excessive heat and sunlight.
- The printer is the most common output device from a computer. There are two categories of printers, impact (dot-matrix) and non-impact (inkjet, thermal and laser).

- Photocopying machines have many computerised parts and, in order to protect the machine and operators, there are several operating precautions that must be observed, including efficient air circulation and space in which to operate.
- Routine maintenance includes replacing copy paper and toner, and keeping equipment clean. Non-routine maintenance includes identifying faults that involve the replacement of parts. It is important to know how to clear paper jams.
- Facsimile machines have revolutionised communication between organisations within Australia and internationally. Electronic mail permits the almost instantaneous transmission of letters, memos and other documents, from one terminal or computer to another, in a different office, city, state or country.
- There are various methods of binding documents. Two types of binding machines are those using spiral and thermal methods.
- Other common items of equipment used in an office include the following:
  —telephone system
  —answering machine
  —overhead projector
  —copy board
  —shredding machine
  —calculator
  —electronic typewriter
  —laminating machine
  —dictation and transcription machine.

Review questions

(These can be used for group discussion or for formal assessment.)

1. List the reasons why it is important that you observe health and safety precautions when using electrical equipment.
2. Outline briefly shutdown procedures for electrical equipment.
3. List the health and safety procedures when using: a computer, a photocopier, a fax machine, a printer.
4. Explain briefly why you should analyse your reasons for buying equipment.
5. What is 'green' technology?
6. List the advantages in renting/leasing office equipment.
7. What is a service maintenance agreement?
8. What are the factors to be considered when choosing a service maintenance agreement?
9. An inventory should be kept of all office equipment. List what you should include in this inventory.

10. Describe the different kinds of printers.
11. List the factors you should take into account when choosing equipment for your office.
12. When you are buying a computer for the first time, why should you 'shop around' and ask a lot of questions? List as many reasons as you can think of. What should you do before you buy and after you buy?
13. What is the Internet?
14. What is a network and how do you join the Net?
15. Why are web browsers and search engines useful when using the Net?
16. Write a paragraph outlining the reasons why it is vital to take security measures when using a computer and the Internet.
17. List the maintenance procedures used for computers, including the storage of floppy disks.
18. Describe copyright legislation and why it is necessary.

Summary — Office supplies

- To enable office workers to complete their tasks efficiently, it is necessary to keep a stock of stationery supplies. In a large organisation there may be a centralised system where all supplies are stored and controlled from one location.
- The range of supplies you need to store will depend on the type and size of your business.
- Different types and sizes of paper include bank, bond, onion-skin, parchment, Cambridge, laser and recycled paper.
- There are different ways of controlling and maintaining stationery supplies. These include the open office supplies room, the closed office supplies room, and the controlled office supplies room.
- Stock levels are controlled using 'lead time level', 'safety stock level', 'basic stock level','desired inventory level','desired order quantity', and 'just-in-time' ordering. A popular method using either computer or cards is the perpetual inventory.
- It is important to store stock in the correct way so that items are easily accessed, are not damaged, and do not cause occupational health and safety hazards.
- Reordering of stock may be done in one of three ways—by purchase order, purchase requisition, or direct purchase.

Review questions

(These can be used for group discussion.)

1. List the different types of paper and explain briefly their uses.
2. Why is it important to use recycled paper?
3. List the different types of envelopes and explain briefly their uses.
4. Describe briefly the three different ways of controlling and maintaining stock, outlining the advantages and disadvantages of each system.
5. Explain briefly the perpetual inventory system.
6. List the terminology used for stock control.
7. List three methods of storing stationery supplies.
8. Explain the rules for storing office supplies.
9. What are a purchase order and a purchase requisition? When are they used?

Competency self-assessment and/or tutor/supervisor assessment

When you have finished reading this chapter, and have completed the activities chosen by your tutor/supervisor (**practical in class, at work or theoretical**), fill in this self-assessment. Then hand it to your tutor/supervisor for official assessment.

Learning outcomes	**Self-assessment**		**Tutor/supervisor assessment**	
I have successfully completed the following learning outcomes	**Yes** (please tick)	**No**	**Yes** (please tick)	**No**
1. Office equipment—routine tasks				
Use relevant occupational health and safety practices				
1.1 demonstrate correct power access procedure	☐	☐	☐	☐
1.2 follow safety procedures recommended in the equipment manuals	☐	☐	☐	☐
2. Operate a range of office equipment to complete routine tasks				
2.1 select appropriate equipment for the task (photocopier, printer, fax etc.)	☐	☐	☐	☐
2.2 locate and check for availability	☐	☐	☐	☐
2.3 use in accordance with manufacturers' instructions	☐	☐	☐	☐

2.4 identify any equipment faults and report to the nominated person	☐	☐	☐	☐
2.5 follow shutdown procedures	☐	☐	☐	☐
2.6 introduction to the Internet				
• joining the net and choosing an ISP	☐	☐	☐	☐
• understanding the Web, web browsers and search engines	☐	☐	☐	☐
• glossary of basic computer and Internet terminology	☐	☐	☐	☐
• security measures on the Web and the Internet	☐	☐	☐	☐
3. Office equipment—non-routine tasks				
Operate a range of office equipment to complete non-routine tasks				
3.1 use various functions of the equipment simultaneously	☐	☐	☐	☐
4. Identify and rectify or report minor faults				
4.1 undertake routine maintenance according to manufacturers' operating instructions	☐	☐	☐	☐
4.2 identify equipment faults accurately and describe to nominated person/section	☐	☐	☐	☐
4.3 rectify minor faults following manufacturers' operating instructions	☐	☐	☐	☐
5. The implications of copyright legislation as applied to an office				
5.1 describe the purpose of the Copyright Act	☐	☐	☐	☐
5.2 explain the implications of the legislation for office copying	☐	☐	☐	☐
6. Apply recycling techniques to minimise paper wastage	☐	☐	☐	☐
7. Bind documents	☐	☐	☐	☐
8. Office supplies—stock control				
• office supplies for a range of specific tasks	☐	☐	☐	☐
• controlling and maintaining office supplies	☐	☐	☐	☐
Date:		Tutor/supervisor signature:		

section two

ENTERPRISE

CHAPTER

4

OFFICE ROLE AND FUNCTIONS

ENT101A, NOS120, 7365J

objectives

The purpose of this chapter is to help you to identify the role and functions of an office as an integral part of a specified organisation, in order to complete the following learning outcomes:

1. Identify the role and functions of the office as an integral part of a nominated enterprise

- 1.1 describe the operations commonly carried out in an office
- 1.2 list the titles of common office positions and give a brief description of the duties of each position
- 1.3 describe the prime function of a nominated enterprise
- 1.4 identify and describe the key functions of the office/departments/key personnel within the nominated enterprise

2. Discuss the role of organisational and office procedures

- 2.1 explain the importance of using a procedures manual and adhering to prescribed office and organisational procedures
- 2.2 use a procedures manual to determine designated person/s to address specified irregularities and/or difficulties for resolution
- 2.3 identify an actual or desirable workplace practice/procedure which assists conservation and the environment

INTRODUCTION

We all have frequent contact with offices. We use the services of a bank regularly; telephone our insurance company regarding our policies; call in to the gas or electricity office to pay our bills; or phone Telstra with a query about our accounts. When we visit a real estate office, we meet people at the reception desk and see people in offices using computers and answering telephones. What are these people actually doing? Why does the business need an office, and what role does the office play in the running of the business?

WHAT IS AN OFFICE?

An office is the area of a building where people—using materials, tools and resources—perform the administrative services required to support the main functions and goals of a business. This could be a well-appointed, luxurious area located in a large organisation in the city, or it could be a very basic area in a corner of a market garden, a school or a club. If you are a salesperson, your office could be your car. Your documents might be stored in the boot and you might use a mobile phone and laptop computer (linked to the computer LAN system at the main office) to communicate with customers.

Part of your home becomes an office (perhaps the spare bedroom or living-room) when you sort and pay your bills and file documents away in a concertina file or a cupboard.

The electronic office

With today's electronic communications systems, an office is not necessarily in one place. More people are working from home, linked to their employer's main office through computer networks. This method of office work may become more common in the future, with the increasing administrative costs of the traditional office and with the development of innovative information technology.

Remote-control software lets users access another PC via dial-up networking (**DUN**), direct dial, or network connections (including the Internet). After remote users connect, they can perform file transfers, troubleshoot systems, run remote applications, and train another user sitting in front of the remote computer. Other features include voice chat, text chat and disk cloning. There are additional remote-control features, such as support for remote printer direction, multiple remote sessions, data encryption, colour scaling to increase performance, and screen blanking the host server for security reasons. Remote access used to be relatively slow, but a recent introduction of universal serial bus (**USB**) means that what used to take half an hour now takes only about one minute.

E-commerce. Electronic data interchange (EDI) (computer-to-computer transactions) means that business-to-business commerce over the Internet will grow rapidly in the near future. This is already big business in the USA, and E-commerce for the consumer is starting to take off in Australia, when we buy goods over the Internet. We already pay telephone and electricity bills with credit cards over the telephone. We will shortly all be able to pay bills over the Internet with the introduction of 'electronic bill presentation and payment (EBPP)', which has already been put in place by one or two corporations with huge numbers of customers to bill.

The electronic remote-access office

All this means that the traditional office of the present day is going to change to a more electronically controlled and remote-access office (see Figure 4.1). Already many businesses have information stored on a computer system at the reception desk, for easy access to customers' records and/or stock control of products. You will need information technology training if you want to obtain a position in an office.

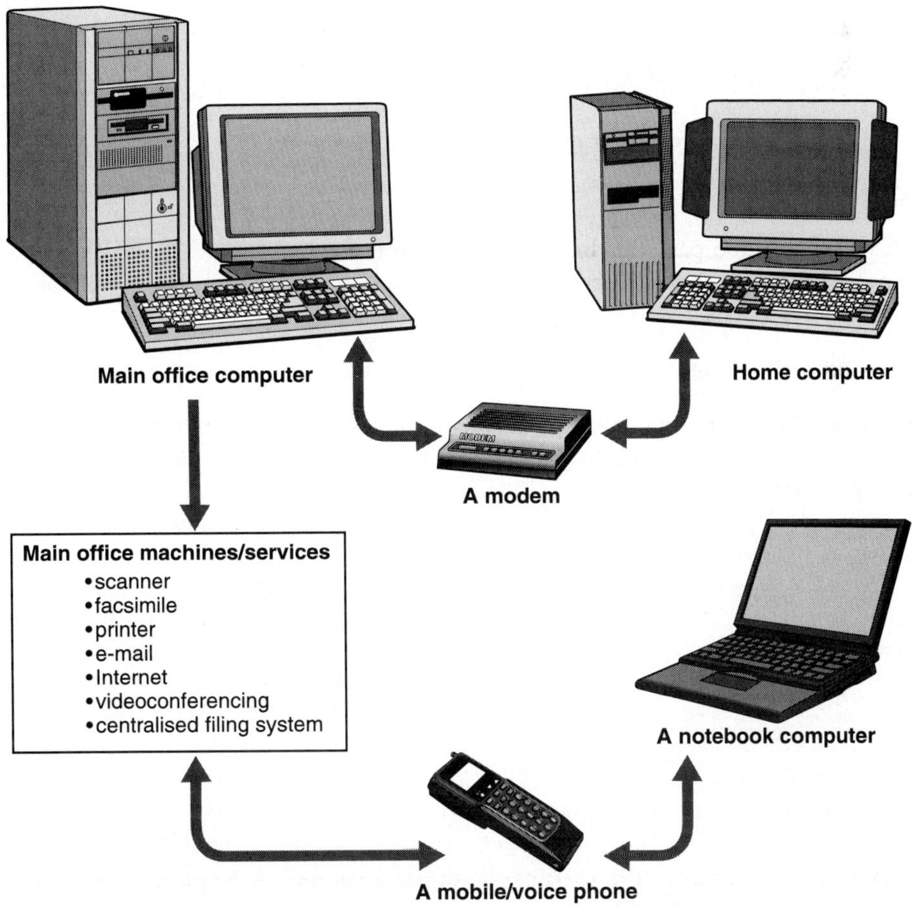

Fig. 4.1 Remote-access computer systems

The components of an office

Wherever an office is located, or whatever the size—whether it is a traditional office or an electronic remote-control office—there are four basic components of all offices:

- **people**, who organise, communicate, solve problems and perform numerous other functions;
- **records/data**, which are created, collected, calculated, analysed, sorted, stored, reproduced, retrieved, and used in a variety of other ways;
- **tools**, such as computers, word processors, cash registers, photocopiers;
- **other resources**, such as management of time and money, office space, office supplies.

Office personnel, using these records, tools and other resources, perform tasks that fulfil the functions of an office.

THE ROLE, FUNCTIONS AND OPERATIONS OF AN OFFICE

The major roles of an office

An office performs three major roles, as follows.

Secondary role

The office functions support the primary concern of the business, which may be either the manufacture of goods or the provision of services, without which there would be no need for an office.

Complementary role

Although the role of the office is secondary, without its services the business could not survive. The office complements the other activities of the business. For example, office services include:

- ordering supplies;
- paying wages;
- the communication process required to provide goods or services to consumers.

Controlling role

An office controls operations connected with budgets, stock, finance and human resources management.

The functions of an office

Office functions support the goals of the organisation and each specific department (Table 4.1 gives examples). Office operations are the tasks that are performed in order to carry out the office functions.

Office functions ⟶ support ⟶	Business functions
Communicating with customers	Marketing goods and services
Managing staff	Production of goods and services
Managing finance	Controlling finance
Processing data	Human resources management
	Administration management

Table 4.1 Office functions and business functions

The operations carried out in an office

The work you perform in your office will depend on your position in the organisation and the size and type of business of that organisation. To carry out the operations allocated to you that support the functions of the office, you will need to be trained in three areas.

1. Operating office equipment correctly, in order to:
   - photocopy documents,
   - use a facsimile (fax) machine,
   - use the telephone system,
   - bind documents,
   - use the computer and word-processing systems,
   - handle, process and retrieve documents.
2. Understanding the processes involved in organisational procedures, which include:
   - identifying and solving routine problems,
   - implementing the correct solutions,
   - completing forms,
   - making out invoices,
   - ordering office supplies.
3. Communicating effectively within the office and with customers and suppliers, which requires you to:
   - know the hierarchical set-up of your organisation,
   - know who is responsible for which section of your business,
   - know the products and services of your organisation,
   - use the correct method of replying to letters and internal memos,
   - know the correct methods of using verbal and non-verbal communication,
   - know how to handle difficult customers,
   - organise meetings and travel arrangements.

Activity 4.1

Office roles and functions

Form into a group and visit an office in your college, school or workplace. Take notes on the role each person holds in the office, the work they perform and the equipment they use. Arrange a brief interview with each person, if this is possible. Your notes should include the following information:

- What role does this office perform and what is the primary concern of the business?
- How many people work in the office?
- What is the role and position of each person?
- What are the duties of each position?
- What are the items of equipment required to complete the tasks for each position?

THE ORGANISATION

Prime function of an organisation

The prime function of an organisation is the purpose for which it is in business, namely:

the manufacture of specific goods or the supply of specific services.

The position you hold, and the functions and the operations carried out in the office, will depend on the level of industry and the size of the organisation you work for. This will be determined by the organisation's prime function and the classification of the industry in which it operates.

Organisation structure

An organisation is a planned arrangement of operations. To achieve the smooth-flowing operation of everyday tasks, preparation is required. Ideally, all employees are doing what they are qualified and trained to do in order to achieve their own goals and ultimately the goals of the business. There are seven areas of operation necessary for an organisation to be effective.

Goal-setting. The goals of an organisation must be made clear to all personnel so that each person is working towards the same goals.

Unity of functions. There are five functions within an organisation:

- production of goods or services,
- marketing goods or services,

- controlling finance,
- human resources management,
- administration management.

These are interrelated, so that people in all departments work towards the same goals.

Work assignment. Everyone will have assigned tasks, with job placement matching the abilities and skills of each person.

Chain of command. There must be a formal statement of who reports to whom within the business.

Unity of command. You, as an office worker, should receive orders from, and be responsible to, only one supervisor; otherwise instructions may become confused, and this will affect the practical aspects of your position and your morale.

Authority commensurate with responsibility. When you are given the responsibility for doing a specific task, you should also be given the authority needed to complete that task.

Span of control. Each supervisor should control a specified number of workers, depending on:

- the nature of the work performed,
- the ability of the supervisor,
- the nature and number of his/her responsibilities,
- the position of the supervisor in the hierarchical structure.

Centralisation or decentralisation of an office

With the growth of computer systems there is a trend towards centralisation of offices. Centralisation means that office records or services operate from one location.

Main advantages of centralisation. These are:

- uniformity in methods, procedures, documents, forms and wage rates;
- minimum duplication of files and work;
- workers assigned from the central location to areas where and when they are needed;
- authority and responsibility concentrated in one supervisor.

Main disadvantages of centralisation. These are:

- delays in access of information;
- no individual choice of system or procedure.

Decentralisation is when all staff members keep records near their work station, creating and storing data and using the system most appropriate to their needs.

Main advantages of decentralisation. These are:

- user control and local decision-making;
- those handling the system being familiar with procedures, so fewer errors are made;
- no delays in handling and processing;
- more flexible system.

Main disadvantages of decentralisation. These are:

- less uniformity in methods, procedures and documentation;
- more duplication of files and work;
- less quality control;

- decrease in security, as no one person or persons have authority over the system;
- higher costs.

Management structure

For you to work effectively, you need to know what levels of management there are in the hierarchical structure of your organisation; who is responsible for which department and its functional operations; and how your position fits into this structure. The management structure of a large organisation is based on three levels: top management, middle management, and operating-level management. These are supported by departmental operation managers and supervisors.

- **Top management** consists of the chairperson, the managing director and other directors, and general managers, who specialise in areas such as finance, production and marketing. They make decisions on company goals and policies.
- **Middle management** consists of managers who manage other managers and supervisors. They develop departmental goals and policies in line with organisational goals.
- **Operating management** consists of managers and supervisors who put management's plans into action and are the link between higher levels of management and office and other workers.
- **Departmental operations** are controlled by the managers of each department. The departmental operations level is where all other personnel perform their functions. This forms the base on which the whole organisation operates and on which top management depends for productivity and support.

The organisation chart. The organisation chart includes the chain of command, from top management down to middle management and departmental operations. It shows the title and position of whoever is responsible for which department and to whom they are responsible.

In order to help you to understand the chain of command and the role, title and duties of personnel in a medium-sized business, a company has been created for use in this book for practical group projects. The company is Business Stationery Supplies Pty Ltd, 163 Collins Street, Melbourne, Victoria 3100. It has branches in other states, and its organisation chart is shown in Figure 4.2. The positions described in each chapter are those in this organisation, and the documents you will use in Chapter 2 carry the letterhead, address and telephone numbers for it. The company name will be used in role-play in Chapter 7 and Chapter 8. This will help to create an environment similar to that of a real office.

Lines of authority

The hierarchical structure of an organisation is shown by the lines of authority on the organisation chart (Figure 4.2), with the name and the position of the person responsible for each department or specific area of the business, and who reports to that person. There are three types of authority within an organisation.

- In **line organisation**, authority flows vertically down the line from top management level to operational level ('top down') and responsibility flows vertically up the line from operational level to top management level ('bottom up'). This permits managers and supervisors to control workers immediately below them.

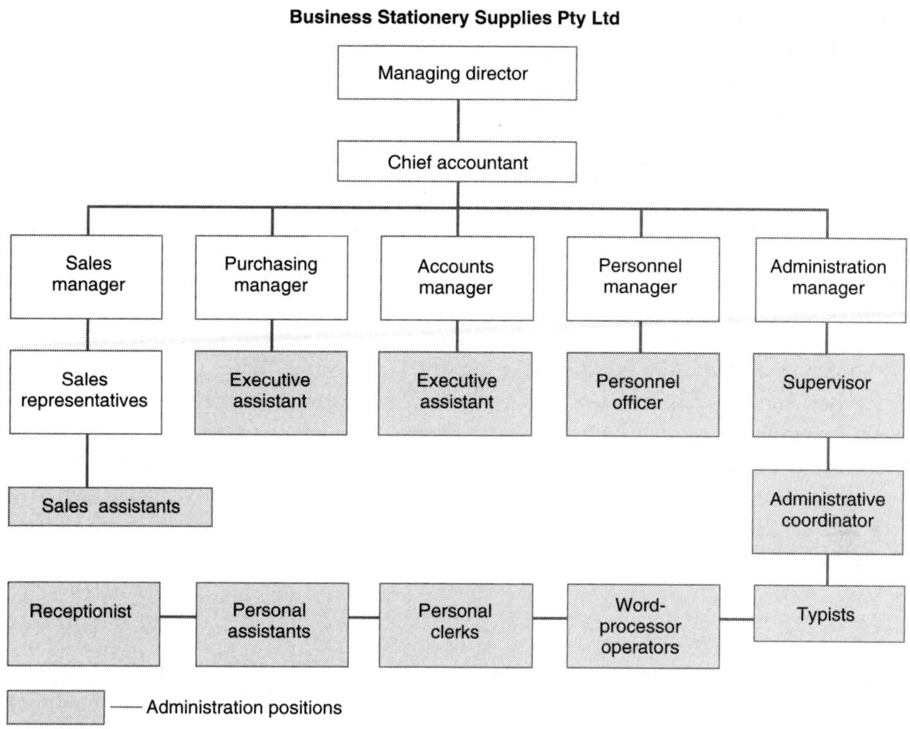

Fig. 4.2 Organisation chart for a medium-sized business (for use in all group projects)

- In **line and staff organisation,** line managers are responsible for key areas and use staff specialists to assist them in specific ways. For example, the accounts manager may use the skills of the personnel department in selecting a new staff member.
- In **functional organisation,** all managers have authority over a pool of workers or human resources (e.g. the typing pool).

The structure within large organisations is usually based on line organisation, with some use of staff organisation and functional organisation.

The structure of any organisation depends on the following:

- the size of the business;
- the form of ownership;
- the industry activity;
- the type of industry classification it belongs to.

These are described in more detail in Chapter 5.

Lines of communication

Lines of communication in an organisation can be either formal or informal.

- **Formal downward communication (top down)** operates in the same direction as the hierarchical structure of the business. Information regarding policies and procedures relating to corporate goals is sent in the form of reports and internal memoranda from top management to middle management down to departmental operations level.

- **Formal upward communication (bottom up)** operates similarly. Staff problems, feedback on product and service problems, and production and sales problems are communicated in the form of reports and internal memoranda from workers and staff through department managers up to top-level management.
- **Horizontal communication** is formal and informal communication between personnel in departments on the same hierarchical level in the organisation. For example, managers of departments communicate with one another regarding production and sales, and administration personnel communicate regarding policies and procedures. The manager of the sales department would need to communicate with the manager of the production department to find out whether the goods are ready for sale, and the chief accountant would need to communicate with the managers of all departments on costs and financial matters.
- **Informal communication** flows on what is called the 'grapevine'. Everyone in an organisation receives some information on what is happening through casual telephone calls or from other personnel in the organisation. This kind of communication spreads rumours that may or may not be correct.

The titles and duties of office positions

The titles and duties of all positions in an organisation will depend on the size and type of business. Table 4.2 shows the titles of key positions, the qualifications and experience, duties and skills required for the position, and the business equipment used in some of the more common positions in a large organisation.

Management decides what operations you carry out in your position through **job analysis**. These duties are described in detail in the **job description** and the **job specification** for each position. (This subject is dealt with in more detail in Chapter 11.)

Title	Qualifications and experience	Duties	Skills	Use of business equipment
Top management level				
Directors Managing director General managers	Higher degree Many years' experience in present position and former similar positions	Determining policy and formulating goals and objectives Controlling capital and expenditure Complying with all statutory requirements (e.g. Factory Act and Occ. Health & Safety) Responsible for the overall management of the business and coordination of all departments	Negotiating Management Interpersonal communication Leadership	Telecommunications Telephone system Computer packages • spreadsheet • diary/calendar • database • electronic mail Dictating machine Fax Calculator Overhead projector

Middle management level				
Department managers	Many years' experience in specialised area of business (e.g. sales, production)	Planning, organising, staffing, leading and controlling the operations of the department towards achieving the organisation's goals	As above	As above
Operating management level				
Office manager Supervisors Administrative coordinators	Business degree Many years' administrative experience	Direction and control of office functions Organisation of office Designing effective systems and procedures Scheduling and coordinating work of section Providing staff training Problem-solving Cooperating with management Supervising staff Ensuring good working environment	Leadership Organisational abilities Good interpersonal communication Good delegation Conflict resolution Good verbal and non-verbal communication	As above
Departmental operations level				
Executive assistant Personal assistant Private secretary	Business degree/diploma Many years' experience in a similar position	Supporting senior management Office management Preparation and production of written communication (internal and external clients) All work of high quality and highly confidential Filing, organising travel, meetings, conferences Making appointments Training new staff	Excellent verbal and non-verbal communication Good English including spelling Good word processing Good conflict resolution	Computer Word-processing packages • spreadsheet • diary/calendar • database Electronic mail, fax Calculator Photocopier Shredder Telephone system Document binder

Receptionist	Good general education Receptionist course/certificate Perhaps some years in similar position	Switchboard duties Filing, clerical Banking, petty cash Making appointments Solving problems Dealing with people (internal/external)	Excellent verbal communication Good conflict resolution Positive, cheerful personality	Switchboard Word processor Calculator Cash register Message pad Appointment book Teledex Word-processing packages Telephone directories Internal directory Fax Franking machine Photocopier
Accounts clerk	Good general education Accounts course certificate Perhaps some years in similar position	Processing invoices Filing, clerical Payroll, banking Bookkeeping Dealing with clients	Good with figures Good verbal/non-verbal communication Computer and spreadsheets Word processing	Computer packages • spreadsheet • diary/calendar • database Photocopier Fax Telephone Shredder

Table 4.2 The titles and duties of office positions in a large organisation

Activity 4.2

Titles and duties of office positions

Reproduced in Figure 4.3 are a number of job vacancy advertisements. Form groups of three and select three vacancies (one from each different organisational level). Using the sheets provided in your workbook for Activity 4.2, fill in the following information for each of the positions (refer to Table 4.2 to help you complete this activity):

- the title of the position;
- the level of the position in the business (upper, middle, operational);
- the prime function of the business;
- the industry classification (primary, secondary, tertiary);
- the operations you think will be carried out in the office for this specific business;
- the qualifications and experience required for the position;
- the duties you think will be performed in this position;
- the skills required for this position;
- the business equipment that will be used in this position.

ADMINISTRATIVE CO-ORDINATOR
($35,000 to $40,000 SALARY PACKAGE)
The School of the Performing Arts is the state centre for professional training and education in the performing arts. The administrative co-ordinator is responsible for providing effective administrative support within a small team environment. The successful applicant will have at least two years corporate or small business experience. Experience with PCs, wordprocessing and spreadsheets is essential. Applicants should be familiar with accounts payable and purchasing procedures and reconciliation processes. Knowledge of and interest in the performing arts would be desirable.

ASSISTANT TO BOOKSHOP MANAGER
Hometown College requires a person to assist the manager of the Campus Bookstore which sells text and general books. Applicants should have experience in cash register sales, customer relations, stock ordering and control, supervision of casual sales staff, and be competent in reconciling sales and banking records. There is some lifting of heavy boxes. Familiarity with Apple Macintosh PC is desirable.

LEGAL SECRETARY
We require an experienced legal secretary to assist one of our partners. The person will be familiar with litigation procedures, have good organisation skills, fast and accurate dicta/ typing and a sound knowledge of Word Perfect 5.1. They will need a mature personality, enjoy working co-operatively with others, and be able to cope in high pressure situations.

FINANCE DIRECTOR c$100,000
Our client has a turnover of $30 million and is in need of an interim Financial Director for an initial period of 9 months. Interested applicants should have relevant experience and qualifications, a 'hands-on' approach, strong people management and a broad commercial background.

• ADMINISTRATIVE ASSISTANT
A vacancy exists in our accounts department for an experienced administrative assistant. The applicant will require: a good telephone manner, accurate, consistent clerical skills, word processing skills. This position should appeal to those who like interacting with customers and a busy environment.

BOOKKEEPER
Part-time
We are a busy management consulting film with a need for a bright and enthusiastic, part-time bookkeeper for 2 days per week. We operate our financial management system using MYOB and experience with that system will be highly favoured. Salary and conditions are negotiable. Please forward your résumé and contact details.

CLERK RECEPTIONIST
Small company has full time junior position for a person with pleasant phone manner. For general office duties, computer/WP skills essential, Microsoft Works an advantage. Job start applicants need only apply to Mrs J. Brown.

MEDICAL SECRETARIES
FULL-TIME & PART-TIME
(25.5 HOURS PER WEEK)
This large medical practice needs two competent secretaries with excellent skills in medical terminology, WP5.1 and audiotyping. Both present ideal opportunities for someone looking for experience in a busy general practice amongst other skilled and experienced staff.

INTERNAL SALES CLERK
$26,500
This position involves a variety of duties in a very busy office, with an emphasis on customer service. Duties include handling and processing customer orders, stock control including computer entries, preparing quotations and reports using Microsoft general credit control and daily invoicing.

State Manager
■ Suit Ambitious Executive
This organisation is commited to excellence in providing quality services that exceed client expectations.

The State Manager will make a significant contribution to the business's ongoing success. The successful applicant will need a good track record in business development, sales and operations management, supported by sound commercial acumen.

His or her communication, interpersonal skills and professional executive presentation will be of the highest order and supported by relevant tertiary qualifications and strong leadership qualities.

RECEPTIONIST - Junior Position
This is an exciting and unique opportunity for a bright and enthusiastic receptionist. Join a growing, friendly corporate consulting film in this front desk role using your excellent communication and presentation skills as you greet clients, distribute and sort mail and undertake general clerical duties. Your initiative, willingness to learn, preferred reception experience and friendly team approach will be welcomed.

Fig. 4.3 Examples of positions in organisations

Organisational policies and procedures

Every business has its own way of carrying out operations within the organisation, and rules and regulations which all employees are required to follow. These are known as organisational policies and procedures.

- **Organisational policies** are decisions made by management on what operations need to be done, and why they need to be done, in order to fulfil the functions and ultimately the goals of the business.
- **Organisational procedures** are the way these operations are performed by the people who have to do them. If your office does not have well-planned procedures, your work will build up to create a 'bottleneck' or barrier, which will create problems such as those listed here:

Stress, frustration and exhaustion Impatience with customers Low job satisfaction High staff turnover	*leading to*	Breakdown of machinery or mistakes resulting from human error	*leading to*	Inability to meet deadlines Missed contracts Late orders Loss of continuing business Failure to keep track of debtors

The result is that you will lose customers, with consequent reduction in business and profits. Therefore it is essential that your work station and your office have well-planned and organised procedures.

Office procedures	=	Incoming work	+	Staff and equipment	=	Outgoing work

The advantages of well-planned and organised procedures are:

- simplicity of operation of tasks;
- the use of minimum supervision;
- utilisation of specialist staff and equipment;
- avoidance of non-essential steps;
- less duplication of work;
- less need for checking work;
- the allocation of work according to ability and essential experience;
- the need for fewer rules;
- reduction in paperwork;
- better revision and updating;
- improved quality and uniformity of work and documentation.

Office procedures manual

To ensure that these procedures are followed by all personnel, a good office supervisor will see that every person working in the office has access to an **office procedures manual,** which outlines clearly and simply:

- the organisational chart of the business and who is responsible for which department or area of specialisation;
- who is responsible for which operation and how this should be carried out;
- the format and layout to be used for all written documents;
- the protocol to be used by everyone who answers the telephone;
- steps to be taken in case of accident or fire;
- how the filing system operates;
- rules for the use of office equipment;
- any other relevant procedural information.

Activity 4.3

Procedures manual

1. Form into groups of three and discuss the reasons why using a procedures manual is important for office workers and management. (Write the reasons individually on Activity sheet 4.3 provided in your workbook.)
2. Refer to your college procedures manual or your office procedures manual (or a manual provided for you by your tutor/supervisor), and describe the procedures you would follow for the following situations.
   (a) You have tripped over the edge of the carpet and damaged your ankle. You are not sure whether it is only bruised or a much more serious injury. What should you do?
   (b) You have started a new job today, and you are not sure what layout you should use for letters. Find out the correct layout from the relevant section in your procedures manual.
   (c) Find out from the procedures manual what you should do if there is a fire in your building. Write down the instruction from the manual.
   (d) You have a problem at work and don't know who to consult. Find the relevant section in the procedures manual and the name and position of the person you should refer to.
   (e) You are the new filing clerk and have been shown the filing system, but you are still not sure exactly how to use it. Refer to the relevant section in the procedures manual and write down the filing system that is outlined for your business.

OFFICE LAYOUT AND WORKFLOW

In order to operate well-planned and organised procedures, your work station and your office will need to be well designed. Examples of good design are:

- flexible office layout (in the event of growth of the business or changing circumstances);

- convenient arrangement of furniture and equipment to allow effective workflow;
- accessibility of documents and equipment;
- privacy (which may be required for managers to interview staff or clients);
- areas that are interrelated in their business functions placed near one another to allow easier interpersonal relationships and better communication.

Office layout

Organisational procedures can be improved by establishing the most effective layout of office work areas. When an office is planned or reorganised, it is important to consider the relationship between people, equipment and the flow of work.

Office workers	+	Equipment	+	Office layout	+	Flow of work	=	Productivity

When you are planning or reorganising your work station or your office, you need to:

- provide effective allocation and use of floor space for office furniture and equipment;
- create a pleasant working environment for yourself and other employees;
- create a positive impact for clients visiting your office;
- organise your work area to provide an efficient and effective workflow;
- provide yourself and other employees with efficient and productive work areas.

Office layout can be organised in two main ways, or using a combination of these two. **Conventional layout** is where work areas are separated by fixed walls.

Advantages

- There is a high degree of privacy.

Disadvantages

- Interpersonal communication is reduced.
- Workflow is ineffective.
- Layout is inflexible in view of any growth of the business or changing circumstances.

Open-plan layout creates work areas by placing acoustic screens around work stations, consisting of desks, chairs, filing cabinets, bookcases or computers.

Advantages

- There is a flexible and economical use of space.
- Interpersonal communication is easy.
- Construction costs are lower.

Disadvantages

- There is reduced privacy.
- There is a high noise level from conversation, telephones and equipment.

Most large organisations use open-plan layouts, with one or two conventional offices for senior management.

Some ways of reducing the disadvantages of an open-plan layout are:

- using acoustic screens and pot plants to provide more privacy;
- using thick carpet and acoustic ceiling tiles to reduce noise levels;

- reducing the level of ringing tone of telephones, and placing photocopiers and other equipment away from work areas to reduce noise and distractions.

A good open-plan office is shown in Figure 4.4.

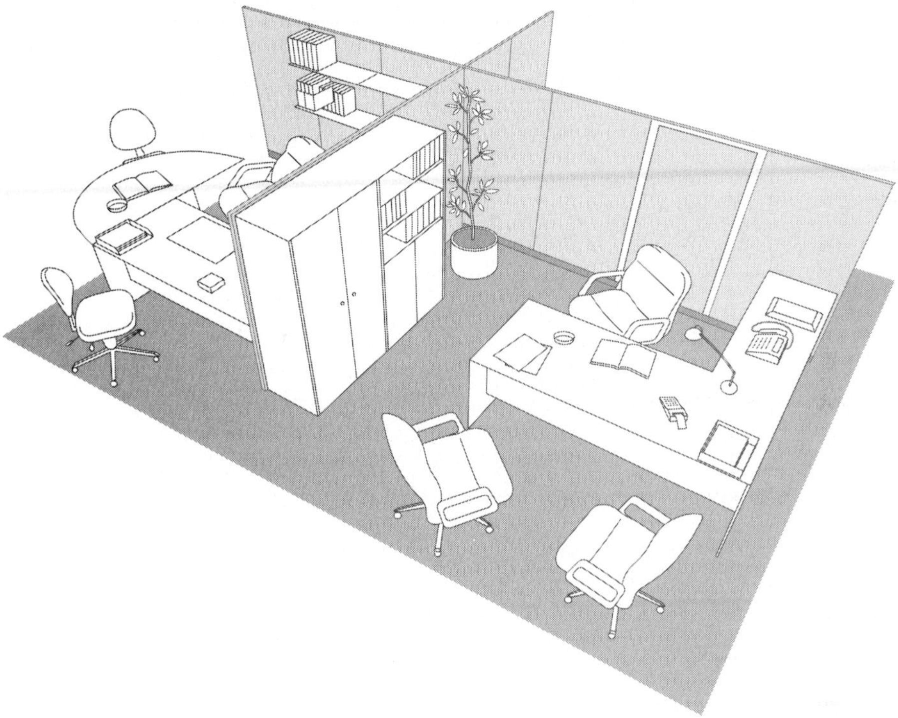

Fig. 4.4 Open-plan office

Workflow

To improve the operations undertaken in your personal work area and the operations in your office, it is essential that procedures flow in the simplest and most effective way. Improved workflow is established by:

- investigating the system as it now exists; and
- making required changes to it.

Two methods are commonly used: an office layout chart, and a flow process chart.

An **office layout chart** shows the location of office workers and the way work moves from person to person. Some rules for better workflow are:

Avoid: • 'criss-crossing and backtracking'.
Remember: • 'once around and out'; and
• follow a circular route of least distance.

Figure 4.5 is a simple example of how to improve the layout of an office workflow arrangement for procedures involving two people, using four pieces of equipment in addition to their work stations.

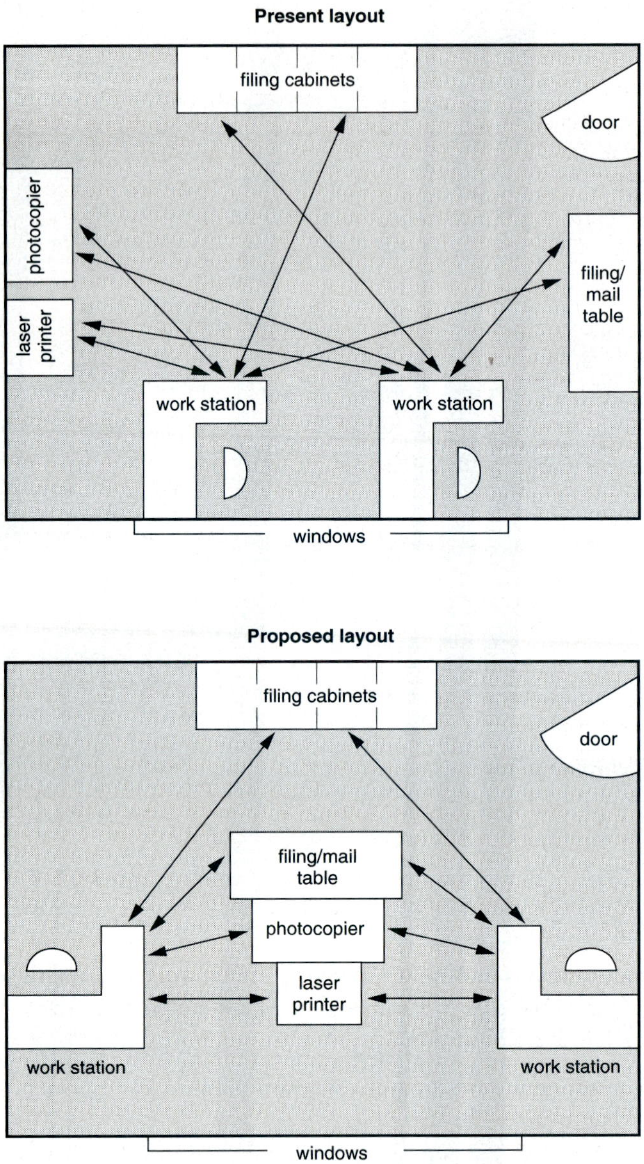

Fig. 4.5 The present office layout with two people, and the proposed improvement

Activity 4.4

Office layout

On Activity sheet 4.4 (1) in your workbook there is an office layout chart showing the outline of the office floor space for the open-plan office of Business Stationery Supplies Pty Ltd.

Arrange the personnel, office furniture and equipment listed below in this open-plan office layout. Take into account which departments should be near the main entrance, and which departments, or personnel, should be placed next to one another. The executives will need a private office.

Departments or personnel

- managing director's office
- managing director's secretary
- sales manager and two sales clerks
- accounts manager and two accounts clerks
- administration manager and secretary
- computer operator
- receptionist

Furniture and equipment

- meeting/conference room
- staff room
- reception couch and coffee table
- compactus/stationery supplies
- central filing system
- printer and photocopier

Discuss this floor layout with other group members. Cut out the diagrams of furniture and equipment that are provided on Activity sheet 4.4 (2) and place these in the most appropriate positions on the office layout chart. Take into account the flow and criss-crossing of people when they are using the equipment. Place the people and areas needing frequent communication close together.

You may like to draw your own acoustic screens or even pot plants around the work stations to separate them. (This may be completed individually.)

Figure 4.6 shows **flow process procedures**. To improve the workflow of procedures in an organisation, you should investigate the present operation, take out unnecessary stages and make any other required changes. Figure 4.6 shows how filing procedures can be improved to save the filing clerk time and make the system easier for everyone involved in records management.

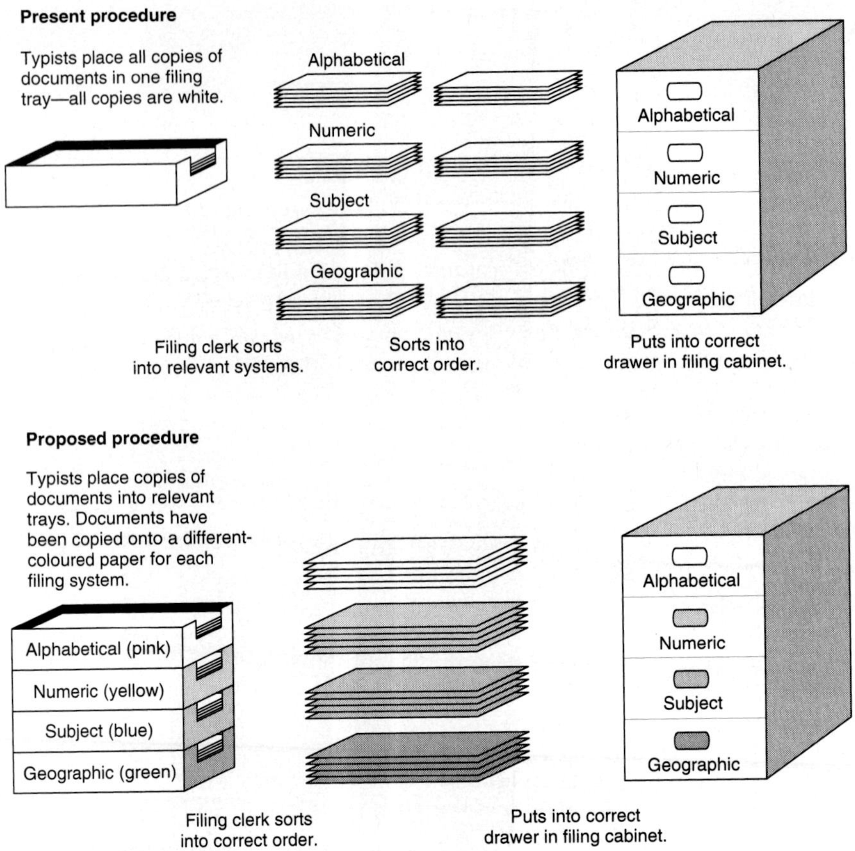

(Note: There are nine operations in this system—four fewer than in the present procedure—which will save time and effort for the filing clerk.)

Fig. 4.6 Improved flow process of records management

A great deal of time in your working day, and in the working day of your office, will be saved by improving the workflow of the operations of all procedures in the office. (This is one of the areas of better time management dealt with in more detail in Chapter 11.)

Activity 4.5

Workflow

Form into groups of three and discuss one procedure you use at college or work. Using brainstorming, think of ways in which you could improve the flow process of this procedure. When you have decided on the best improved flow process, draw a simple diagram of it with notes explaining the advantages of the changes. (Take advantage of the skills of the most artistic member of your group.)

CONSERVATION OF THE ENVIRONMENT

A vital part of office procedures is to try to conserve as many natural resources as possible to help preserve our environment. Some ways in which you could help are as follows:

- Switch off light and power when it is not required.
- Recycle paper in the following ways:
  —reuse envelopes wherever possible;
  —use internal envelopes, which can be readdressed up to 40 times;
  —place any used or damaged paper in recycling bins;
  —reuse manila folders from dead files in the filing system;
  —cut paper from dead files into squares and use these for taking notes.
- Take your own cup to work in order to avoid using plastic or polystyrene cups, which are not biodegradable.

Biodegradable waste products naturally decompose and do not cause an environmental problem. Goods made of plastic (such as disposable nappies) will not decompose for decades, and will cause problems for future generations.

Case study 4.1

Conservation of the environment

You are the administrative coordinator for Business Stationery Supplies Pty Ltd. The chief accountant has informed you that there has been a directive from management on a policy to improve the conservation of natural resources that must be implemented in all departments of the organisation. You have been given the responsibility to put this policy into operation and to make sure that staff are using procedures to conserve natural resources.

- How would you inform the staff in all departments of the new directive, and how would you check that the procedures were being implemented?
- What suggestions would you make to the following departments of ways in which they could conserve natural resources?
  —All departments: How could they reduce the amount of electricity used? How could they recycle paper? Think of as many ideas as you can.
  —The mailroom: How could they reuse envelopes and any other materials used or received in the mailroom?
  —The centralised filing system: How could they recycle files and other materials used in the filing system?
  —The office equipment area: How could they recycle paper? Save electricity? What else can you think of, especially when using computers and printers?
- Make suggestions for any other areas of the organisation in which you think procedures for conservation of natural resources could be put into operation.

Summary

- An office performs three major roles: a secondary role, a complementary role, and a controlling role.
- An office is an area of a building where people, using materials, tools and resources, perform the administrative services required to support the main function and goals of a business.
- The functions of an office (communication, managing staff and finances, and processing data) support the functions of a business (marketing and producing goods and services).
- The operations carried out in an office include operating office equipment and communicating effectively within the office and with customers and suppliers outside the office.
- The prime function of an organisation is the manufacture of specific goods or the provision of specific services.
- Organisation structure is necessary for the effective operation of a business. This includes goal-setting, unity of functions, work assignment, a chain of command, unity of command, authority commensurate with responsibility, and a specified span of control.
- The structure of key personnel includes lines of authority, and lines of communication—formal downwards (top down), formal upwards (bottom up), and horizontal.
- The chain of command is shown in the organisation chart. This includes the levels of management (top, middle and operational), and the job titles and positions of responsibility. These have been decided through job analysis and are written down in the job descriptions and job specifications.
- Organisation policies are decisions made by management on what operations need to be done. Organisational procedures are how these operations are performed and are outlined clearly and simply in an office procedures manual. A vital part of office procedures is to try to conserve as many natural resources as possible.
- Office layout should be well planned and well designed to provide the most effective methods of workflow.
- Open-plan layout uses the space available in the most efficient way, by dispensing with permanent walls and using screens and plants to divide work areas.

Review questions

(These may be used for group discussion.)

1. Write a short paragraph to explain what an office is.
2. Name the four components of an office.
3. Describe the functions of an office.
4. List the five main functions of a business.
5. An office performs three major roles for a business. Describe each one briefly.
6. What is the prime function of an organisation?
7. There are three classifications of industry. What are they?
8. In your role as an office worker, you will need to be trained in three different areas. What are they?
9. There are seven areas of operation necessary for an organisation to be effective. What are these?
10. An office can be centralised or decentralised. List briefly the advantages and disadvantages of centralisation.
11. Name the three types of authority structure within an organisation.
12. Describe briefly the three levels of management.
13. What is an organisation chart? Write a brief paragraph to explain what it is.
14. What are organisational policies and procedures and why are they so important within an office and an organisation?
15. How would you improve workflow in an office? Write a short paragraph on the methods you would use.
16. Explain how you would improve office layout.

Competency self-assessment and/or tutor/supervisor assessment

When you have finished reading this chapter, and have completed the activities chosen by your tutor/supervisor (**practical in class, at work or theoretical**), fill in this self-assessment. Then hand it to your tutor/supervisor for official assessment.

Learning outcomes	**Self-assessment**		**Tutor/supervisor assessment**	
I have successfully completed the following learning outcomes	**Yes** (please tick)	**No**	**Yes** (please tick)	**No**
1. Identify the role and functions of the office as an integral part of a nominated enterprise				
1.1 describe the operations commonly carried out in an office	☐	☐	☐	☐
1.2 list the titles of common office positions and give a brief description of the duties of each position	☐	☐	☐	☐
1.3 describe the prime function of a nominated enterprise	☐	☐	☐	☐
1.4 identify and describe the key functions of the office/department/key personnel within the nominated enterprise	☐	☐	☐	☐
2. Discuss the role of organisational and office procedures				
2.1 explain the importance of using a procedures manual and adhering to prescribed office and organisational procedures	☐	☐	☐	☐
2.2 use a procedures manual to determine designated person/s to address specified irregularities and/or difficulties for resolution	☐	☐	☐	☐
2.3 identify an actual desirable workplace practice/procedure which assists conservation and the environment	☐	☐	☐	☐
Date:	Tutor/supervisor signature:			

CHAPTER 5

WORK ENVIRONMENT

NOS119

objectives

The purpose of this chapter is to help you to understand aspects of the work environment that will affect your working conditions, in order to complete the following learning outcomes:

1. The function and structure of public and private organisations

- 1.1 primary, secondary and tertiary sectors in relation to business activity
- 1.2 the importance of goals and objectives to an organisation
- 1.3 public and private organisations, and their roles and source of funding
- 1.4 the three levels of public service in Australia
- 1.5 the different forms of ownership of private business organisations
- 1.6 the need for structure within an organisation

2. The rights and responsibilities of employers and employees, reciprocal nature, and the importance of cooperation in the workplace

- 2.1 the need for cooperation, mutual respect and reciprocal effort in the workplace
- 2.2 the rights and responsibilities of employers and employees
- 2.3 the purpose of a 'statement of duties'
- 2.4 the nature and role of 'performance feedback' and its relationship to work performance and levels of conduct

3. The principles and implementation of equal opportunity and antidiscrimination as they apply to the workplace

- 3.1 the concept of equal opportunity, and issues arising from it
- 3.2 Commonwealth and state Acts relating to equal opportunity and equal employment opportunity, and their purpose
- 3.3 the kinds of discrimination that might occur in the workplace and sources of information/assistance

4. The industrial relations system in Australia

- 4.1 the nature, role and function of unions
- 4.2 the nature, role and function of employer associations
- 4.3 the nature and purpose of Awards
- 4.4 the parties to the industrial relations system, and the roles within it
- 4.5 the steps in the process of resolving disputes between employers and employees
- 4.6 the nature and purpose of Award restructuring in the context of workplace restructuring

INTRODUCTION

In Chapter 11 we discuss various procedures that could help you to work more effectively, such as: knowing your duties as outlined in your job specification; using your aptitudes, abilities and skills; putting in place the correct procedures; and having a positive attitude. In addition to this, the environment in which you work can have a positive or negative effect on your attitudes towards work and the people with whom you work. It can affect how you are treated in your specific position; your conditions of employment; and your promotional prospects. The environment in which you work will depend on the size, type and structure of the particular organisation.

THE FUNCTION AND STRUCTURE OF ORGANISATIONS

The structure of an organisation is determined by the type of ownership, its business activities, and its size. In a small business, the contact between owner and staff is simple and direct. In a large business, the organisation structure is centred on chains of command, authority and responsibility, and the formal communication network. (This is described in more detail in Chapter 4.) Structure within an organisation is necessary in order for the business to achieve its specific goals and objectives.

Goals and objectives are essential for an organisation to operate effectively. The primary goal of a business, in the private sector of industry, is to make a profit. If it is not making a profit it will go out of business. The objectives of the staff in each department of a business organisation are set to reach this goal. In an effective organisation:

- clear goals and objectives have been outlined to all personnel in the organisation, so that everyone in the business is working towards the same goals;
- the functions necessary to achieve the goals have been determined from the objectives;
- the staff and resources necessary to perform the functions have been established.

The public and private sectors

There are two sectors of industry in Australia. In the **public sector**, business is operated by three levels of government: Commonwealth, state, and local. Each of these levels of government provides essential services for the public, funded from taxes, revenue and grants. Many of the services provided to the public are slowly being privatised, such as the provision of electricity, gas and fuel.

In the **private sector** of industry, business is operated by private individuals. Businesses in the private sector may be large organisations employing hundreds of staff, small companies employing two or three staff, or individuals working alone. The businesses are financed with personal funds or by shareholders in the company. A comparison of the two sectors is given in Table 5.1.

Public service		Private enterprise	
Owners: the electorate (voters)		Owners: shareholders of the company	
Parliament		Company	
Prime Minister		Chair of the board	
Cabinet		Directors	
Minister for . . .		Managing director	
Government departments		Company departments	
Head of department		General manager	
Divisional directors		Executive managers of sections	
Aims:	welfare of community	Aims:	profit
Accountability to:	taxpayers	Accountability to:	shareholders board of directors government
Promotion:	based on merit or, when two people have same qualifications, on seniority	Promotion:	based on demonstrated ability
Administration by:	minister department heads	Administration by:	one or two individuals (small organisation) board of directors administration department (large organisation)
Security of employment—used to be the case, but changing at present		Insecurity of employment—depends on profits	

Table 5.1 Comparison of the structure of public and private sector

Business is split over three main areas of production:

- **primary industry**—the extraction of raw materials such as minerals, fuels, agriculture, forestry and fishing;
- **secondary industry**—the manufacture of products from the raw materials;
- **tertiary industry**, which provides services to the public such as medical clinics, accounting and law firms, educational institutions and the distribution and sale of manufactured goods (this is the largest classification of industry).

TYPES OF BUSINESS OWNERSHIP

Business ownership in the private sector is divided into five main categories: sole proprietors; partnerships; private companies; public companies; and non-profit organisations. Features of each are listed below.

Sole proprietors

Examples: a butcher's shop, a fashion boutique, a pharmacy.
Owner: one person.
Capital invested in the business: by the owner, who is liable for any debts incurred by the business; a sole trader could lose everything he/she owns if the business fails.
Profits: received by the owner.
Regulating documentation: no regulating documentation if the owner operates under his/her own name; if the business has a business name (e.g. 'Celine's Flowershop'), then the business must register the name under the *Business Names Act.*

Partnerships

Examples: a medical centre, a small accounting business, a small law business.
Owners: two or more people (to a maximum of 20).
Capital invested in the business: by partners, who could lose everything if the business is not successful.
Regulating documentation: partnership agreement controlled by the *Partnership Act*; registration of name under the *Business Names Act.*
Profits: distributed between partners.

Private companies

Example: Business Stationery Supplies Pty Ltd.
Owners: between one and 50 (previously a minimum of two). Public company must be formed after the members exceed 50 and the board of directors are elected by shareholders.
Capital invested in business: by shareholders in the company (shares are not bought and sold on the stock exchange and cannot be bought by the public).
Regulating documentation: registered under the Corporations Law and controlled by its own Memorandum and Articles of Association.
Profits: a dividend distributed to shareholders according to profits earned.

Public companies

Examples: Business Enterprises Ltd.
Owners: minimum of five (no upper limit); board of directors elected by shareholders.
Capital invested in business: by shareholders (there is a public invitation to invest and a prospectus is issued); shares are bought and sold on the stock exchange.
Regulating documentation: registered under the Corporations Law and controlled by its Memorandum and Articles of Association.
Profits: a dividend distributed to shareholders according to profits earned.

Non-profit organisations

Various types of organisation such as credit unions, building societies, cooperatives, charity organisations, and private clubs not run for profit.
Capital invested in business: by members of cooperatives and non-profit-making clubs; by donations from the public to charitable organisations.
Profits: being a non-profit organisation does not mean the business makes no profit; any profits made by the business are distributed between the members, or used for administrative purposes, or used to help needy people in the case of charitable organisations.

Public sector

Capital invested in business: taxation and revenue from public and business.
Profits: capital and revenue used for public services.
Regulating documentation: laws passed by parliament.
This refers to the three levels of the Australian public service:

1. the Commonwealth public service such as Australia Post, the Australian Customs Service and the Department of Social Security;
2. the state public service, such as the state police force, state schools and hospitals;
3. local government.

Activity 5.1

Organisations

Form a group of three. With the other members of your group discuss the different types of business ownership, then complete the following activities individually.

1. Photocopy an advertisement from the *Yellow Pages* of one organisation for each of the following forms of business ownership:
   - a sole proprietor
   - a partnership
   - a private company (Pty Ltd)
   - a public company in private sector (Ltd)
   - a non-profit organisation
   - a public company in public sector (state which level of government).
2. Attach each advertisement to the top of a separate sheet of paper and type (or print neatly by hand), underneath each organisation, the following details:
   - the name of the organisation
   - the form of ownership
   - the type of industry classification (primary, secondary or tertiary)
   - the prime function of the business (the goods and services it produces)
   - the source of capital invested in the business
   - the number of people owning the business
   - the regulating documentation.
3. On a separate sheet of paper, write a brief paragraph on the following:
   (a) Why are goals and objectives important to an organisation?
   (b) The structure of any organisation is determined by the type of ownership. Describe the structure in a large organisation and how it operates. State why this type of structure is necessary in a large organisation.
4. Draw an organisation chart for the private company you chose in number 1 above (refer to the organisation chart for Business Stationery Supplies Pty Ltd in Chapter 4).

COOPERATION, MUTUAL RESPECT AND RECIPROCAL EFFORT

Cooperation is essential for developing the ability to relate effectively to management, customers/clients and colleagues, to work as a team and to communicate well at an interpersonal level. For example, when one section of the business is working to complete a large order, staff from other sections will help out. Reciprocal effort at a personal level means that if you have a quiet period you will help someone who has a deadline to meet, and they will then help you when you have a large volume of work that must be completed in a specified time period.

Mutual respect means treating other people as you would like to be treated yourself, including acknowledging other people's cultural and religious beliefs.

RIGHTS AND RESPONSIBILITIES

Your responsibilities as an employee

When you agree to accept a position in an organisation, whether this is a signed agreement or a verbal agreement, you have entered into a contract of employment. This contract of employment sets out terms and conditions, which include:

- arriving at the office on time and working the agreed number of hours;
- carrying out the work with care and skill, to a high quality, and in a responsible manner;
- obeying lawful orders from your employer in the context of your work;
- following management's policies and procedures in order to achieve your work goals and the organisation's goals;
- maintaining the confidentiality of the work carried out;
- following the safety rules according to the occupational health and safety procedures put in place by management;
- being courteous and helpful to other employees and visitors from outside the organisation, in a manner that helps to promote a positive corporate image;
- respecting colleagues' and customers' cultural and social differences;
- behaving in an ethical manner and giving disclosable information only;
- dressing in a suitable manner according to what is expected by the organisation;
- appearing neat, tidy and well-groomed during working hours;
- being honest and trustworthy.

Your rights as an employee

According to various Acts of Parliament and standards set by the Federal Industrial Relations Commission, as an employee you have a right to expect:

- a safe physical working environment (see Chapter 6);
- fair practices in the workplace in respect of equal employment opportunity and equal promotional opportunity (see later in this chapter);
- a working environment free from sexual harassment;
- a working environment free from any kind of discrimination;
- a fair rate of pay;
- a choice as to whether to join a union or not;
- training and guidance in the correct methods of performing your duties.

Full-time workers are entitled to four weeks' leave a year on full pay. Some employees receive 17.5 per cent above full pay as a 'holiday' loading. Long-service leave may be granted to people who remain with one employer for a long period of time.

Rights of an employer

According to standards set by the Federal Industrial Relations Commission, your employer has the right to dismiss you if:

- you do not adhere to your duties and responsibilities in accordance with your job description, or do not comply with instructions;
- you commit a criminal offence;
- you are negligent or careless and cause an accident;
- you commit acts of disloyalty, such as divulging your employer's trade secrets, revealing confidential information to a competitor, or making profit from information obtained at work.

Unfair dismissal

You have the right, if you think you have been dismissed unfairly, to take your case to the Commission for Reconciliation with the Industrial Relations Commission.

Notice of termination

Employers usually give the following notice of termination:

Length of employment	*Notice*
1 year or less	1 week
1 to 3 years	2 weeks
3 to 5 years	3 weeks

Employees should give the same period of notice, but this will be negotiated with your employer through your individual contract of employment, which is becoming the more common method of employment.

STATEMENT OF DUTIES (JOB DESCRIPTION/JOB SPECIFICATION)

The purpose of a statement of duties (job description or job specification) is to set out clearly the duties of the position, the responsibilities and the level of authority. Knowing which duties you are responsible for and which you are not enables you to perform your duties more effectively. It informs you of who is in charge of your section and who you should report to. It outlines the qualifications and skills required to fulfil the requirements of the position (see Chapter 11).

Performance feedback

At the end of the year, you may be asked to complete a 'staff performance review'. There is likely to be a section in which you state how you think you have performed your duties throughout the year, and a section which is completed by your manager, who states how she/he thinks you have performed throughout the year. This review is used by management to assess your performance, and helps you to assess your own performance. The results of the performance review help to identify areas where you may need training, or whether you may be promoted to a higher position.

Case study 5.1

Rights and responsibilities

You have been promoted from your position as receptionist with Business Stationery Supplies Pty Ltd, and someone else is taking over your position when you leave it next week. Write an internal memorandum to Judy Chan welcoming her to your company and giving her advice on the following aspects, which will help her in her new position. (Refer to Chapter 9 for memorandum format.)

1. Using the heading 'Cooperation, mutual respect, and reciprocal effort', advise Judy why these attitudes are essential in her position as receptionist. Give her at least three examples of each.
2. Using the heading 'Rights and responsibilities of employers and employees', advise Judy of what rights she can expect as an employee of the company, and what her responsibilities are. Give her at least six examples of each.
3. Explain to her the purpose of a 'statement of duties' and why it is important that she should know the information it contains.
4. Tell her that at the end of the year she will have to complete a performance review, which will provide feedback on how she has performed her duties during the year. Explain to her in what ways this can be useful to her and useful to her employer.

EQUAL EMPLOYMENT OPPORTUNITY

There is now a range of legislation in Australia, at both Commonwealth and state level, which is designed to prevent discrimination and promote equal opportunity. It is unlawful to discriminate against a person in the following areas:

- race or ethnic origin;
- physical, intellectual or psychological impairment (including an impairment you previously suffered from, or having or being thought to have AIDS or being HIV-positive);
- sex;
- pregnancy;
- having or not having children;
- marital status, including being a de-facto spouse;
- religious belief or activity;
- age
- parental status
- industrial activity.

Sexual harassment is also unlawful.

What is discrimination?

Discrimination is:

- treating a person less favourably than someone else in the same or similar circumstances on one or more of the grounds listed above;
- treating a person less favourably because of characteristics that apply to a person of a particular group as defined by one of the grounds listed above;
- imposing an unreasonable condition which some people of a particular group are less able to meet than others.

The Victorian *Equal Opportunity Act 1984* makes discrimination unlawful in the following areas:

- employment
- education
- provision of goods and services
- accommodation
- government and council services
- membership of clubs and associations on crown land or which receive government funding
- sport
- disposal of land.

Discrimination against employees

As an employee, you may also suffer from discrimination if you are inhibited or prevented, because of your gender, race, disability etc., from:

- joining a training program;
- applying for promotion or transfer;
- taking part in meetings;
- having equal access to facilities.

Discrimination in staff recruitment

In the past there have been advertisements that specified the sex or race of the person required to fill the position. This is now illegal. It is also illegal to discriminate against people at job interviews by asking questions about their beliefs, their sexual preferences, whether or not they intend to have children, and so on.

Interviewers should not ask personal questions like the following:

- Have you completed your family?
- What are your child care arrangements?
- Are you still under treatment for X? or How many sick days did you take last year?
- Were you born in Australia? Do you speak English at home?

Instead, they should focus on the job's requirements, as follows:

- This job requires interstate travel. How flexible are you with regard to overtime or out-of-hours activities?
- This job requires heavy lifting/sitting at a computer terminal for long periods of time. Are you able to do this?
- In this job you will need to write detailed reports. Would you please spend the next 10 minutes completing this writing exercise?

Job advertisements should show that the vacancy is open to every qualified person. Unacceptable words are 'waitress', 'manageress', 'he' or 'she', 'him' or 'her', 'woman' or 'man', 'boy' or 'girl', 'single' or 'married'. Acceptable alternatives are worker, personal office assistant, principal, cleaner, bar attendant, employee, supervisor, sales assistant, clerk, 'he/she'.

Indirect discrimination

Discrimination is sometimes not directed against a specific person or persons but is built into the system. An example is the setting of a condition with which one group can comply more easily than another, such as making unbroken service a requirement for promotion. This discriminates against mothers who take maternity leave.

Sexual harassment in employment

Under the *Sex Discrimination Act 1984* (Cwlth), it is unlawful for an employer or a potential employer to sexually harass any worker. Sexual harassment is unlawful if a person has reason to believe that she or he will be disadvantaged in any way at work by refusing to accept or tolerate the sexual harassment. It is also unlawful for an employer or supervisor knowingly to permit a person to be sexually harassed by a fellow worker.

Sexual harassment is behaviour of a sexual nature that is unwelcome, unsolicited and unreciprocated. It can be physical, verbal or written and can consist of any or all of the following:

- unwelcome comments about a person's sex life or physical appearance;
- suggestive behaviour such as leering or ogling;
- unnecessary familiarity such as brushing up against a person;
- sexual jokes, offensive telephone calls, photographs, reading matter or objects;
- sexual propositions or continual requests for dates;
- physical contact such as touching or fondling;
- indecent assault or rape (which are also criminal offences).

Affirmative action

In 1986, Commonwealth legislation introduced the *Affirmative Action (Equal Employment Opportunity for Women) Act.* This is positive action, as it sets up processes in the workplace that will help to remove the problems and barriers women face in employment.

Affirmative action is about getting the best person for the job, whether that person is male or female. The goal of affirmative action is equal employment opportunity for women, and affirmative action provides the means to achieve this.

- Employers are required to issue a policy statement to all employees, making it clear that the organisation is committed to equal opportunity for its female employees.
- Responsibility for affirmative action should be given to a person with authority in the organisation.
- Consultation between management, employees and (where appropriate) unions should take place in order to see if the distribution of work, training and promotion opportunities is made on a basis of sexual equality.

Activity 5.2

Equal opportunity and discrimination

Research this topic to obtain up-to-date information on the Regulations covering each Act in your state. Obtain articles from national newspapers and journals on equal opportunity issues and write a short report under the following main headings, including examples from the articles:

1. *Equal opportunity (EO)*
   Explain what EO means.
   Give examples from the articles you have obtained.
   State the Acts relating to EO.

2. *Equal employment opportunity (EEO)*
   Explain what EEO means.
   Give examples from the articles you have obtained.
   State the Acts relating to EEO.

3. *Discrimination*
   Describe the types of discrimination that might occur in the workplace. Give sources of information and assistance for dealing with these types of discrimination.

THE INDUSTRIAL RELATIONS SYSTEM

The main parties involved in the industrial relations system are the following:

- the Australian Council of Trade Unions (ACTU);
- the state labour councils;
- the unions;
- the federal and state governments;
- the Australian Conciliation and Arbitration Commission;
- the state Industrial Relations Commissions;
- employer associations.

The nature, role and function of unions

A union is an association of people with a common interest who are in the same employment category, such as the Metals and Engineering Workers Union. It is the policy of a union to ensure that members get a good and fair return for their labour. Members pay a fee, and in return their union provides support and protection in areas of:

- awards in relation to grading of positions;
- better salaries and conditions;
- occupational health and safety in the workplace;
- workers' compensation accident claims;
- unfair dismissal or retrenchment;
- work-related problems;
- leave (sick, maternity, bereavement, special, short, holiday, long-service);
- representation of employees' interests in cases before the various industrial tribunals;
- affiliation with the Australian Council of Trade Unions (ACTU), which processes wage or productivity cases before the Arbitration Commission.

The benefits for members are voting rights on issues being considered; eligibility to attend union meetings, which are held regularly; and up-to-date information on issues concerning their employment. For office workers it is an employee's right to join or not to join a union.

Employer associations

Employer associations in Australia are usually active political lobbyists, making representations and submissions to government concerning industrial relations legislation and government policy at both Commonwealth and state levels. Their role is to maintain and create a legal and political environment beneficial to their membership and to keep their members informed of wage movements, prepare cases for various industrial tribunals, and interpret the complexities of changes in Awards and legislation that will affect their members' operations. The functions of employer associations include:

- providing information, research and advice;
- representing members;
- education, training and other specialised services;
- representing employers in collective bargaining and associated activities.

The nature and purpose of Awards

The Industrial Relations Commission in each state sets out the wages and conditions that employers are obliged by law to provide for their employees. This is set out in documents called Awards. The provisions that make up a typical Award include such things as sick leave, overtime pay, shiftwork, allowances for dirty or hazardous work, grievance procedures, as well as rates of pay for particular skill levels and jobs. One of the most important responsibilities of union representatives is to ensure that Award conditions are enforced. (However, this situation is changing as enterprise bargaining and contracts of employment are becoming more widespread.)

Activity 5.3

Industrial disputes

The industrial relations situation in Australia is changing at present. Obtain recent articles regarding disputes between employers and union representatives of the workers. Bring these articles to a group discussion. After the discussion, write a summary of each article. It should include the following:

- the reasons for the dispute;
- the parties taking part in the dispute;
- the area of industry;
- the process of any discussions that took place;
- the parties to the discussions;
- the results of the discussions.

The steps in the process of resolving disputes

Step one. Employee takes up dispute with supervisor. Supervisor resolves dispute to employee's satisfaction; or does not resolve dispute but employee accepts decision; or supervisor does not resolve dispute and employee is dissatisfied.

Step two. Dispute is discussed by union job representative, grievance and management representatives. If not resolved, the dispute is taken to the Industrial Relations Commission to try to resolve it by either conciliation or arbitration.

Conciliation means discussion between disputing parties to resolve a dispute. **Arbitration** is when a third party resolves a dispute.

Award restructuring

Award restructuring is when employers and unions agree to make changes to the industrial Awards covering the way that work is organised, the range of duties employees carry out, and the training that is provided for them.

The key ingredients of Award restructuring are:

- training employees to acquire new skills so that they can move to higher grades or classifications;
- providing employees with the opportunity to learn a wider range of skills;
- giving employees the chance to become multi-skilled (skilled in more than one area and kind of job);
- setting up a framework for unions and employers to work together to change jobs for the benefit of all.

TECHNOLOGICAL CHANGE

Many more tasks are now performed by machines, and while there are advantages resulting from this change there are also quite a few disadvantages. Some **advantages** in an automated office are as follows.

Improved quality. Documents produced on a word processor are of a much higher quality than when they were produced on a typewriter. There are fewer errors (or at least these are easier to correct), and presentation of letters and reports is to a much higher standard. Desktop publishing enables organisations to produce their own publications, which look like professionally printed documents.

Competition. Automation is necessary to keep up with competition. For example, replies to requests for quotations can now be sent by fax or by e-mail, and if you do not possess this equipment you could lose business.

Service. Customers/clients can be given a better service through improved communication equipment and skills.

International business. There has been an increase in international business in recent years, and if your business wants to keep up with the competition it must have the necessary equipment and skills to communicate with international organisations.

The **disadvantages** of office automation include the following.

Cost. The cost of installing new equipment can be high and it may be many years before the business recoups this initial cost.

Training. Costs for training staff to operate the new equipment can be high.

Change. The effects of change can disrupt the operation of a business through the training process and through the fears of employees of the change situation.

Effects of change in the workplace

We live in a rapidly changing world, and in the work environment change is occurring constantly. A supervisor or manager should create a working environment where change is viewed as normal.

Office automation has changed both the way work is done and the way workers are viewed by management. Changing organisational structure resulting in changing roles in the workforce has caused a great deal of stress to employees, and has met with considerable resistance.

A supervisor or manager should work to overcome resistance to change and encourage the adoption of new attitudes in the following ways:

- by informing all staff of the need for change;
- by representing change in a positive way (helping staff to realise what they will gain from the changes);
- by asking staff involved in the change to put forward ideas to help the changeover (as they are the people who work with the equipment and know the procedures involved, if their ideas are accepted they are more likely to accept the changes);
- by communicating often with staff involved in the change;
- by holding frequent meetings to encourage staff to voice their opinions;
- by keeping staff informed of each step of the change process;
- by instituting training sessions where necessary;
- by conducting orientation sessions to explain the change;
- by communicating a clear picture of the part staff will have in the new arrangements and how change will affect each specific role and the work in general;
- by arranging for feedback on the effects of the change from those involved after the change has been made.

Activity 5.4

The industrial relations system

Write a short paragraph on each of the following:

1. the nature, role and functions of unions;
2. the nature, role and function of employer associations;
3. the nature and purpose of Awards;
4. the main parties to the industrial relations system, outlining their roles within it;
5. the steps in the process of resolving disputes between employer and employees;
6. Award restructuring in the context of the workplace.

Case study 5.2

Technological change

You are the administrative coordinator for Business Stationery Supplies Pty Ltd. The managing director, Edward Gerrard, is installing a new computer system in all areas of the business. He has asked you to report to him on the factors that will affect the workplace through this changed situation.

Write an internal memorandum to the managing director outlining your views on the following:

1. the advantages and disadvantages of the technological change;
2. how the change will affect the roles of, and work performed by, people in the workplace;
3. the methods he should put into operation to make sure this change will take place as smoothly as possible.

Summary

The function and structure of organisations

- The structure of an organisation is determined by the type of ownership and its size.
- Goals and objectives are essential for an organisation to operate effectively. The primary goal of a business, in the private sector of industry, is to make a profit, and the objectives of the staff within each department are set to reach this goal.
- There are two sectors of industry in Australia, the public and the private. The private sector is operated by private individuals, and organisations may be small or large. The business is financed with personal funds or by shareholders of the company.
- Business is split over three main areas of production: primary industry—the extraction of raw materials; secondary industry—the manufacture of products from the raw materials extracted; and tertiary industry, which provides services to the public.
- Business ownership in the private sector is divided into five main categories: (1) sole proprietorship, which is owned by one person; (2) partnership, which is owned by 2–20 people; (3) private company (Pty Ltd), operated by a board of directors elected by shareholders within the company; (4) public company (Ltd), operated by board of directors elected by shareholders, with shares

being bought and sold on the stock exchange; (5) non-profit organisations, such as credit unions, building societies, cooperatives and charity organisations.

- Public organisations in the public sector of industry consist of the three levels of the Australian public service: (1) the Commonwealth public service; (2) the state public service; and (3) local government.
- Cooperation is essential to be able to work as a team, and reciprocal effort is helping someone with work when you have a quiet period and they helping you when they have a quiet period. Mutual respect means treating other people as you would like to be treated yourself, and acknowledging other people's cultural and religious beliefs.
- When you agree to accept a position in an organisation, whether this is a signed or a verbal agreement, you have entered into a contract of employment. This sets out terms and conditions with which you have to comply.
- According to the standards set by the Federal Industrial Relations Commission, both employers and employees have rights and responsibilities within the contract of employment.
- The purpose of a statement of duties is to set out clearly the duties and responsibilities of a position. Performance feedback may involve a 'staff performance review' by you and your manager or supervisor to determine your performance and identify areas in which you may need training.
- Cooperation is essential for developing the ability to relate effectively with management, customers/clients and colleagues, to work as a team and to communicate well at an interpersonal level.

Equal employment opportunity

- The equal opportunity Acts of the Commonwealth and the states make it unlawful to discriminate against a person because of race or ethnic origin; physical, intellectual or psychological impairment; sex; pregnancy; marital status; or religious belief.
- Discrimination is treating a person less favourably than someone else in the same or similar circumstances on one or more of the grounds listed above.
- As an employee you may also suffer discrimination if you are inhibited or prevented from joining a training program; applying for promotion or transfer; taking part in meetings; or having equal access to facilities.
- It is also illegal to discriminate against people when advertising a job vacancy, and at job interviews by asking questions about a person's beliefs, sexual preferences, whether or not they intend to have children, and so on.
- Indirect discrimination is built into the system by, for example, the setting of a condition with which one group can comply more easily than another.
- Sexual harassment is behaviour of a sexual nature that is unwelcome, unsolicited and unreciprocated. It can be physical, verbal or written.
- Affirmative action is positive action to help remove the problems and barriers women face in employment. It is about getting the best person for the job, whether that person is male or female.

The nature, role and functions of unions and technological change

- A union is an association of people in the same employment category. It is the policy of a union to ensure that members get a good and fair return for their labour, and it provides support and protection to employees in other areas of employment.
- Employer associations in Australia are usually active political lobbyists making representations and submissions to government concerning industrial relations issues on behalf of employers.
- The Industrial Relations Commission in each state sets out the wages and conditions that employers are obliged by law to provide for their employees. This is set out in documents called Awards.
- The parties involved in the industrial relations system are the Australian Council of Trade Unions (ACTU); state labour councils; the Commonwealth and state governments.
- Award restructuring is when employers and unions agree to make changes to the industrial awards covering the way that work is organised, the range of duties employees carry out, and the training that is provided for them.
- We live in a rapidly changing world, and in the work environment change is occurring constantly. Office automation has changed both the way work is done and the way workers are viewed by management. A supervisor or manager should work to overcome resistance to change and encourage the adoption of new attitudes to change in the workplace.

Review questions

(These can be used for group discussion.)

1. Why are goals and objectives essential for an organisation to operate effectively?
2. Name the two sectors of industry in Australia.
3. List some of the structural differences between the public service and private enterprise.
4. Name the three main classifications of business and give a brief description of each.
5. There are six main categories of business ownership in Australia. Name these and indicate how many people own the business in each category.
6. Why are cooperation, mutual respect and reciprocal effort between people important in a work environment?
7. List your responsibilities as an employee.
8. List your rights as an employee.
9. What are some of the rights of an employer?
10. What can you do if you think you have been dismissed from your work unfairly?

11. Explain the purpose of a statement of duties (job description) and the ways this can be useful to you.
12. Explain the meaning of performance feedback.
13. In what ways is it unlawful to discriminate against a person?
14. What is discrimination?
15. Describe some of the ways in which you can discriminate against people in staff recruitment.
16. How can you discriminate indirectly?
17. What is sexual harassment?
18. Explain briefly the reasons why affirmative action was introduced in the workplace.
19. Write a paragraph on the nature, role and function of unions.
20. What are employer associations?
21. What are the provisions that make up an Award?
22. Name the parties involved in the industrial relations system.
23. List the steps involved in the process of resolving disputes.
24. What is Award restructuring?
25. List some advantages and disadvantages of technological change.

Competency self-assessment and/or tutor/supervisor assessment

When you have finished reading this chapter, and have completed the activities chosen by your tutor/supervisor (**practical in class, at work or theoretical**), fill in this self-assessment. Then hand it to your tutor/supervisor for official assessment.

Learning outcomes	**Self-assessment**		**Tutor/supervisor assessment**	
I have successfully completed the following learning outcomes	**Yes** (please tick)	**No**	**Yes** (please tick)	**No**
1. The function and structure of public and private organisations				
1.1 primary, secondary and tertiary sectors in relation to business activity	☐	☐	☐	☐
1.2 the importance of goals and objectives to an organisation	☐	☐	☐	☐
1.3 public and private organisations, and their roles and source of funding	☐	☐	☐	☐
1.4 the three levels of public service in Australia	☐	☐	☐	☐

1.5 the different forms of ownership of private business organisations	☐	☐	☐	☐
1.6 the need for structure within an organisation	☐	☐	☐	☐
2. The rights and responsibilities of employers and employees, reciprocal nature, and the importance of cooperation in the workplace				
2.1 the need for cooperation, mutual respect and reciprocal effort in the workplace	☐	☐	☐	☐
2.2 the rights and responsibilities of employers and employees	☐	☐	☐	☐
2.3 the purpose of a 'statement of duties'	☐	☐	☐	☐
2.4 the nature and role of 'performance feedback' and its relationship to work performance and levels of conduct	☐	☐	☐	☐
3. The principles and implementation of equal opportunity and antidiscrimination as they apply to the workplace				
3.1 the concept of equal opportunity and issues arising from it	☐	☐	☐	☐
3.2 Commonwealth and state Acts relating to equal employment opportunity, and their purpose	☐	☐	☐	☐
3.3 the kinds of discrimination that might occur in the workplace and sources of information/assistance	☐	☐	☐	☐
4. The industrial relations system in Australia				
4.1 the nature, role and function of unions	☐	☐	☐	☐
4.2 the nature, role and function of employer associations	☐	☐	☐	☐
4.3 the nature and purpose of Awards	☐	☐	☐	☐
4.4 the parties to the industrial relations system, and the roles within it	☐	☐	☐	☐
4.5 the steps in the process of resolving disputes between employers and employees	☐	☐	☐	☐
4.6 the nature and purpose of Award restructuring in the context of workplace restructuring	☐	☐	☐	☐
Date:	Tutor/supervisor signature:			

CHAPTER

6

OCCUPATIONAL HEALTH AND SAFETY

ENT202A, NOS110, 7365G

objectives

The purpose of this chapter is to help you to understand the general aims and objectives of legislation for health and safety in the workplace as required by Commonwealth and state governments, in order to complete the following learning outcomes:

1. **Occupational health and safety responsibilities as required by Commonwealth and state legislation and their implementation in the workplace**
   - 1.1 general aims and objectives of existing Commonwealth and state occupational health and safety legislation
   - 1.2 employer and employee responsibilities, rights and obligations relating to occupational health and safety
   - 1.3 the major functions of safety committees and representatives
   - 1.4 the role of WorkSafe Australia

2. **Potential hazards in an office environment and remedial action**
   - 2.1 physical, chemical and behavioural hazards that may cause injury or health problems in an office environment
   - 2.2 remedial action

3. **The causes of occupational overuse injuries of office workers and preventive methods**
   - 3.1 work methods and conditions that can lead to occupational overuse injury, and preventive and remedial actions
   - 3.2 the basic principles of correct posture
   - 3.3 correct lifting and handling techniques

4. **The accident/incident and reporting procedures in the workplace**
   - 4.1 the procedures for reporting an accident/incident/injury in the workplace
   - 4.2 complete an accident/incident report form for a real or simulated accident/incident
   - 4.3 the purpose of workers' compensation
   - 4.4 the procedure for claiming workers' compensation
   - 4.5 the first aid facilities that should be found in the workplace and their location

5. **Factors affecting work performance**
   - 5.1 substances that are subject to misuse
   - 5.2 stressors
   - 5.3 the impact of stress and substance abuse on work performance

INTRODUCTION

Work can be dangerous. On an average day, somewhere in Australia at least one worker will be killed, 300 will injure their backs, 20 will be diagnosed with hearing loss, and a dozen will break at least one bone. According to the Bureau of Statistics, work-related illness and injuries consistently cause the loss of three to six times the number of working days lost in industrial action. In each state, occupational injuries cost almost $5 billion a year in WorkCover, medical bills, lost working time, and compensation.

Few people are fatally injured in the office, but many injuries and illnesses occur through:

- lifting objects that are too heavy, too awkward, or positioned too low or too high (e.g. lifting office machinery, too many reams of photocopy paper, or moving furniture);
- sitting at the computer screen incorrectly;
- sitting at the computer for too long;
- not using ergonomic furniture;
- excessive noise from equipment;
- incorrect lighting;
- chemical hazards, such as fumes released from photocopiers causing lung irritation and skin rashes;
- slips, trips and falls.

The object of this chapter is to draw to your attention to the hazards that occur in an office environment and the steps you and your colleagues should take to avoid injury and/or illness. Effective management of health and safety hazards can lead to lower costs, improved productivity, higher morale in the workplace and better industrial relations.

Employers have a duty of care to provide a healthy and safe working environment under occupational health and safety legislation and common law. The *Occupational Health and Safety (Commonwealth Employment) Act 1991* (Section 16) states that:

> **an employer must take all reasonable practicable steps to protect the health and safety at work of the employer's employees.**

The following are occupational health and safety (OH&S) Acts of the Commonwealth and all states.

• *Occupational Health and Safety (Commonwealth Employment) Act 1991*	Commonwealth
• *Occupational Health and Safety Act 1989*	Australian Capital Territory
• *Occupational Health and Safety Act 1983*	New South Wales
• *Work Health Act 1986*	Northern Territory
• *Workplace Health and Safety Act 1995*	Queensland
• *Occupational Safety, Health and Welfare Act 1977*	Tasmania
• *Occupational Health and Safety Act 1985*	Victoria
• *Occupational Health, Safety and Welfare Act 1984*	Western Australia
• *Occupational Safety, Health and Welfare Act 1986*	South Australia

Legislation under the Act for each state can vary, but in general all the Acts state that:

> **employers should do all that is reasonably practicable to set up a working environment that is safe and without risks to the health of employees, contractors and other persons in or near the workplace.**

The hazard management approach

Managers and supervisors are not only responsible but are also accountable for this duty of care.

Hazard means the potential to cause injury or illness.

Risk means the likelihood of injury or illness arising from exposure to any hazards. For example, cleaning fluids or detergents are a potential hazard but may not be a risk unless they are incorrectly stored or handled.

Managing health and safety in the office requires a process of hazard identification, risk assessment, risk control and evaluation of control measures. Effective management of health and safety hazards also involves training, consultation, documentation of health and safety activities and regular review of the management system. Figure 6.1 outlines the hazard management approach.

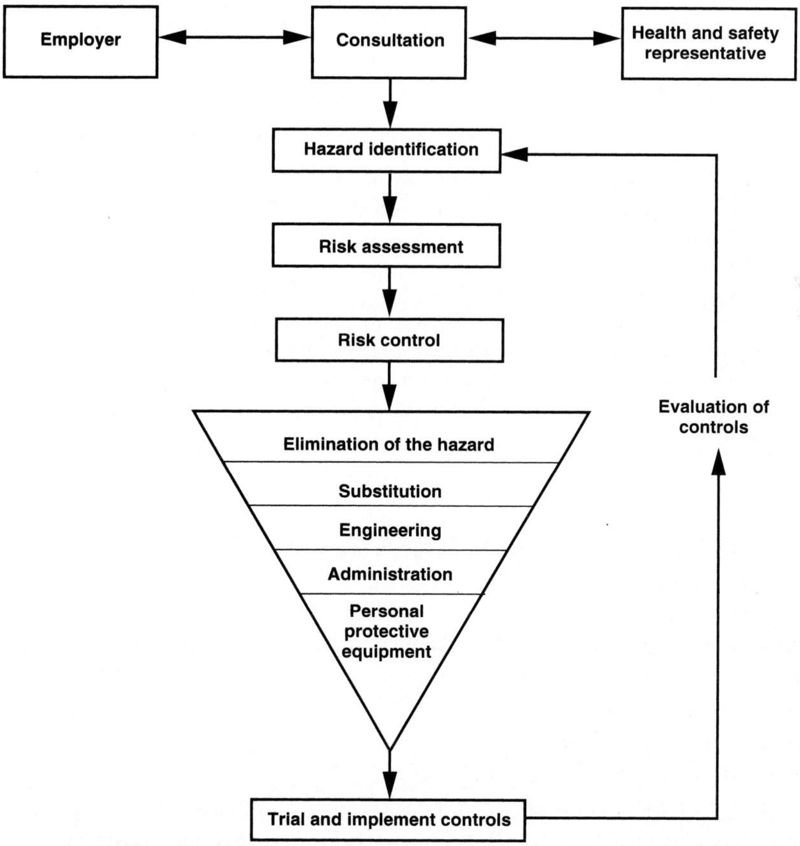

Fig. 6.1 The hazard management approach

Hazard management is a continuous, ongoing process, as technology changes and further options for the control of risks become available. It requires consultation between employers, employees and health and safety representatives when determining the approach and methods to be used.

The Commonwealth occupational health and safety legislation has three major parts. These are:

1. the Act—the parent Act is the *Occupational Health and Safety (Commonwealth Employment) Act 1991*;
2. Regulations made under the Act; and
3. Codes of Practice approved under the Act.

REGULATIONS AND CODES OF PRACTICE

OH&S Regulations

Regulations are laws written under the authority of an Act. The Act often gives broad terms to describe what is required. Regulations are used to give detailed effect to the Act. A person who does not comply with a Regulation can be prosecuted.

OH&S Codes of Practice

A Code of Practice is written to give people practical guidance under the Regulations. A person cannot be prosecuted for not complying with a Code of Practice, but a Code of Practice can be used as evidence that a person did not fulfil his/her general duty under the Act or Regulations, which can lead to prosecution.

Australian Standards

There are many Australian Standards that relate to work in offices. Standards set some basic requirements in the design, development and use of equipment, furniture, plant and work practices in both office and industrial work environments. Australian Standards are not law unless incorporated in Regulations.

ORGANISATIONAL POLICIES AND PROCEDURES

Policies and procedures are set by management and recorded in the corporate procedures manual.

Employer responsibilities

Under the occupational health and safety Acts, the employer has a 'duty of care' to provide a safe and healthy work environment. Below are the steps involved in working out the responsibilities for health and safety in a workplace.

- Top management takes overall responsibility.
- Senior management works out the health and safety implications for your workplace.
- Senior management allocates the responsibility and resources for each activity for each position or person in the workplace.
- Management must work out how to check whether the responsibilities are being carried out.
- The responsibilities and accountabilities are written into a health and safety procedures manual, so that it is clear what the company promises to do to prevent injury and disease and comply with legislation.

Safety committee

A joint employer–employee health and safety committee is appointed by management to check and advise on Regulations and Codes of Practice relating to all health and safety matters in the organisation. The OH&S Act outlines the role of the committee, which is to establish any new OH&S policies and review and amend existing policies. A designated workgroup will be informed about all health and safety aspects of the workplace, inspect the workplace, consult with management and employees, and issue provisional improvement notices.

Training

Employers are required to provide information, training and supervision so that employees can perform their work in a safe manner.

Supervisors should be trained in the knowledge of health hazards in the workplace, and the agreed duties should include policies and procedures on how to respond to such hazards. Training should include the following:

- *specific hazards*—identification, inspections, control and reporting;
- *investigation*—accidents, incidents and issues, with corrective actions;
- *reactive and response activities*—first aid and medical emergencies; reporting illness, injury; reporting incidents, issues and unsafe acts;
- *administrative activities*—record-keeping, buying of first aid medical supplies, issue resolution, consultation on change;
- *legal responsibilities*—to contractors, to visitors, keeping injury records, monitoring new regulations that apply to the workplace.

Training should also provide employees with an understanding of:

- health and safety legal responsibilities;
- the nature of the hazards in the workplace;
- the process of hazard identification, risk assessment and risk control;
- the arrangements for reporting circumstances likely to cause hazards;
- the reasons for and safe use of the risk control measures in place in the workforce;
- safe work practices.

Employee responsibilities

One of your rights as an employee is for your employer to provide a safe environment for you to work in. But equally you, as an employee, have a legal responsibility to observe the health and safety rules outlined in the corporate procedures manual.

- Use equipment correctly and for the purposes intended.
- Take all reasonable care to avoid accidents.
- Be responsible for your own health and safety and do not endanger the health and safety of others.

THE ROLE OF WORKSAFE AUSTRALIA

The role of WorkSafe is to facilitate the adoption of uniform OH&S standards by the Commonwealth, the states and territories; to improve OH&S performance and reduce the incidence, severity and costs of occupational injury and disease in selected national industries; to provide a national focus for Australian OH&S research, responsive to industry needs, and new knowledge that will improve OH&S performance in industry. WorkSafe concentrates on the most serious injury and disease problems confronting workers, which are:

- hearing defects caused by excessive noise;
- workplace exposure to hazardous substances;
- incorrect manual handling, resulting in back injuries;
- injuries caused by mechanical equipment;
- certification of operators of industrial equipment;
- control of major hazardous facilities.

THE ROLE OF UNIONS

The role of unions is to represent the workers if cases of unsafe working conditions are not resolved in the workplace; to act as mediators in claims for sickness and accident workers' compensation. Union representatives have the responsibility for checking working conditions and reporting any potentially hazardous situations.

Comcare Australia

Comcare has offices in all capital cities in Australia. Under the Commonwealth Safety Rehabilitation and Compensation Commission (SRCC) Comcare controls the legislation regulating compensation and rehabilitation claims made by workers in the Commonwealth public service.

Activity 6.1

Work injuries and illness

Form into groups of three. Each student should obtain at least one article from the newspapers regarding a case of work injury or illness, and write a summary of the article. Bring the articles and summaries to a group discussion at a time and location organised by your tutor/supervisor. Aspects for group discussion are:

- details of the injury or illness;
- the type of industry in which it occurred (primary, secondary, tertiary);
- when and where the injury occurred;
- why you think the injury occurred (details to be taken from the article to help you form your opinion);
- what you think should have been done to prevent the accident or illness;
- who you think was responsible for the accident or illness;
- what action was taken after the accident or illness, and who was involved in that action (WorkCover, employer, employee, union etc.).

The leader of the group should appoint one of the members to take notes and another member to write a short report on the results.

PHYSICAL, CHEMICAL AND BEHAVIOURAL HAZARDS IN THE OFFICE

The Commonwealth and state OH&S Acts in general define a hazard as:

> **… anything that could cause harm—a moving forklift, a chemical solvent, a repetitive job, a badly designed workplace.**

Most of the deaths and serious injuries occur in primary and secondary industry in occupations such as: **farming**, where accidents occur with equipment; **building**, where workers fall from high levels; **mining**, where accidents occur underground; and **manufacturing**, through accidents with machinery.

Death or serious injury does not often occur in an office environment, but wrong practices can result in serious long-term illness and pain, causing absence from work for many months, or even years. Frequent visits to doctors and hospitals, and pain and suffering, can lead eventually to the loss of your job, and could affect you for the rest of your life.

Preventive and remedial action

In general, preventive measures under Commonwealth and state OH&S Acts include the following:

- *Provide a safe and healthy workplace.* Use the hazard management approach outlined in this chapter.
- *Recognise the hazards in your workplace.* Check your workplace and your particular work area for hazards. Make a checklist of the present layout of furniture and machines, of what is hazardous and of how you can change it. Use your checklist so that you do not miss new hazards.
- *Assess the hazards in your workplace.* Make sure you know what a hazard is in relation to furniture and machines. Work out how serious the hazards are in your list. Decide which hazards you need to fix first.
- *Control the hazards in your workplace.* Find out about different ways of controlling hazards and decide on the best method of fixing each hazard on your list.

Physical hazards

Use of the computer

Occupational overuse syndrome (OOS). OOS (or, as it used to be called, RSI) may be linked to certain types of physical discomfort or injury. It is caused by performing repetitive tasks that require awkward body positions or forceful exertions, without allowing the body a chance to recover. These include carpal tunnel syndrome (CTS), tendinitis, and tenosynovitis.

Ergonomics. Ergonomics is a recent scientific discipline, which links the use of equipment (particularly computers), the design of furniture, and the work environment, to human performance and wellbeing. Ergonomists have developed methods to prevent work-related illness and injury.

To prevent back and neck problems, **ergonomic furniture** has been designed to enable adjustments to be made to suit individual body height and length of arms and legs while using a computer. If special ergonomic furniture is not provided by your employer, you should adjust the existing furniture to make yourself as comfortable as possible in the following ways:

- Ensure that your chair, work surface and keyboard are in the most comfortable position for you. (We are all different shapes and sizes.)
- Adjust the back of the chair so that there is contact between it and the most curved part of your back and waist level.
- If you have to adjust the height of the seat to enable you to keep your wrists straight, use a footrest to prevent thigh pressure.
- Armrests should not be used when you are typing for long periods of time on a computer.

Figure 6.2 shows the ergonomic position for using a computer.

Correct posture. The position in Figure 6.2 is typically shown as the 'correct posture', but sitting even in this position for any length of time will make your hip joints ache. You should take breaks from sitting at a computer after every 50-minute interval. Work on some other task, in order to give you exercise from sitting in one position for too long.

Maintaining a comfortable body posture can not only lessen your risk of injury but also help you work more comfortably and effectively. A lot of the discomfort, and excessive fatigue, is experienced when using a computer:

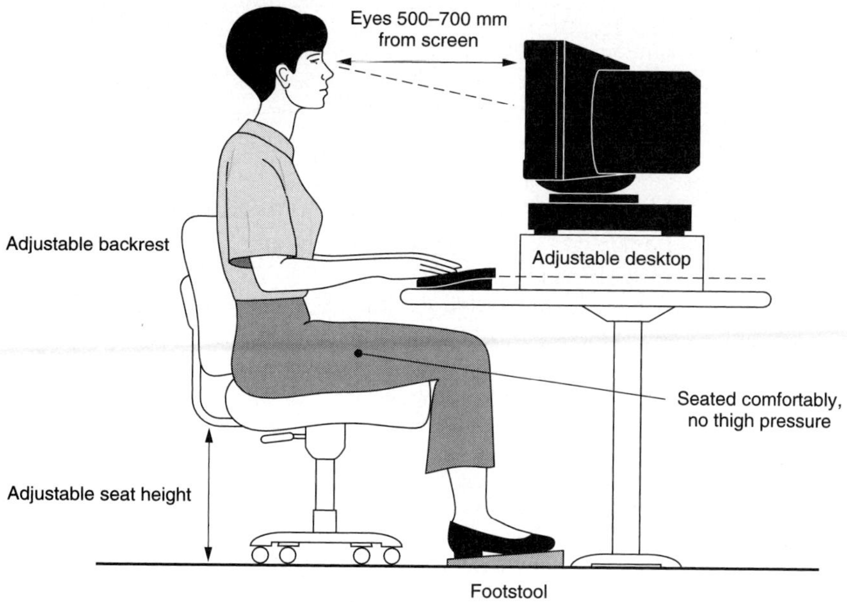

Fig. 6.2 Ergonomic position for using a computer

- from sitting too long and typing too fast;
- sitting in awkward or twisted postures or maintaining the posture for too long;
- reaching often or too far, or using too much muscular effort for relatively easy tasks.

To prevent twisting and excessive amounts of reaching, which could damage your back, arrange the tools around your desk into primary, secondary and reference zones (as described in Chapter 11).

As shown in Figure 6.3, place the keyboard as close to the front edge of the desk as is comfortable. Do not place documents between the keyboard and the front edge of the desk while using the keyboard, as this lengthens the reach distance to the keyboard and may result in excessive bending of the neck to look at the documents. Ensure that there is room to put the keyboard to one side when it is not in use. Note the following points:

- *VDU (screen)*—The screen should be positioned once the chair and desk height has been established.
- *VDU stands*—A variety of VDU stands is available to raise screens to your preferred level.
- *Height*—The screen should be positioned so that the top of the screen is level with or slightly lower than your eyes.
- *Distance from the eye*—First place the screen so that it is approximately an arm's length away from your usual seated position. Move it further away or closer as required.
- *Positioning the screen*—As shown in Figure 6.4, the screen should be placed in front of you so that you avoid glare or reflections from windows. It should also be tilted to stop reflection from overhead light and windows.

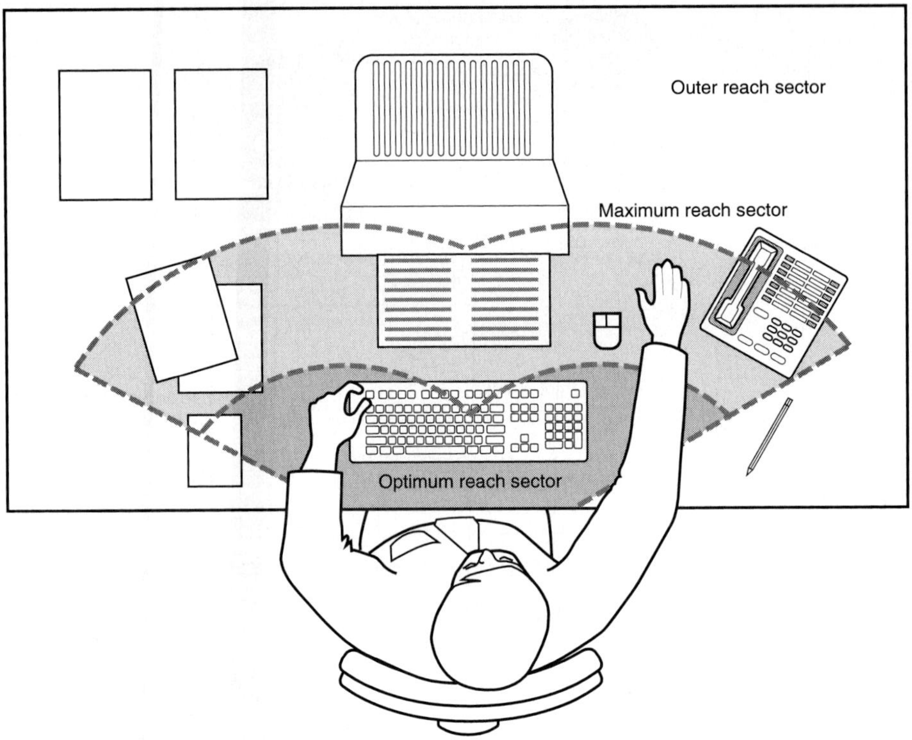

Fig. 6.3 Position of computer and keyboard on the desk

Figure 6.5 shows the position of screen and document holder, if you look evenly between the two. If you use one more than the other, a slight adjustment should be made.

Ergonomic guidelines for keyboarding. It is important when you are using your computer to do the following:

- Position your hands over the keyboard with the elbow at a 90-degree angle (not less). Do not raise your shoulders to keep your arms at right angles, but let your arms relax at your sides.
- Make sure that your wrists are straight when typing. If necessary, adjust the wrist leveller under the keyboard to promote a comfortable and straight-wrist position.
- Move your keyboard well in from the edge of the desk to avoid resting your palms on a hard desk edge.
- Place your monitor at eye level, minimising the bending or twisting of your neck while typing.
- Use a document holder to place documents in such a position that you are not twisting your neck to read them.
- Alternate between using the mouse and the keyboard.
- Use a light touch when pressing the keyboard or holding the mouse.
- Keep the mouse close to the keyboard so that you do not have to reach far.
- If you do not use the keyboard as much as you use the mouse, place the keyboard slightly to one side and the mouse in front of you.

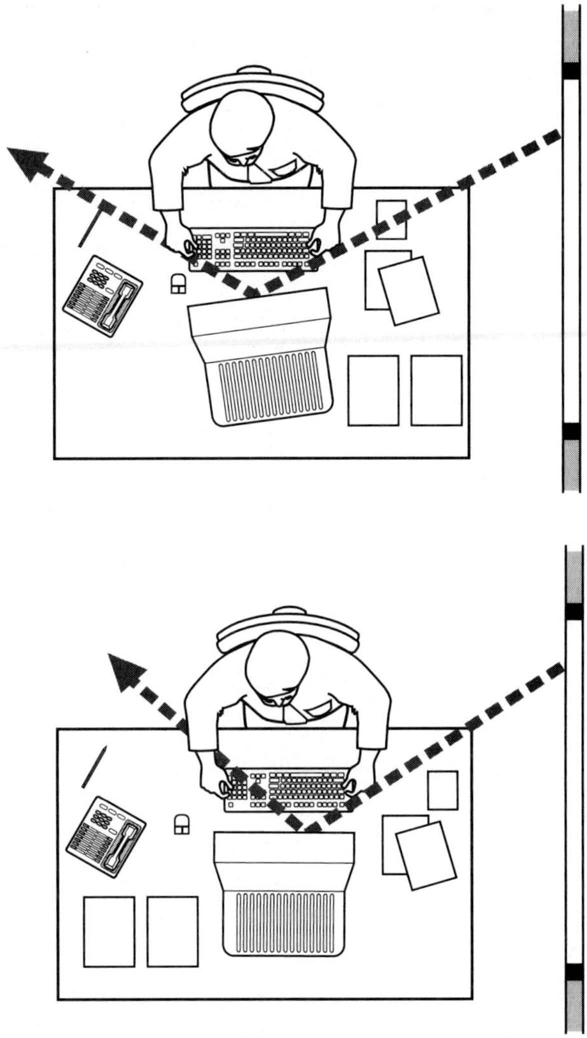

Fig. 6.4 Suitable screen locations in relation to windows

To avoid long-term injury when using a keyboard, vary your tasks in order to alternate your posture. Computer companies have developed special ergonomic keyboards to help you work more comfortably, with hands, wrists and forearms in a more relaxed position. An example of an ergonomic keyboard in given in Figure 6.6.

Ergonomic desk design. A good desk should have:

- rounded corners with no sharp edges;
- good access for legs with no obstacles under the desk to cause discomfort or possible injury (see Figure 6.7);
- a flat, smooth surface for ease of writing, or the correct height for computer use, and of a neutral colour with non-reflective finish.

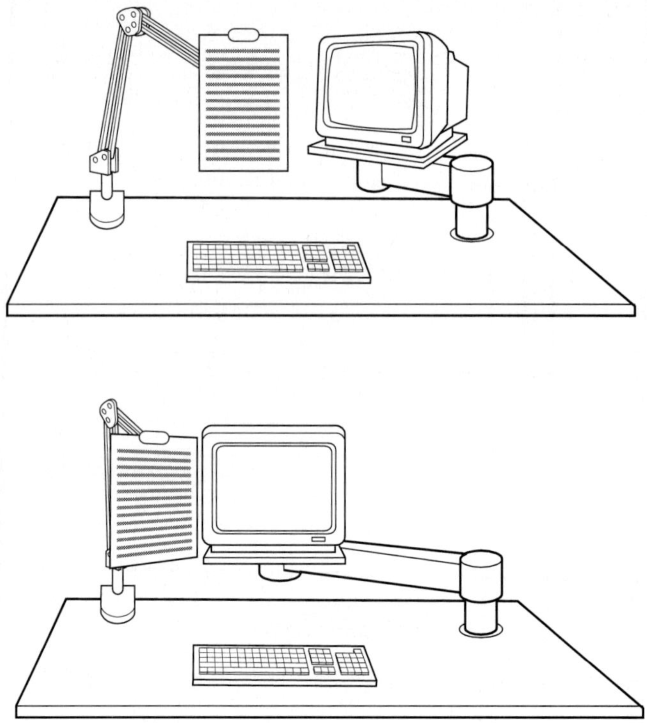

Fig. 6.5 Suitable screen locations when using a document holder

Fig. 6.6 Example of an ergonomic keyboard

When choosing a desk, consider:

- the tasks to be performed and the type of equipment and materials to be used;
- modular work stations that permit flexibility in design and layout.

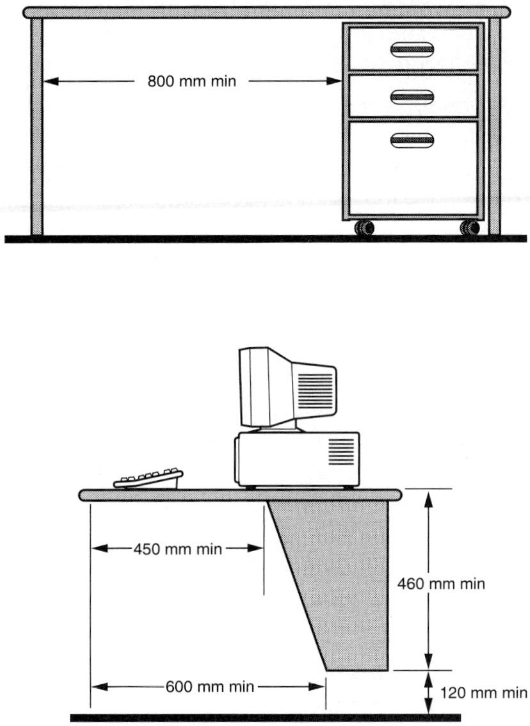

Fig. 6.7 Knee-clearance dimensions for office desks (Source: WorkSafe)

Exercises. Eye strain may be a problem when you are doing any form of close work. When you are using a computer it is a good idea to pause often to rest your eyes. You need only pause for a short time—between about 30 seconds and two minutes.

While you are resting your eyes you can do some exercises, such as these ones suggested by Microsoft:

- Look up and away from the screen. Focus on a distant object (more than 3 m away). For example, look out of the window, or at a picture on a far wall.
- Yawn and blink to keep your eyes lubricated.
- Changing focus and covering your eyes with your hands can help to relieve eyestrain.
- Sit up straight, face forward. Look up, then down. Look left, then right. Repeat this sequence several times without moving your head.
- Make sure there is enough light and that the type of light is correct (see Figures 6.4 and 6.8).
- Make sure there is no glare from the window.
- Use a screen with a light background when working with text.

- Deep breathing can help relieve tension in your muscles.
- Massage your hands and fingers to keep them warm and reduce stiffness.
- Stretching is good exercise for the entire body. Clasp your hands behind your head and gently stretch your elbows back while taking in a deep breath.
- Shrugging your shoulders gently up to your ears and relaxing gently downwards can help to reduce tension or stiffness in the upper back and neck.
- Move out of your seat regularly—at least once every 50 minutes.

Notebook and laptop computers. The portable nature of laptops and notebooks results in their being used in a wide variety of situations and settings where there is limited capacity to adjust a desk. This can result in the work level being too high or too low. Docking laptops or notebooks into a desktop computer at an adjustable work station can avoid these problems. The advantage of docking stations is the capacity easily to connect the laptop to a desktop computer back at the main office, and to use other peripheral devices.

Excessive noise

Occupational health and safety Regulations in all states require employers to assess employees' exposure to noise, which should not exceed exposure standards.

Noise levels are measured in terms of decibels (dB). The recommended levels of sound in an office are: up to 50 dB for a private office; up to 60 dB for a general office; and up to 80 dB for an equipment room. If the level of noise were to rise to between 100 and 130 decibels, the noise would cause pain. The duration of sound will also affect a person's ability to tolerate it. Long periods of high levels of noise—even if the noise does not reach 80 decibels—over 60 decibels can result in:

- decreased efficiency of your work;
- more stress, leading to headaches and other stress symptoms;
- misinterpretation of verbal communication;
- permanent hearing loss.

Preventive and remedial action. In order to reduce noise levels, the following remedial methods should be put in place.

To eliminate as much **external noise** as possible, an office should be located away from a major roadway, a factory or industrial areas. If this is not possible, windows should be double-glazed or constructed of industrial-quality glazing.

Internal noise may be controlled in a number of ways. **Noisy equipment** such as photocopiers, electronic staplers, guillotines, shredders and binders should be located in an area away from the main office and from areas that require concentration or conversation. Equipment that cannot be isolated, such as dot-matrix printers, should have acoustic covers fitted. Machines that produce noise through vibration should be placed on a surface that will absorb the sound and the vibration. (Chapter 3 gives information about the health and safety Regulations for the use of specified equipment, as outlined in the manufacturer's manual for each item.)

Sound absorption should be used to reduce the noise of loud voices, telephones and equipment in an open-plan office. Sound-deadening materials include:

- acoustic ceiling tiles
- acoustic screens
- pot plants

- window blinds and curtains
- carpets
- bookshelves between work stations.

(Chapter 4 gives more information about open-plan offices.)

Poor lighting

Poor lighting at work can affect health and safety by:

- making it difficult to see and recognise hazards in the workplace;
- causing and contributing to eyestrain;
- causing people at work to adopt stressful postures to see properly;
- causing a range of other problems that can directly, or indirectly, affect health and safety.

Remedial action includes:

- increasing or decreasing the number of lights;
- changing the type of light—swapping light globes (incandescent lights) for fluorescent tubes or tungsten lights;
- changing the fittings or housings—providing diffusers to shield bare light globes or tubes to decrease glare and reflection;
- changing the colouring of the light used;
- changing the position of your desk and your computer;
- using blinds to reduce external glare.

The correct positions for artificial lighting are illustrated in Figure 6.8. (The correct positions for natural light from windows are illustrated in Figure 6.4.)

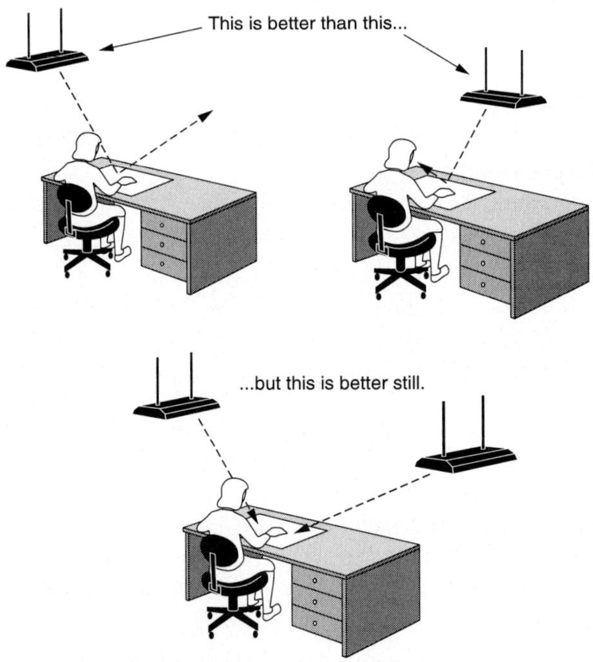

Fig. 6.8 The correct positions for artificial lighting *(continues)*

Fig. 6.8 (cont.)

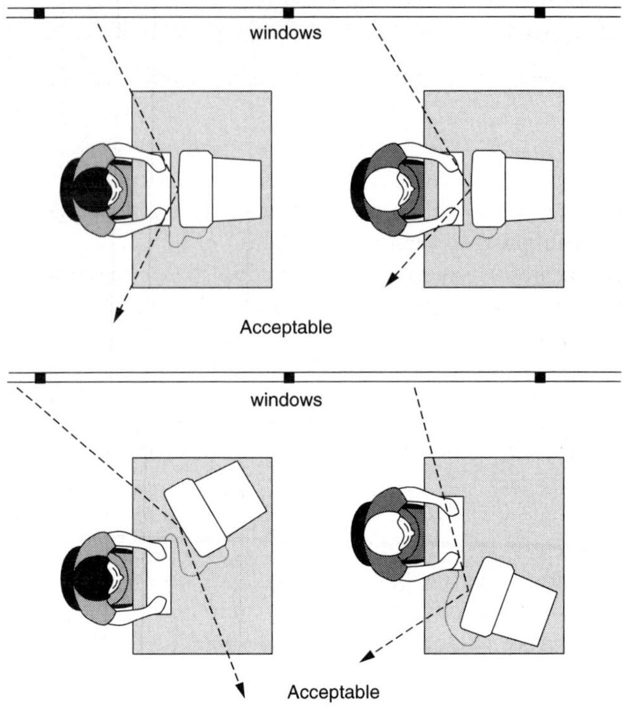

Fig. 6.8 The correct positions for natural lighting

Activity 6.2

Work conditions

With other members of your group, check the furniture in the offices in your building. Describe the desks and chairs. Are they suitable for the work they will be used for? Check the lighting provided for desk surfaces. Is there any natural light? If not, what kind of artificial lighting is provided? Is this the correct kind of light? If not, what should be done to improve this? Write a short report answering these questions.

Manual handling

Under the Occupation Health and Safety (Manual Handling) Regulations in each state, and the Code of Practice for Manual Handling, this is defined as any activity requiring the use of force exerted by a person to lift, push, pull, carry or otherwise move or restrain any animate or inanimate object.

Remedial action. The Regulations require employers to control risk, using the following order of control options.

- *Most preferred*—redesigning to eliminate or reduce the risk, as the first step.
- *Second preference*—provide mechanical aids to reduce risk, and training in their use.
- *Least preferred*—particular training and education appropriate to the task.

Where redesign and mechanical aids are not available, training and education in the correct lifting and handling techniques must be incorporated in health and safety procedures.

Correct lifting and handling techniques. If a heavy object needs to be lifted and moved to another position, the activity should be performed by at least two people. This means that the weight is being shared and there is less risk of back strain or the object being dropped. The use of a trolley to carry materials to and from a central storage area may be required to minimise the demands of this task.

Storage and shelving systems. The general principles of storage areas are as follows.

- Large or heavy items should be stored to minimise the demands of handling. Frequently handled items should be placed within easy reach. Items carried on a trolley should remain on the trolley while in storage.
- Smaller, lightweight and infrequently handled items may be stored in the lower or higher areas of a storage system.
- It should be easy to place items in the storage unit and take them out.
- The storage system should accommodate the size and shape of the item being stored. For example, dividers will secure files stored in shelving and improve access to them. Documents or small publications may be stored in suspension files or folders, making them easier to handle.
- Climbing shelves, or standing on a chair with wheels, to access higher shelves is an unsafe practice. A small platform on rollers or small sets of stepladders should be used, as shown in Figure 6.9.

Fig. 6.9 Incorrect (left) and correct (right) lifting practices

Other physical hazards

Physical conditions of the workplace, such as the condition of the floor surface, poor housekeeping and extreme temperatures, may increase the risk of injury. If a floor has been cleaned and is still wet, a sign should be placed in a prominent position to warn people to take care when walking over the surface.

Fig. 6.10 Signs like the one on the left prevent accidents

Filing cabinets. Never open more than one filing cabinet drawer at the same time, particularly with three- and four-drawer cabinets, as they could tip over onto you, or you might bang your head on the drawer above you, causing serious injury.

Fig. 6.11 Another physical hazard in the office

Compactus or mobile storage. The compactus is a very efficient way to use storage space, but one or two features can present hazards.

- They can become heavy when full of stored files and equipment, and may be difficult to control.
- Large ones, with many openings, can trap people or fingers between the units if more than one person is trying to use the compactus.

- The raised platform or rails can create a tripping hazard.
- It should not require great effort to operate the handle and open the compactus. If the opposite is the case, proper installation and regular maintenance of the unit has not been carried out.

Office layout remedial action. Keep your office clean and tidy. Remove all objects from walkways and from the tops of cupboards or filing cabinets. Make sure that frayed power cables and carpet edges are repaired. Carry out regular hazard management practices as outlined in Figure 6.1 and use regular preventive measures listed under the heading *Preventive and remedial action* above.

Activity 6.3

Physical hazards in the office

Figure 6.12 gives many examples of physical hazards in a poor office working environment. On a separate sheet of paper, list as many hazards as you can find in the picture.

Fig. 6.12 A poor office working environment

Chemical hazards

The following substances, if not used with caution, can have an adverse effect on your health and wellbeing.

- *Correction fluids.* These can cause headaches, eye irritation, inflammation of the skin and other ailments.
- *Thinners for use with correction fluids.* These contain powerful solvents, which can depress the central nervous system, cause fatigue, nausea and lack of muscular coordination.
- *Cleaning fluids or detergents for use on office equipment.* Serious effects on the eyes and skin may result from careless use.
- *Dust, fumes and synthetic mineral fibres.* Allergies are growing more common through increased exposure to dust and synthetic mineral fibres. Over an extended period this exposure may lead to respiratory diseases and ultimately lung cancer, as was the case with exposure to asbestos fibres.
- *Cigarette smoke.* There is now medical evidence to support the dangers of 'passive smoking' (inhaling the fumes from other people smoking cigarettes). Many offices now insist on non-smoking areas.
- *Air conditioning.* Poor or badly maintained air-conditioning systems (particularly water-controlled systems) have been blamed for the spread of disease. In one or two very serious cases of 'Legionnaires' disease', people have died. Hazardous substances have been transmitted through air-conditioning systems, and chills and influenza and various symptoms connected to the 'humidifier effect'.

Remedial action

To prevent injury or illness when using hazardous chemical substances, the following preventive measures should be taken:

- adequate ventilation provided in areas where hazardous chemicals are used;
- training given to all personnel in the use of these materials;
- instructions written down in the procedures manual;
- your office area kept clean and the carpets vacuumed and steam-cleaned regularly to reduce the amount of dust in the atmosphere;
- the air-conditioning system maintained regularly;
- any dangerous chemical materials stored as listed below.

The National Safety Council of Australia Limited provides multilingual signs and posters to warn employees of dangerous hazards. These are particularly useful in manufacturing businesses, warehouses, corridors and rooms in colleges and schools, and all public buildings where people do not speak English.

Storage of hazardous chemical materials

In all states there are OH&S Regulations to control the handling of dangerous goods. This legislation is known as '**HAZCHEM**', and includes a system of warning signs for buildings that store or handle dangerous goods.

An up-to-date 'material safety data sheet' (**MSDS**) must be held for each hazardous substance used at the workplace. Guidance on what should be included in a good MSDS may be found in the approved *Code of Practice for the Preparation of Material Safety Data Sheets (1994)*. A copy of an MSDS must be displayed at one or more accessible points as a register.

Behavioural hazards

How people behave at work sometimes creates hazardous situations. Examples are:

- not operating equipment according to the operating manual and not observing health and safety rules;
- behaving in an irresponsible manner, such as playing practical jokes;
- leaving bags and other objects in walkways;
- not using electrical equipment with due care;
- running, or rushing about quickly, which can cause an accident;
- using chemicals or solvents in a careless or dangerous manner;
- poor management practices;
- bullying by other members of staff or managers and supervisors.

Electrical hazards

As mentioned in Chapter 3, one of the most dangerous hazards is the incorrect use of electrical equipment. To avoid causing electric shocks, burns and fires which could damage the building, the following procedures should always be followed.

- Never let anyone who is not a qualified electrician carry out any repairs to electrical equipment.
- Carry out a regular check of all electrical equipment in your hazard management control procedures.
- Regular maintenance should be carried out on all electrical equipment.
- Check for frayed power cords, as these are also a fire risk.
- Always turn off the power at the wall switch before taking out or putting in a power plug.
- Never run too many items of equipment off one power plug. An electrician should provide you with one power point for each item of equipment.
- Protect electrical equipment when you know a thunderstorm is approaching by turning off power switches and unplugging power cords from outlets. If this is not possible, a power-surge protector should be permanently connected to all power points using electrical equipment.
- At the end of the day, you should get into the habit of switching off all electrical equipment where possible.

Not carrying out any of the above procedures could cause a fire, which may destroy not only your office but the whole building. We see items of this nature in the news all the time. Standard fire procedures should also be carried out regularly and a fire officer appointed to supervise these procedures.

Standard fire orders

In all public buildings there are standard procedures in case of fire hazard. Procedures for the evacuation of the building must be outlined in the procedures manual. A chief warden is appointed to control any emergency situation. **Fire extinguishers** are placed at strategic points in the building, and are colour-coded and labelled for use with different categories of fire hazards. Check that you or other members of staff know how to operate fire extinguishers. Standard Fire Orders on what to do and where to go in case of a fire (see below) must be placed on the wall of every room.

Standard Fire Orders

ACTIONS TO BE CONSIDERED ON DISCOVERING A FIRE

R 'Rescue' any person/s in immediate danger **ONLY** if safe to do so.
A **'Alarm'**—Raise the alarm and follow your emergency procedures.
C **'Contain'**—Close doors to contain the fire.
E **'Extinguish'**—Attempt to extinguish the fire only if you are trained and it is safe to do so.

Assembly point: In large public and private institutions, you are asked to go to your assembly point if there is a fire alarm or drill.

Activity 6.4

Health and safety notices and equipment

With other members of your group, arrange a tour of your college building or work building. Look for the following:

- notices of hazardous chemical storage;
- the different types of fire extinguishers, and how to use them;
- signs for Standard Fire Orders and evacuation procedures (Are they in every room?);
- any other occupational health and safety signs.

Write a short report listing the notices and equipment, and indicating whether the building is using the correct procedure for health and safety precautions or whether it should be improved.

Other factors affecting work performance

All the factors already discussed in this chapter can affect your work performance, but there are others. Behavioural factors include:

- taking prescribed or unprescribed drugs;
- excessive alcohol intake, which can affect your performance at work the following day;
- drinking excessive amounts of tea or coffee, which can make you hyperactive, with resulting loss of sleep and damage to your health;
- staying out late on many evenings during the week, which can make you tired and listless the following day;
- disturbed patterns of sleep occasioned by shiftwork or small children, which can cause a lack of tolerance of others.

Some other environmental factors that can affect your work performance are: interpersonal relationships with other members of staff or your supervisor; and poor working conditions, including the layout and decor of your office. Even colours can affect how you feel. The choice of colour can make your work area light, welcoming and stimulating, or dark and depressing. The best colours for an office are usually light cream, or light grey with a matt finish to prevent reflective glare.

Stress

Too much stress is one of the major contributors to illness in an office environment. **Stressors** are factors that cause stress in the workplace and in your personal life. Examples are:

- unrealistic work requirements (work overload, resulting in working long hours to complete tasks);
- poor time management skills;
- poor management communication skills;
- changing situations in the workplace;
- working irregular hours (shiftwork);
- repetitive work with poor promotional prospects;
- productivity-linked wage systems;
- poor interpersonal relationships with colleagues and/or supervisor/manager;
- increased workload due to reduced staff levels.

Common behavioural signs of excess stress. These include:

- insomnia and poor sleep patterns;
- inability to concentrate;
- intolerance to noise;
- irritability and short temper;
- reduction in willpower;
- poor memory;
- overreaction to situations at work or at home;
- increased use of tobacco, alcohol and drugs.

Common physical symptoms of excess stress. These include:

- feeling of tension in the back or neck;
- increasing incidence of headaches linked with tension;
- excessive tiredness;
- dizziness, and feelings of faintness;
- difficulty in breathing;
- unexplained rashes or skin irritations;
- excessive perspiration, especially of the hands, when the environment is not warm;
- indigestion and stomach upsets;
- palpitations or irregular heartbeat.

Managing stress (remedial action). Some methods of managing stress are:

- using better time management (as described in Chapter 11);
- better workplace and work station design;
- having regular breaks when using a computer, by varying your work tasks;
- not skipping your morning tea breaks and lunch breaks;
- enriching your job and making it more interesting;
- keeping in better health through regular exercise and regular relaxation;
- reducing exposure to physical and chemical hazards.

There are many types of stress management relaxation techniques. Some involve yoga, massage therapy, flotation tanks and meditation. Just relaxing at home with a good book or your favourite video can take your mind off your problems at work.

Many people find that going to the gym after work (particularly if you have a sedentary job) helps to provide relaxation and fitness. Aerobics, swimming, playing tennis, walking or cycling stretch muscles and help to relax the body and the mind.

Performing some exercises while you are at work can help to reduce tension and stress. Some have already been mentioned in this chapter. The following are a few more:

- stop, stretch and check
- head roll
- chin tuck
- shoulder roll
- shoulder stretch
- wrist stretch
- upper and lower back stretch
- back arch
- foot rotation
- deep breathing.

Activity 6.5

Stress

(You may complete this activity as part of a team or on your own.)

There are many articles that have been written about stress in the workplace, as this is the most common form of illness in the work environment today. Search in the library, or on the Internet, for one article on stress for each student in the group. Bring the articles to your next class and hold a discussion on their contents. After the discussion, each student should write a short report on stress management.

When completed, hand the report to your tutor/supervisor for assessment.

FIRST AID FACILITIES

The Occupational Health and Safety (First Aid) Act in your state and the Workplace Code of Practice provide guidance in the provision of first aid facilities. Organisations must provide persons suitably trained to give first aid in a working environment.

First aid officers (other than medical practitioners or trained nurses) who are responsible for first aid in an organisation should:

- hold a current occupational first aid certificate;
- obtain recertification after three years;
- attend training or retraining in an accredited course during normal working hours;
- be paid a first aid allowance by the organisation.

To comply with legal requirements, when an accident occurs, or someone is taken ill, you are required to call a registered first aid officer to deal with the situation.

In a large organisation a **first aid room** or area should be provided for the treatment of illness and accidents. This should be fitted with:

- a couch, where someone who is feeling ill can lie down or an accident victim can be treated;
- facilities such as blankets, pillows, a screen, a workbench, a sink with hot and cold running water;
- a first aid kit.

First aid kit

There are three sizes of first aid kit (A, B and C). The size to be held in the workplace depends on the total number of employees. First aid kits contain items such as:

adhesive sterile plastic dressing strips
disposable gloves
dressings (sterile)
safety pins
scissors
adhesive strips and tape

eye pads (sterile)
eyewash solution
gauze bandages
splinter forceps
wound dressings, bandages and swabs

A first aid pamphlet approved by the OH&S should also be enclosed.

Activity 6.6

First aid

With other members of your group, obtain an OH&S first aid pamphlet from the first aid officer in your organisation. If this is not available, find the correct source in the telephone directory and ask for one to be sent to you. From the first aid pamphlet, check the facilities that should be provided in the first aid room and list items that should be stored in the first aid kit, some of which may not be listed above.

Injuries and illness in the office

Strains, sprains, cuts and bruises are the most common injuries occurring in offices. Policies and procedures for first aid in the office should be put in place by management, which should ensure the implementation of an effective approach to the management of injuries. The Victorian Code of Practice for First Aid in the Workplace 1995 provides guidance for establishing procedures to deal with minor injuries and illness at work (see Figure 6.13). Currently there is no Commonwealth Code of Practice on first aid.

WORKCOVER (WORKERS' COMPENSATION)

WorkCover is the Commonwealth government scheme for workers' compensation in all states of Australia. Any person who receives a payment for providing a service within a worker/employer relationship is covered by WorkCover. However, sole proprietors, the self-employed, trustees, partners and some contractors are not covered.

Comcare controls the legislation regulating compensation and rehabilitation claims made by Commonwealth public servants.

Types of eligible claims

For you to claim compensation, your injury or illness must arise out of or in the course of your employment, and employment must be a significant contributing factor. This includes:

- when temporarily absent during authorised recess;
- while travelling for the purpose of employment;
- where an injury or illness has occurred over time.

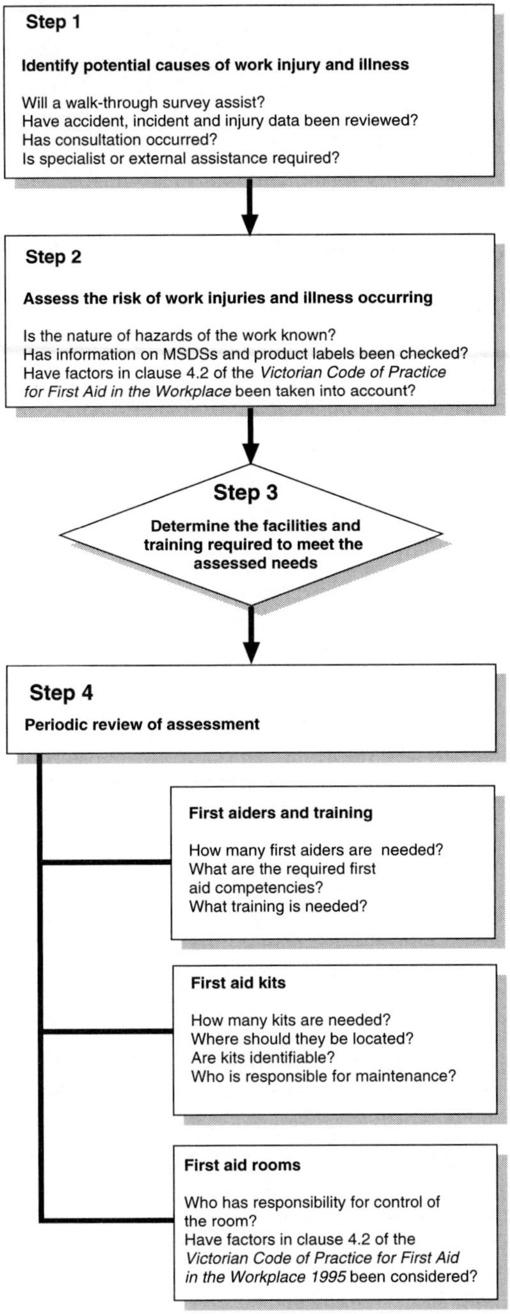

Fig. 6.13 Process for assessing first aid facilities and training in the workplace (Source: Victorian Code of Practice for First Aid in the Workplace 1995)

Claims may be made for:

- medical expenses
- weekly benefits
- permanent disability
- death.

How to report injury or illness

The steps involved if an injury or illness occurs are:

- reporting the injury or illness to the supervisor;
- initial treatment;
- completing a notification of accident form;
- obtaining a medical certificate;
- completing a certificate of incapacity to work;
- claiming for compensation (see Figure 6.15);
- returning to work.

In the case of an accident, the employee must complete a notification of accident form, detailing his or her name, address and date of birth, and the time, date, location and nature of the accident. A witness to the accident is required to enter details in the space provided. State legislation requires an employer to record the details of each accident or incident in an accident record register (see Figure 6.14). This is kept for up to five years in case of workers' compensation claims.

FIRST AID REGISTER OF INJURIES AND TREATMENT		
Name of injured person:	Date of birth:	Male ☐ Female ☐
Address:		Occupation:
Location accident occurred:		
Industry:	Activity:	
Date of injury:	Time of injury:	
Brief description of the type, cause and location of injury (e.g. hand, arm, foot):		
Treatment given:		
Referred to:		
Staff ....Visitor ....Contract ....Other		
Signed ....(first aid officer) Date.................................		

Fig. 6.14 First aid register of injuries and treatment

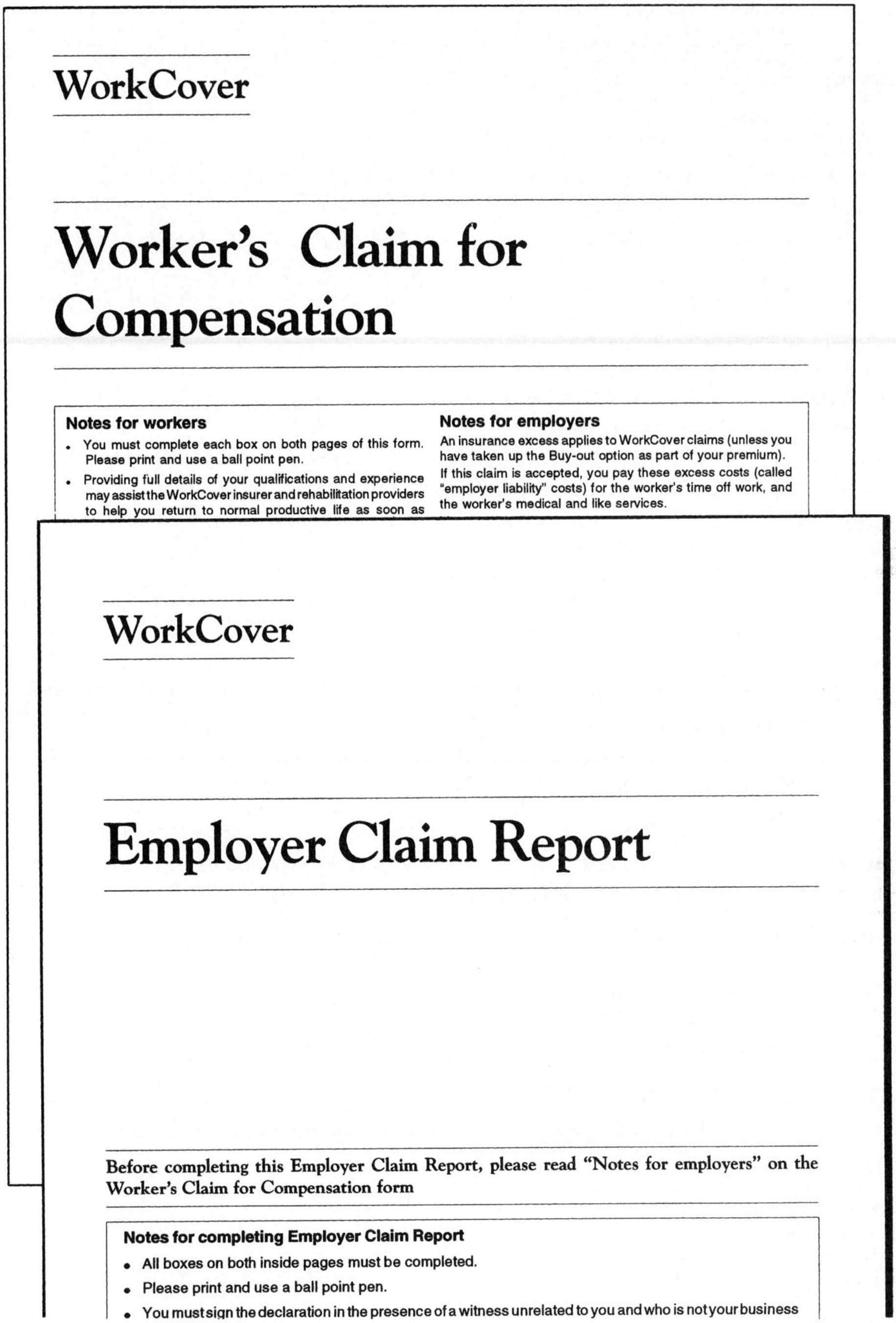

WorkCover

Worker's Claim for Compensation

Notes for workers

- You must complete each box on both pages of this form. Please print and use a ball point pen.
- Providing full details of your qualifications and experience may assist the WorkCover insurer and rehabilitation providers to help you return to normal productive life as soon as

Notes for employers

An insurance excess applies to WorkCover claims (unless you have taken up the Buy-out option as part of your premium).

If this claim is accepted, you pay these excess costs (called "employer liability" costs) for the worker's time off work, and the worker's medical and like services.

WorkCover

Employer Claim Report

Before completing this Employer Claim Report, please read "Notes for employers" on the Worker's Claim for Compensation form

Notes for completing Employer Claim Report

- All boxes on both inside pages must be completed.
- Please print and use a ball point pen.
- You must sign the declaration in the presence of a witness unrelated to you and who is not your business

Fig. 6.15 Forms for worker's claim for compensation and employer claim report

Activity 6.7

Reporting an injury

The following is a conversation between you, the receptionist, and Pam Jones, the junior typist. Pam cries out:

'Come quickly. Jane has fallen. She's hit her head on the table and she's knocked herself out. There's blood everywhere. What's she going to do? She's got three small children and can't afford to be off work. John, her husband, is in Canada. Oh, what can I do?'

'Pam, show me where Jane is.'

Pam is gesticulating and crying. You follow her and survey the scene. Jane has indeed fallen and it looks as though she has banged her head on the desk, as there is blood on the corner of it. There is blood over her left ear. Files are scattered about. There is a swivel office chair but no steps, and it looks as though Jane has climbed first onto the chair, then onto the table to retrieve some files from the top shelf. It also looks as though she has missed her footing and fallen. She must have leaned over too far, judging by the position of the swivel chair and table. Jane's arm looks broken. She has fallen on her side and appears to be unconscious, but you see that she is breathing. Take charge.

1. How would you care for Jane? **(Because she is unconscious, and bleeding, your first priority is to ring for an ambulance. Leave Jane where she is. You have noted that she is breathing. There is nothing more you can do.)**
2. Complete the accident report form on Activity sheet 6.7 in your workbook. **(From the conversation, it is obvious that there is a lot of irrelevant information. It is your job to give the facts only.)**
3. Indicate on the diagram of the human body just where the injury is.
4. As you know the reason Jane fell was because she did not have the right equipment to reach the files (i.e. a stepladder), write a memorandum to your safety officer and ask her if she will talk to the staff about safety in the office.
5. Write a memorandum to the manager of your company, explain what has happened to Jane, and ask if you can buy a suitable stepladder immediately.

Case study 6.1

Dealing with an injury

You are the administrative coordinator for Business Stationery Supplies Pty Ltd. At 10 am, the receptionist telephones you to say that Julie Campitelli, one of the clerks, has tripped over a box left in the passage. She has fallen heavily and injured her arm and hand.

1. What are the implications of this accident that may affect Julie?
2. What are the implications of this accident that may affect the organisation?
3. When you attend Julie, what would you do to help her?
4. What steps would you advise Julie to take?
5. What steps would you take to report this accident?
6. What steps would Julie take after the accident happened?

Summary

- Work can be dangerous. On an average day in Australia, at least one worker will be killed, 300 will injure their backs, 20 will be diagnosed with hearing loss and a dozen will break at least one bone.
- Few people are fatally injured in the office, but many injuries and illnesses occur through poor management practice; lifting; sitting at the computer for too long; excessive noise; incorrect lighting; and chemical hazards.
- Under the occupational health and safety Acts of the Commonwealth and all states, employers are required to 'provide and maintain so far as is practicable for employees a working environment that is safe and without risk to health'.
- Occupation health and safety Regulations are laws written under the authority of an Act. A Code of Practice is written to give people practical guidance under the Regulations.
- Top management takes overall responsibility for occupational health and safety activities in the workplace.
- A joint employer–employee health and safety committee is appointed by management to check and advise on Regulations and Codes of Practice.
- Supervisors should be trained in the knowledge of health hazards in the workplace, and the agreed duties should include policies and procedures on how to respond to such hazards.
- You, as an employee, have a legal responsibility to observe health and safety rules in the workplace, by using equipment correctly and taking all reasonable care to avoid accidents.

- WorkSafe Australia is a tripartite body with representatives of the Commonwealth, state and territory governments, the Australian Chamber of Commerce and Industry (ACCI) and the Australian Council of Trade Unions (ACTU).
- Comcare controls the legislation regulating compensation and rehabilitation claims made by Commonwealth government workers.
- Physical, chemical and behavioural hazards in the office environment are defined as anything that could cause harm, such as a chemical solvent, a repetitive job or a badly designed workplace.
- Ergonomics is a theory that links the use of equipment, the design of furniture, and the work environment, to human performance and wellbeing. Ergonomic furniture has been designed to prevent back and neck problems.
- Arrange your workspace into primary, secondary and reference zones to prevent twisting and excessive reaching.
- A lot of discomfort experienced when using a computer occurs from pushing the body beyond its natural limits. This can result in occupational overuse syndrome (OOS). To prevent injury, breaks and exercise should be taken at regular intervals.
- Excessive noise levels can cause ear damage. Steps should be taken, such as placing noisy equipment away from work areas and using sound absorption materials in an office, to prevent injury.
- Poor lighting in a work environment can make it difficult to see hazards and cause eyestrain, poor posture and a range of other problems, which can affect the health of a worker.
- Correct lifting and handling techniques should be used to prevent injury.
- Factors that affect work performance include taking drugs and excessive amounts of alcohol, coffee and tea.
- Stress is one of the major contributors to illness in an office environment. Stressors are factors that cause stress, including poor time management, poor management communication skills, working irregular hours, performing repetitive work, and poor interpersonal relationships.
- The OH&S First Aid in the Workplace Code of Practice provides guidance in the provision of first aid facilities in a work environment.
- WorkCover is the government scheme for Commonwealth government workers' compensation in all states of Australia.
- The steps involved in reporting injury or illness are:
  —reporting the injury/illness to the supervisor
  —initial treatment
  —completing a notification of accident form
  —obtaining a medical certificate
  —completing a certificate of incapacity to work
  —claiming for compensation
  —returning to work.

Review questions

(These may be used for group discussion.)

1. Write a paragraph on why work can be dangerous.
2. List the kind of injuries and illnesses that can occur in an office.
3. State the name of the occupational health and safety Act in your state.
4. What are the employer's obligations under this Act?
5. Describe briefly an OH&S Regulation.
6. Describe briefly an OH&S Code of Practice.
7. List employer responsibilities under the OH&S Act.
8. List employee responsibilities under the OH&S Act.
9. What is the role of WorkSafe Australia?
10. What is the role of unions in relation to OH&S?
11. List possible physical hazards in an office.
12. Write a paragraph about 'ergonomics', giving examples.
13. Write a paragraph on 'occupational overuse syndrome' (OOS).
14. List some of the steps you can take to avoid these injuries.
15. What remedial action would you take to reduce noise in your office?
16. What steps could you take to improve lighting in your office?
17. List some of the hazardous chemical materials in an office that can cause illness.
18. List some of the factors that cause stress, and some of the common behavioural and physical symptoms of excess stress.
19. What steps can you take to manage stress?
20. Write a paragraph about workers' compensation.
21. Why should you report an accident immediately after it happens?

Competency self-assessment and/or tutor/supervisor assessment

When you have finished reading this chapter, and have completed the activities chosen by your tutor/supervisor (**practical in class, at work or theoretical**), fill in this self-assessment. Then hand it to your tutor/supervisor for official assessment.

Learning outcomes	**Self-assessment**		**Tutor/supervisor assessment**	
I have successfully completed the following learning outcomes	**Yes** (please tick)	**No**	**Yes** (please tick)	**No**
1. Occupational health and safety responsibilities required by Commonwealth and state legislation and their implementation in the workplace				
1.1 general aims and objectives of existing Commonwealth and state occupational health and safety legislation	☐	☐	☐	☐
1.2 employer and employee responsibilities, rights and obligations relating to occupational health and safety	☐	☐	☐	☐
1.3 the major functions of safety committees and representatives	☐	☐	☐	☐
1.4 the role of WorkSafe Australia	☐	☐	☐	☐
2. Potential hazards in an office environment and remedial action				
2.1 physical, chemical and behavioural hazards that may cause injury or health problems in an office environment	☐	☐	☐	☐
2.2 remedial action	☐	☐	☐	☐
3. The cause of occupational overuse injuries of office workers and preventive methods				
3.1 work methods and conditions that can lead to occupational overuse injury, and preventive and remedial actions	☐	☐	☐	☐
3.2 the basic principles of correct posture	☐	☐	☐	☐
3.3 correct lifting and handling techniques	☐	☐	☐	☐

4. The accident/incident and reporting procedures in the workplace				
4.1 the procedures for reporting an accident/ incident/injury in the workplace	☐	☐	☐	☐
4.2 complete an accident/incident report form for a real or simulated accident/incident	☐	☐	☐	☐
4.3 the purpose of workers' compensation	☐	☐	☐	☐
4.4 the procedure for claiming workers' compensation	☐	☐	☐	☐
4.5 the first aid facilities that should be found in the workplace and their location	☐	☐	☐	☐
5. Factors affecting work performance				
5.1 substances that are subject to misuse	☐	☐	☐	☐
5.2 stressors	☐	☐	☐	☐
5.3 the impact of stress and substance abuse on work performance	☐	☐	☐	☐
Date:	Tutor/supervisor signature:			

CHAPTER

7

RECEPTION PROTOCOL AND CUSTOMER SERVICE

ENT201A, NOS137, 7365W,
ENT301A, NOS144, 7370J,
ENT302A

objectives

The purpose of this chapter is to help you to understand the procedures necessary for receiving and attending to visitors' needs at the reception desk, and projecting a positive corporate image, in order to complete the following learning outcomes:

1. **Receive visitors and attend to visitors' needs to ensure a positive image of the organisation**
   - 1.1 receive visitors promptly and courteously
   - 1.2 establish visitors' requirements through appropriate and polite questioning and active listening
   - 1.3 interact courteously with clients and project a positive corporate image
   - 1.4 direct or escort visitors to destination as appropriate
   - 1.5 explain courteously reasons for delay/non-availability
   - 1.6 identify situations outside area of responsibility and obtain assistance from appropriate person promptly
2. **Respond to people in a variety of situations**
   - 2.1 recognise existence of individual and cultural differences
   - 2.2 adjust responses appropriately to meet needs of others
   - 2.3 identify and respond appropriately to those aspects of a foreign culture which affect workplace relationships
3. **Respond to and act on internal/external enquiries to promote the products and services of the organisation**
   - 3.1 deal promptly and courteously with enquiries
   - 3.2 provide relevant information to meet enquirers' needs
   - 3.3 refer enquiries outside area of responsibility to nominated person/section for resolution
   - 3.4 take follow-up action where required
4. **Customer service, corporate presentation, customer relations**

INTRODUCTION

Many people regard the position of a receptionist as being low in the hierarchical structure of the organisation. In fact, the receptionist is one of the most important people. Why?

The receptionist is the first person who is encountered by someone from outside the business, either over the phone or in the office. The clients or customers are crucial to your organisation: without them there would be no business. Therefore, the top priorities of a receptionist, and of all other personnel in the organisation, are to keep present customers/clients happy and to attract new ones. You may have heard the saying 'The customer is always right'. Obviously the customer is not always right, but if you don't keep customers satisfied with your service they will take their business to another organisation.

The other reason why the receptionist is so important is because she or he is the link between personnel in all internal departments and the customers/clients outside the business (see Figure 7.1). How the receptionist carries out specified duties, and receives visitors, can have a significant impact on the public relations of the organisation, and can create either a positive or negative corporate image.

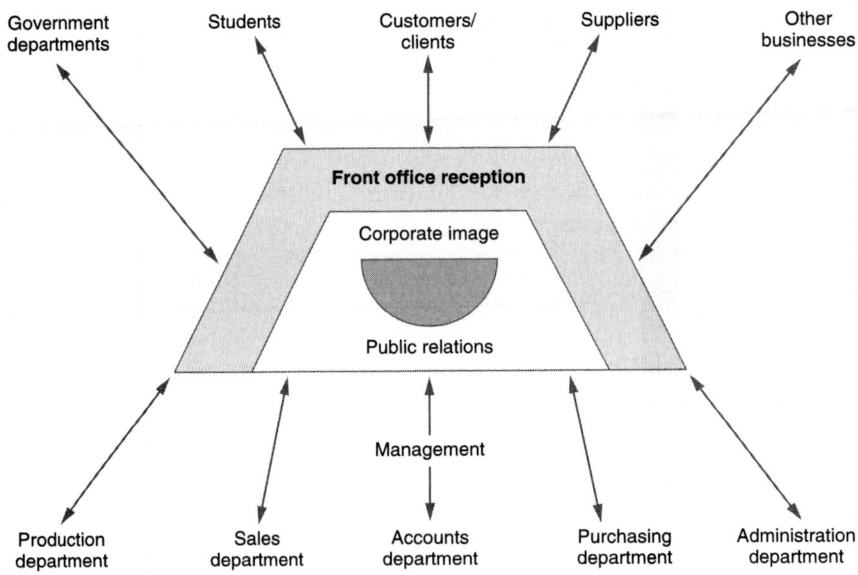

Fig. 7.1 Importance of receptionist to corporate image

RECEPTION DUTIES

There are many tasks a receptionist performs, apart from answering the telephone and greeting callers at the reception desk. Some of these are listed in Table 7.1. In order to perform all these duties effectively, you will need a number of resources to assist you in scheduling appointments, taking messages, and recording mail items and the movement of internal personnel. Some of these are described below.

Main duties	*Probable other duties*
Dealing with visitors • customers/clients • sales representatives • business associates • senior management Operating switchboard • receiving and screening incoming calls • handling enquiries • dealing with complaints • transferring incoming calls • making outgoing calls • taking and passing on messages	Scheduling appointments Typing correspondence Controlling incoming and outgoing mail Controlling office supplies Typing envelopes and labels Receiving and sending documents by courier service

Table 7.1 Reception duties

Scheduling appointments

Scheduling appointments involves the use of a desk diary or an electronic diary system, as shown in Figures 7.2 and 7.3.

Internal telephone directory

You will need to know who is responsible for specialist areas of the business, so you should have a copy of the organisation chart and an internal telephone directory, which records the names and telephone numbers of all personnel who work in the business. If you do not have an internal telephone directory, you can easily make one by keeping a list of names and telephone numbers in a paper or computer file. When people leave or new people start you can keep your records updated by changing your list and printing out a new one for your folder (see Chapter 8).

In/out indicator

You can keep track of the time key personnel leave and return to the building by using an 'in/out board' (see Figure 7.4) fixed to the wall, or a book in which they record the times they leave and return.

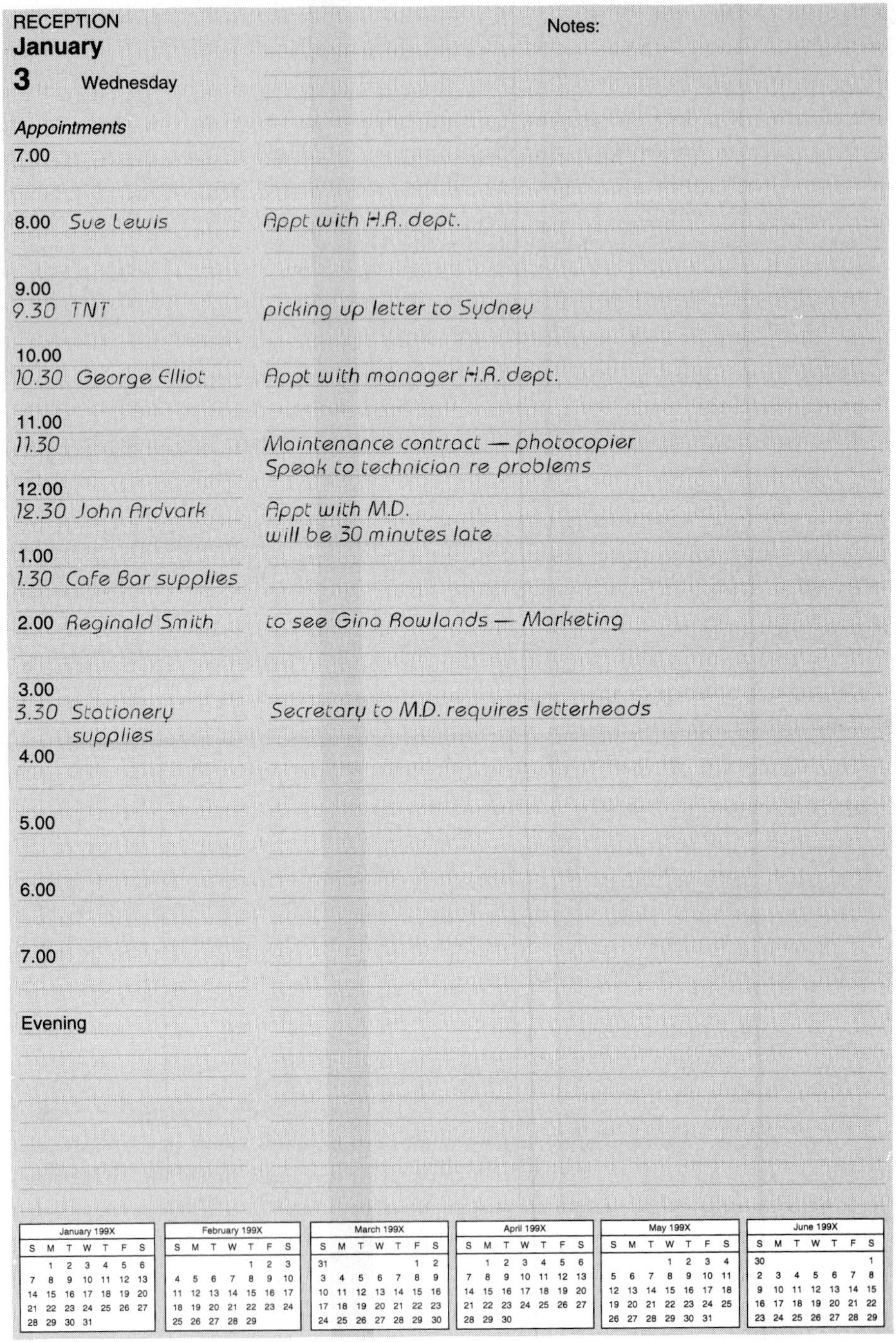

RECEPTION
January
3 Wednesday

Notes:

Appointments

		Notes
7.00		
8.00	Sue Lewis	Appt with H.R. dept.
9.00		
9.30	TNT	picking up letter to Sydney
10.00		
10.30	George Elliot	Appt with manager H.R. dept.
11.00		
11.30		Maintenance contract — photocopier Speak to technician re problems
12.00		
12.30	John Ardvark	Appt with M.D. will be 30 minutes late
1.00		
1.30	Cafe Bar supplies	
2.00	Reginald Smith	to see Gina Rowlands — Marketing
3.00		
3.30	Stationery supplies	Secretary to M.D. requires letterheads
4.00		
5.00		
6.00		
7.00		

Evening

January 199X

S	M	T	W	T	F	S
	1	2	3	4	5	6
7	8	9	10	11	12	13
14	15	16	17	18	19	20
21	22	23	24	25	26	27
28	29	30	31			

February 199X

S	M	T	W	T	F	S
				1	2	3
4	5	6	7	8	9	10
11	12	13	14	15	16	17
18	19	20	21	22	23	24
25	26	27	28	29		

March 199X

S	M	T	W	T	F	S
31					1	2
3	4	5	6	7	8	9
10	11	12	13	14	15	16
17	18	19	20	21	22	23
24	25	26	27	28	29	30

April 199X

S	M	T	W	T	F	S
	1	2	3	4	5	6
7	8	9	10	11	12	13
14	15	16	17	18	19	20
21	22	23	24	25	26	27
28	29	30				

May 199X

S	M	T	W	T	F	S
			1	2	3	4
5	6	7	8	9	10	11
12	13	14	15	16	17	18
19	20	21	22	23	24	25
26	27	28	29	30	31	

June 199X

S	M	T	W	T	F	S
30						1
2	3	4	5	6	7	8
9	10	11	12	13	14	15
16	17	18	19	20	21	22
23	24	25	26	27	28	29

Fig. 7.2 Paper diary system

Reception appointment diary

October 199X						
S	M	T	W	T	F	S
1	2	3	4	5	6	7
8	9	10	11	12	13	14
15	16	17	18	19	20	21
22	23	24	25	26	27	28
29	30	31				

Tuesday, 10 October 199X

		Notes:
8:00	Sue Lewis appointment with H.R. dept	
8:30		
9:00		
9:30	TNT picking up letter to Sydney	
10:00		
10:30	George Elliot appointment with Manager H.R. dept	
11:00		
11:30	Maintenance contract for photocopier	Speak to photocopy technician re problems
12:00		
12:30	John Ardvark—appointment with managing director	John Ardvark will be 30 minutes late
1:00		
1:30	Cafe Bar delivering supplies	
2:00	Reginald Smith to see Gina Rowlands—Marketing	
2:30		
3:00		
3:30	Stationery supplies being delivered	Secretary to MD needs letterheads as soon as they are delivered
4:00		
4:30		
5:00		
5:30		
6:00		

Fig. 7.3 Computer diary system

NAMES	IN	9 am	10 am	11 am	12 noon	1 pm	2 pm	3 pm	4 pm	5 pm	OUT
Aslim							•				•
Barker	•										
Chan	•										
Gerrard				•							•

Fig. 7.4 In/out board

Activity 7.1

A visit to the reception desk

Form a group and visit the main reception desk of your college, school or work. Take note of the following details and activities, and fill in Activity sheet 7.1 in your workbook.

- Is the atmosphere of the reception area pleasant and welcoming?
- Is the decor attractive?
- Is there an area where visitors can sit down and wait?
- Are there magazines, and facilities for refreshments?
- Are there pot plants?
- Are visitors welcomed promptly and in a friendly manner?
- Are the people at the reception desk dressed appropriately?
- Is there an 'in/out board' or a record book for staff leaving and returning?

Reception protocol for visitors

There are three different types of visitors to the office. These are listed in Table 7.2.

1. Those expected (with appointments)
   - customers/clients transacting business
   - representatives from other businesses
   - applicants for jobs
   - visitors from different countries and cultures
2. Those expected without appointments who provide regular services
   - couriers making deliveries and collections
   - maintenance and service workers
3. Those not expected (without appointments)
   - people making enquiries about products/services
   - customers making complaints
   - sales representatives from other businesses
   - people from charity organisations asking for donations

Table 7.2 Types of visitors

Dealing with these different types of visitors involves the communication process (dealt with in more detail in Chapter 11). The protocol for receiving all types of visitors must always be the same. Receive and welcome visitors promptly, pleasantly and courteously, and put them at ease by giving them your undivided attention and using their name often. This is not as easy as it sounds, particularly if you have a busy reception desk. The question is often asked:

> **What do I do if I am already on the telephone dealing with a client, and a visitor walks up to the reception desk?**

The correct procedure is to make sure your visitor does not wait for long (which annoys most people). If your telephone conversation is not nearly finished, you courteously and conveniently break it off *very briefly*, asking the caller to hold the line for just a moment. Put the call 'on hold', then ask your visitor or visitors to take a seat and explain that you will attend to them in one moment. Finish your telephone conversation as courteously but as quickly as possible, then deal with your visitor/s. This is a procedure that will need practice to make it work effectively.

PUBLIC RELATIONS AND CORPORATE IMAGE

Public relations

Public relations is the relationship between the people in your organisation and people outside the organisation. You, and all other employees in your business, are part of the 'image' of the organisation. How the people in your organisation deal with customers/clients, suppliers and employees in other organisations can give your organisation a positive or negative corporate image. Examples are the way telephone enquiries are handled, appointments are scheduled, and visitors are treated at reception through to highest management.

Corporate image

Areas of importance required to generate a positive corporate image include:

- the appearance of your reception area;
- the personal appearance of people at the reception desk;
- customer/client service;
- positive attitudes;
- how problems are identified and resolved.

The reception area

The first impression visitors receive when entering the reception area reflects on the rest of the business. Reception is the 'shop window' of the business. Therefore it should have:

- a welcoming atmosphere with attractive decor;
- modern furniture and comfortable chairs for visitors to use while waiting;
- a coffee table with magazines and pot plants;
- tea and coffee facilities.

All this creates a pleasant environment and enhances the image of your business. A well-organised and pleasant, welcoming reception area is shown in Figure 7.5.

Personal appearance of the receptionist

Included in the corporate image is your personal appearance at reception. In today's climate of equal opportunity and affirmative action, if women want to be considered on an equal footing with men it is important that they follow the corporate rules and regulations that men have to acknowledge. In the service sector of industry, men are expected to dress smartly and in a businesslike style. Particularly in sales, details of personal presentation, such as the colour and style of business suits, shirts and ties, and the length of hair, are strictly controlled. How you dress outside the office is your own business, but how you dress in the office, and in particular in the reception area, is part of the corporate image.

Dress

There is often a formal code (or an unspecified one) in organisations as to how personnel should dress. This will depend on the type and size of business you are working for. In some organisations, such as banks, airports, hospitals and pharmaceutical companies, you may be required to wear a uniform. In other businesses, you should follow how other people dress. There may be more formal dress in the city than in the suburbs or country areas. Some businesses like to project a modern image and will be more casual with their dress code than conservative businesses, such as law and accounting firms. Some hints on dressing to suit your organisation are given below.

Fig 7.5 Reception area

You should be smart, neat and tidy. Classic suits or jackets are appropriate for most offices. Women's skirts should be fashionable but not too short. Men's suits should be businesslike, well-presented and well-fitting. You do not have to pay hundreds of dollars for items of clothing (and if you have just started work you won't be able to afford it). Wait for the sales and build up your wardrobe slowly. Black shoes, stockings and a skirt, with a different-coloured shirt or top and a plain jacket, is almost a uniform for women working in the city. This combination is popular because of its flexibility (through mixing and matching) to meet different seasons and occasions. For men, black shoes and socks, a dark pair of pants with a plain white shirt and a blazer-type jacket is all that is needed to start with. Later, men should see what others are wearing to work, and buy accordingly. Men's business attire is generally very sober.

Hygiene and grooming

Hygiene includes clean, tidy hair, careful make-up for women, and discreet jewellery.

Hair. Whatever the style preferred, hair should be washed often and be neat and tidy.

Women who wear their hair long should make sure it does not hang over their faces. It is probably better to tie it back or wear it up for work. Hair that hangs down over the face obscures expression. Facial expressions are part of non-verbal communication or body language, and are often more important when dealing with people than verbal communication. For men, long hair should definitely be tied back. However, it is better to have hair cut in a manageable and businesslike style. Again, look at the haircuts of the other men in the organisation, and take your cue from them.

Make-up. This should be discreet. Women wearing heavy lipstick and eye make-up may look attractive when going out in the evening, but this can give the wrong impression during the day at work. While it is frowned on for men to wear make-up in the office, they should be aware that skin must be clean.

Jewellery. Women's jewellery should be businesslike, not flamboyant. Heavy, dangling earrings and bracelets can be distracting for other people. Men should be discreet with earring ornaments, rings and necklaces. Some clients have preconceived ideas about how men should dress and, as with women, too much jewellery can be distracting.

Customer service

As mentioned earlier, customers and clients are very important, and the first priority of a receptionist is to provide them with satisfactory service. There are several negative attitudes you should *never* use when dealing with visitors, as this will damage the corporate image of your business.

Negative attitudes

- *Couldn't-care-less.* **Do not** stand at the reception desk and pay little attention to the customer.
- *Brush-off.* **Do not** try to get rid of the customer because it is almost lunchtime or the end of the day.
- *Coldness.* **Do not** be impatient, unfriendly and inconsiderate, leaving the customer with the feeling that he or she is a nuisance.
- *Disrespect.* **Do not** patronise or talk down to customers because you do not expect them to understand what you are doing or saying.
- *Robotism.* **Do not** use the same actions and comments for all customers, showing no feeling or individuality.
- *Run-around.* **Do not** give customers the impression that you are not interested in solving their problems or get rid of them as quickly as possible.
- *Company rules.* **Do not** say to a customer that you cannot do something because it is against company rules, instead of trying to help him/her in some other way.

Instead, you should *always* use positive attitudes when dealing with customers/clients.

Positive attitudes

Listen very carefully to what customers and internal personnel say to you, because this:

- prevents misinformation;
- saves time;

- avoids confusion;
- improves the attitudes and morale of the workgroup;
- enables the needs of customers to be accurately processed.

There are five types of active listening you should use:

- non-verbal
- encouraging
- showing empathy
- clarifying
- silence.

Non-verbal. Non-verbal listening is giving the customer your whole attention, smiling and nodding to indicate that you understand what he or she is saying.

Encouraging. You should use verbal 'encouragers'. Use the customer's name often. Let the customer finish what he or she is saying.

Showing empathy. Empathising is putting yourself in the other person's position and understanding his/her point of view.

Clarifying. In order to clarify a situation, you should ask questions. Use both **open** and **closed questions** (see Chapter 8).

Dealing with difficult customers

This is also outlined in Chapter 8.

Dealing with visitors from different cultures

When receiving visitors from another country, or visitors who have only recently come to Australia and do not understand English well, you should remember the following.

- If English is not the first language of your visitor, you should try to speak a little more slowly and clearly, emphasising key words. (Australians tend to run words together and to speak too quickly.) But do not speak too slowly, too loudly, or shout at your visitor.
- People from Japan and other Asian countries communicate in a much more formal and polite manner than we sometimes do in Australia. You should behave similarly.
- People from some other cultures have different religious beliefs that should be respected.
- Try to be helpful, friendly and understanding.

Activity 7.2

Memo to the new receptionist

You have obtained a new position within Business Stationery Supplies Pty Ltd, and someone else (Bill) is taking over your position at the reception desk tomorrow. Write a memo to Bill outlining how he can perform his duties more effectively, and how he can promote a positive corporate image for the organisation. The memo should be approximately one page in length and should outline each point under headings (see Chapter 9 for correct memo layout).

Identifying and resolving problems (conflict resolution)

Sometimes you may not be able to resolve a problem despite using the above techniques, and you then have a conflict situation. Five types of conflict situations in an organisation have been identified:

- conflict between individuals;
- conflict between individuals and groups;
- conflict between individuals and the organisation;
- conflict between groups in the same organisation;
- conflict between organisations.

In your role as a receptionist you could experience any of the five types of conflict, but the most common one would be between you and a customer. The steps used to resolve this conflict on a win/win basis (in which both sides win) are to:

- explore the problem;
- search for mutually acceptable solutions;
- evaluate the solution with the customer;
- decide together whether the solution is mutually acceptable;
- follow up and evaluate the solution after it has been put into operation.

Some points to remember when dealing with a conflict situation are:

- If the customer is aggressive, try not to be aggressive in return.
- Do not take criticism personally and try not to get upset.
- Try to stay calm and in control of the situation.
- Use the 'boomerang principle' (what you send out you receive back). If you are pleasant, calm, empathise with the customer, apologise and offer solutions, the customer will also calm down, and usually appreciates what you are trying to do. Customers who are pleased with what you have done for them become ambassadors for you and your organisation by telling potential new customers how you have helped them and solved their problems.

Body language

Body language comprises messages conveyed by facial expression, eyes, posture and hand gestures. More than 50 per cent of human interaction is through body language and you can, through your body language, make a visitor either feel welcome or feel that he/she is being treated badly. You should use body language to create positive feelings between you and visitors and you and your colleagues in the following ways.

- Sit in an attentive posture (face the visitor).
- Make good eye contact and nod in acknowledgement of what the visitor is saying.
- Give your visitor your complete attention, and try to ignore other distractions.
- Be interested in what the visitor has to say.
- Smile, and be cheerful and cooperative.

Activity 7.3

Reception role-play

In your workbook there are two scripts for role-playing situations dealing with visitors to the reception desk, and a number of situations for which you will need to work out your own scripts. Form into groups and role-play these situations, observed by your tutor/supervisor. A diary sheet is also provided in your workbook to schedule these performances. Enter the date and time when your group will perform reception protocol role-play. This will be arranged to comply with other groups and with the approval of your tutor/supervisor.

Case study 7.1

Reception situations

You are the new receptionist for Business Stationery Supplies Pty Ltd. For the first week before you take up your duties, you are training and observing the duties which Gina, the present receptionist, performs so that you can take over the position at the beginning of next week. You have observed the following situations and feel that Gina is not handling these correctly.

1. Gina has a very friendly personality and has come to know many of the customers who telephone staff in Business Stationery Supplies, which does give a positive corporate image to the business. But she spends a lot of time chatting to these callers while other incoming calls are ringing for a long time and visitors wait at the reception desk.
2. On another occasion you notice that, while she is talking on the telephone, the personal assistant to the sales manager (who is away on business today) leaves the building and does not indicate on the 'in/out board' when she is returning.
3. You notice that a visitor, who has an appointment with the manager of the accounts department, has been sitting waiting for 15 minutes and is looking annoyed and restless. You know that the accounts manager is not in his office and appears either to have forgotten the appointment or to be running late.
4. In the afternoon a customer calls at the reception desk and appears to be very angry. He explains that he has been trying to contact a sales representative on three previous occasions and has not been able to do so. In desperation he has called in person, but again she is not in the office. Gina is reacting aggressively and you can see that the situation is getting out of hand.

On a separate sheet of paper, explain what Gina is doing wrong in each situation above, and what the outcome could be.

Then describe in detail what your attitude would be, and what action you would take, to solve the problem in each situation.

Summary

- The receptionist is one of the most important people in a business. The way in which she/he carries out specific duties and receives visitors can create either a positive or a negative corporate image.
- The duties of a receptionist include receiving and screening incoming calls, making outgoing calls, receiving visitors, scheduling appointments, typing correspondence, dealing with mail and controlling office supplies.
- There are three types of visitors: those expected (with appointments), those not expected (without appointments), and those expected without appointments who provide regular services.
- The protocol for receiving all types of visitors includes welcoming them promptly, pleasantly and courteously.
- Public relations (which is the way in which personnel in your organisation deal with customers/clients, suppliers and employees in other organisations) can give your business either a positive or a negative corporate image.
- The reception area is the 'shop window' of the business. It should therefore have a welcoming, pleasant, cheerful environment, which reflects the image of your business.
- The personal appearance of the receptionist is also important in reflecting the image of the business. She/he should be clean, neat and tidy at all times.
- Customers' and clients' requirements are the first priority of a receptionist, and positive attitudes should always be adopted. These include listening effectively and actively, using non-verbal communication, encouraging and empathising.
- When dealing with visitors from different cultures, the receptionist should be helpful, friendly and understanding, speak a little more slowly and clearly, and respect cultural and religious beliefs.
- When solving problems involving conflict (conflict resolution), he/she should try to resolve this conflict on a win/win basis by exploring the problem and searching for a mutually acceptable solution. If the customer is aggressive, he/she should try not to be aggressive in return, but try to stay calm and not to take criticism personally.
- Body language (non-verbal communication) is a message conveyed by facial expression, eyes, posture and hand gestures.

Review questions

(These can be used for group discussion.)

1. Why is the receptonist one of the most important people in the business?
2. List the main duties of a receptionist.
3. There are three different types of visitors to the office. What are they?
4. Write a short paragraph on the public relations of an organisation.
5. In the reception area, what aspects help to create a positive corporate image? Write a short sentence on each of these.
6. There are a number of positive attitudes that will help public relations and the corporate image. What are they?
7. How would you help to identify and resolve problems when dealing with a customer?
8. If the customer becomes very aggressive, how would you deal with this situation?

CUSTOMER SERVICE

Australia has approximately 19 million people, who are potential customers in the domestic market. This represents the total client-based market which all organisations in Australia must share. There is also unlimited potential in the international market, where communication between people can be more formal than it is in Australia.

Customers have a need, and they will buy from you only if they are satisfied with your service and your product. If they are dissatisfied they will take their custom to another company. If your organisation has a total commitment to quality service it has a very good chance of business success. This commitment could increase the number of customers by persuading them to leave other companies and buy from you—with resulting increased profits.

Total quality service

Customer service is the process of being of help or assistance to customers. All employees in an organisation should work to achieve a total service package addressing the needs of customers. This can be achieved only where there is:

- a knowledge of market demand;
- an understanding of the customer's needs;
- a commitment by all employees to the organisation's goals and strategies;
- a capacity to put into place workable practices to establish a quality customer service culture;

- the provision of the necessary resources to maintain the customer service culture;
- a willingness to decide when changes should be made;
- a capacity to review and improve customer service;
- the ability to advise customers about the improved customer service.

It is now accepted all over the world that quality service to the customer is the most important aspect of business. There are international standards for quality systems which include customer service.

ISO 9001/9002/9003

These are a series of three international standards for quality systems that can be used for external quality assurance purposes. The quality systems standard ISO9000 was first published in 1987. The quality system as defined by the International Organization for Standardization (ISO) in its 9000 series of standards was developed in response to the challenge of increasing market globalisation. Since then thousands of organisations around the world have adopted these standards to show that they are following accepted quality practices. One of the features of the ISO9000 standard is third-party certification of an organisation's quality system.

ISO9001 is a model for quality assurance systems in design, development, production, installation and servicing.

Activity 7.4

Quality service

(This activity may be completed by an individual person.)

Discuss with members of your group instances when you have enquired about a product or have bought a product. After the discussion, complete the questions on a separate sheet of paper.

1. Did you have sufficient details about the product?
2. Were the staff polite and helpful?
3. Were you happy with the price and the after-sales service?
4. Were you happy with the quality of the product and the warranty?
5. Did you have to complain about a product? If so, how did the company staff treat your complaints?

Identifying specific needs of customers

In a recent survey of 40 respondents in a series of interviews that were identical for both suppliers and customers, nine customer services were selected as the most important:

1. availability of the item;
2. after-sales service and backup;
3. efficient telephone handling of orders and queries;

4. ordering convenience;
5. competent technical representative;
6. delivery time;
7. reliability of delivery;
8. demonstrations of equipment;
9. availability of published material.

Some other features of successful customer service are:

- providing a pleasant environment;
- making sure that staff are:
  —friendly and well-groomed,
  —well-informed about products and services,
  —providing willing assistance,
  —polite and helpful.

Types of customers

There are basically three different types of customer—regular, chance and visiting.

Regular customers

These are people who are happy with your service package and like your corporate image. They are repeat customers and are the basis of your business. Never take them for granted or you may lose them. If you upset one of your regulars, you will have a dissatisfied customer who may take her/his business to another company.

To keep regular customers happy you should list their details, and their particular requirements, in a computer or paper file, and service them regularly by:

- making regular telephone calls to enquire whether they have any present or future needs;
- sending them regular promotional material;
- sending them cards at Christmas or on special occasions;
- returning telephone calls or letters promptly, using the 'you principle' outlined in Chapter 9;
- showing them that you care about their opinions and requirements (empathising);
- keeping up-to-date customer records.

Customer records. A large part of quality service to customers is the ability to greet them by name and answer any queries promptly and efficiently. Having a record of all their details and requirements filed away in a location where you can retrieve them immediately and efficiently, and give quick, accurate answers, adds to the positive image of your company.

In service industries such as a dentist's surgery, a doctor's surgery, or hairdressing business, customer records are often kept on a preprinted card system or in a paper file. At the same time, computer records are also kept at the reception desk for easy and quick reference of personal details.

In large organisations such as Telstra, electricity and gas corporations, and insurance companies, customer records are stored in a computer database system and information is retrieved very quickly. Figure 2.1 in Chapter 2 is an example of a computer database customer-record filing system.

Chance customers (passing trade)

Chance customers, or passing trade, are equally important to your business. They usually live in the area. Chance customers have entered your business because they may have been attracted by your window display or your advertisement in the *Yellow Pages*, or you have been recommended by a regular customer. These customers are vital to your business because you increase your business by making chance customers into regular customers. Your quality of service will decide whether this happens.

Visiting customers

Visiting customers may be people from another town, another state or another country. These customers are also important to build up your business. If you work in a travel agency, your visiting customers may be tourists who may return some time in the future. If your service is excellent, visiting customers are ambassadors for your business by recommending you to other customers.

When dealing with all of the above types of customers, your service should be the same—excellent quality.

How to achieve excellent quality customer service

There are many ways in which an organisation can achieve excellent quality customer service:

- Put out a survey to gather information on customers' needs and expectations of service.
- Provide customers with information about the organisation's activities.
- Develop a cultural awareness towards total quality service within the organisation through good communication, staff training and development.
- Meet customers' needs and expectations.
- Create a set of customer service performance standards to be placed in the corporate procedures manual which all staff are made aware of.
- Obtain feedback from your customers. Regularly review the customer service results and be willing to change and improve the service where necessary.

Activity 7.5

Product survey

(You may complete this activity as part of a group or on your own.)

Complete Activity sheet 7.5 in your workbook. Choose three products that you have used recently and enter each one at the top of the columns numbered 1, 2,

3. Answer the questions in each column and, at the bottom of each column (after you have answered the questions), decide whether you were very satisfied, fairly satisfied, or dissatisfied with the product.

If you were dissatisfied with one of the products, list at the bottom of the survey (where indicated) the steps you would take and what action you would expect the organisation to take.

The organisation's role

Your organisation's role, and your role within the organisation, is to focus on the customer and deliver a quality service that meets the needs of customers. How staff deal with customers, clients, suppliers and employees in other organisations—in the reception area, on the telephone, in written communication and computer communication, and face to face—can have a significant effect on the success or failure of your business.

To operate a total quality service scheme effectively, good communication skills are vital. Continuous evaluation of the quality of the provision of service will reduce the possibility of customer dissatisfaction and complaints. Identification of the different communication skills is required and staff must be trained to use these skills effectively.

Providing customers with information

The most effective way of informing prospective customers of your products is through advertising. Advertising influences the customer's perception of your organisation by:

- informing them of what you can do for them and how your product or service is of value to them;
- demonstrating why your organisation is ahead of others;
- illustrating how your product is better than others.

Billions of dollars are spent each year on advertising in the following ways:

- We are confronted with it in all walks of life: on television, in magazines, on the Internet and on billboards by the roadside etc. etc.
- We are sent advertising material in the post by large organisations such as Telstra, local government, gas and electricity boards and so on.
- You receive in your mailbox mountains of 'junk mail' advertising all kinds of products.
- Agents and sales representatives in the insurance industry, and other industries, send out letters with enclosed flyers informing customers of products and services.

You should keep in touch with existing customers and clients to inform them of any changes in products or services, as this is all part of quality customer service.

Activity 7.6

Advertising

(You may discuss this activity with members of your group or you may complete it on your own.)

You have decided to open your own business and want to advertise for new customers. One way you are going to do this is by hand-delivering to all the houses in your district a persuasive one-page folded form letter, which you will drop into all mailboxes.

Choose your new business, e.g. hairdressing, a dress boutique, a computer shop, a car mechanic business, a gift shop. Write a form letter using the steps for persuasive or sales letters—the AIDA principle, as outlined in Chapter 9.

- Use few words and make them bold to stand out and persuade.
- Use graphics to attract attention (you can draw these or use computer graphics).
- Think of ways this product or service would persuade you to buy.
- Discuss with other members of your group what would persuade them to buy your product.

(Examples are the form letters you have received in your mailbox advertising products and services.)

Hand your completed form letter to your tutor/supervisor for assessment.

Your role in customer service

A customer's first impression is a lasting one, and your role within the organisation is to provide a service of consistently high quality in as short a time as possible. This may be achieved in the following ways:

- Acknowledge the customer and show empathy.
- Listen effectively and provide feedback.
- Use appropriate verbal and non-verbal communication.
- Use conflict resolution techniques to solve problems.
- Make sure you follow up the measures you have taken to solve conflict to ascertain whether your customer is satisfied.
- Keep a record of complaints in a 'complaints register'.

Activity 7.7

The organisation's role and your role in customer service

(You may complete this activity as part of a group or on your own.)

Each member of your group should photocopy an advertisement of a business from the *Yellow Pages*. Then discuss with the other members of your group how the organisation could provide excellent quality service to customers for this specific type of business. After the discussion, list these on a separate piece of paper.

Also list the ways in which you, as an employee, could improve service to customers in the particular business you have chosen.

Handling customer complaints

In general, there are two types of customer complaints—justified and unjustified.

A justified customer complaint

Deal with a justified complaint by using negotiation and conflict resolution skills. (These are described in more detail early in this chapter.) You can also attempt to solve a customer's complaints by trying one of the two following methods.

The PAIR approved method

Placate: listen, empathise, respond with concern.
Attend: to the complaint.
Investigate: the circumstances and details of the incident.
Resolve: decide on the action to be taken, with the approval of the customer.

The five-step method

1. *Listen:* Be open-minded. Do not take the complaint personally.
2. *Respond:* Show concern and empathy, and apologise for any inconvenience. Put yourself in their position.
3. *Decide on action:* What action will influence you here? Is the complaint justified? What is company policy? You may need to seek advice from your supervisor.
4. *Take action:* Whatever action you decide to take (with the approval of the customer), make sure you act promptly.
5. *Follow up:* Always follow up the action you have taken and confirm that the customer is happy with the result.

Follow-up. When you take any action for a customer (whether it is a request or a complaint), check that the course of action to achieve the solution suits the customer. Clearly explain each step. As you do this, ask questions and receive feedback to clarify that the customer understands what is happening, agrees to the solution, and will be happy with the results.

If possible, it is a good idea to put everything down in writing to avoid any misunderstanding at a later date. Follow up the action taken to make sure it has been carried out. This will ensure that you end up with a happy customer who will stay with your organisation.

An unjustified customer complaint

Not all complaints are justified. When you think the complaint is not justified you should deal with the situation using negotiation or conflict resolution (explained early in this chapter).

Activity 7.8

Customer complaints

(You may complete this activity on your own.)

Below are some customers' complaints. Choose one of these and decide whether the complaint is justified or unjustified, and list on a separate sheet of paper how you would deal with that *one* situation. Make sure you analyse the situation carefully and write at least two or three paragraphs.

1. You work in a dress boutique and a customer has called into the shop with a dress she bought four weeks ago in your January sale. She says she has only worn the dress once and, after washing, it has shrunk and the colours have run.
2. You are the manager of a white goods store and a customer has telephoned complaining that the washing machine she bought two years ago has broken down and she is unable to do any washing. The machine has a two-year warranty which has just run out. She is very agitated and says she has had nothing but trouble with this machine since she bought it and she absolutely hates it. She says she will never recommend it to her friends and relatives.
3. You are the owner of a hairdressing business in a small suburban shopping centre, with residential houses surrounding your business. A regular client has telephoned to complain about her hair. Two days ago you permed her hair and she insisted that you also applied a blond permanent rinse, even though you advised her that it was unwise to colour hair immediately after a perm as this was bad for the hair. She still wanted you to do it as she was going out to a function that evening. Now her hair is dry as straw and tinged with green.

Summary

- There are international standards for quality systems which include customer service—ISO9001/9002/9003.
- In a recent survey nine customer services were selected as the most important, including availability of item, after-sales service and backup, efficient telephone handling of orders and queries, reliability of delivery, and a pleasant environment and willing staff.
- There are basically three different types of customers—regular, chance and visiting.
- Keeping efficient customer records, whether computer-based or paper-based, is part of quality service.
- There are many ways in which to achieve excellent customer service: put out a survey, provide information, meet customers' needs, create a set of customer service performance standards, and obtain feedback.
- The organisation's role and your role is to focus on the customer and deliver a quality service that meets the needs of the customer.
- There are two types of customer complaint—justified and unjustified.

Review questions

(These may be used for group discussion or formal assessment.)

1. What will give your organisation a good chance of business success?
2. List the ways in which an organisation can achieve total quality service.
3. List the features of successful customer service.
4. Describe the three types of customer.
5. Describe different ways of keeping customer records and state why these are important.
6. List the ways in which your organisation can achieve excellent quality customer service.
7. Write a short paragraph about the organisation's role in customer service.
8. Write a short paragraph about your role in customer service.
9. Describe how you would deal with an unjustified customer complaint.
10. How would you deal with a justified customer complaint?

Competency self-assessment and/or tutor/supervisor assessment

When you have finished reading this chapter, and have completed the activities chosen by your tutor/supervisor (**practical in class, at work or theoretical**), fill in this self-assessment. Then hand it to your tutor/supervisor for official assessment.

Learning outcomes	**Self-assessment**		**Tutor/supervisor assessment**	
I have successfully completed the following learning outcomes	**Yes** (please tick)	**No**	**Yes** (please tick)	**No**
1. Receive visitors and attend to visitors' needs to ensure a positive image of the organisation				
1.1 receive visitors promptly and courteously	☐	☐	☐	☐
1.2 establish visitors' requirements through appropriate and polite questioning and active listening	☐	☐	☐	☐
1.3 interact courteously with clients and project a positive corporate image	☐	☐	☐	☐
1.4 direct or escort visitors to destination as appropriate	☐	☐	☐	☐
1.5 explain courteously reasons for delay/non-availability	☐	☐	☐	☐
1.6 identify situations outside area of responsibility and obtain assistance from appropriate person promptly	☐	☐	☐	☐
2. Respond to people in a variety of situations				
2.1 recognise existence of individual and cultural differences	☐	☐	☐	☐
2.2 adjust responses appropriately to meet needs of others	☐	☐	☐	☐
2.3 identify and respond appropriately to those aspects of a foreign culture which affect workplace relationships	☐	☐	☐	☐
3. Respond to and act on internal/external enquiries to promote the products and services of the organisation				
3.1 deal promptly and courteously with enquiries	☐	☐	☐	☐
3.2 provide relevant information to meet enquirers' needs	☐	☐	☐	☐

3.3 refer enquiries outside area of responsibility to nominated person/section for resolution	☐	☐	☐	☐
3.4 take follow-up action where required	☐	☐	☐	☐
4. Customer service, corporate presentation, customer relations	☐	☐	☐	☐
Date:	Tutor/supervisor signature:			

section three

COMMUNICATION

CHAPTER

8

TELEPHONE OPERATIONS

COM201A, NOS126, 7365L

objectives

The purpose of this chapter is to help you to operate a telephone system effectively in a workplace environment using 'hands-on' methods of practice and assessment, in order to complete the following learning outcomes:

1. Operate a telephone system effectively in the workplace

1.1 operate equipment to receive and generate calls, using features such as:
- picking up group
- transferring calls
- putting calls on hold and redialling, according to procedures outlined in the operating manual

1.2 operate a telephone answering system according to procedures outlined in the operating manual

2. Receive and respond to incoming telephone calls in the workplace

2.1 answer calls promptly and clearly, using designated business protocol procedures

2.2 identify callers correctly and establish their requirements accurately

2.3 give disclosable information only

2.4 transfer calls to appropriate location/person promptly

2.5 record messages effectively using clear and concise language

2.6 distribute messages according to organisational procedures

3. Make telephone calls in the workplace

3.1 obtain accurate telephone numbers from an appropriate source

3.2 establish contact using designated business protocol procedures

3.3 convey purpose of call clearly and concisely

This chapter is written in the form of a telephone procedures section from the procedures manual for Business Stationery Supplies Pty Ltd.

If your class or group does not have access to an operating telephone system connected to Telstra, the practical activities in this chapter may be completed successfully by using two unconnected handsets.

INTRODUCTION

The way in which all personnel in the organisation answer telephone calls and deal with people on the telephone can have a positive or negative effect on the corporate image of the organisation. This includes:

- how quickly you answer the call;
- how you greet the caller;
- the non-verbal communication in your voice;
- how clearly you speak;
- how effectively you listen to other people.

In the business communication process, the telephone is used more than any other method of communication. Therefore, one of the first things you should do when you commence a job is to find out how to operate the specific system used in your new business.

OPERATING TELEPHONE EQUIPMENT

Many different types of telephone systems are used in Australian businesses. A lot of these are automatic systems with no switchboard operator, so that when the telephone rings anyone can answer the call and transfer it to the person requested. This is why uniform corporate procedures are essential for all personnel who answer the telephone and transfer calls. Check the manufacturer's user guide on all the operating facilities of your system so that you can confidently hold and transfer calls to another extension. See Figures 8.1(a) and 8.1(b) for some of the functions of an automatic network telephone system.

Many different items of telephone equipment are available today for business organisations. Some of these are shown in Figure 8.2.

Internet telephone

You can make telephone calls using the same technology as computers use to exchange information. Several packages have been released that allow you (with the help of a soundcard, microphone and speakers) to talk to your friends overseas at the cost of connecting to the Internet. At present calls can be made only to someone who is using similar software and hardware, but this technology will be used more often in the future. To find out more about Internet telephone calls, check the following sites:

www.freetel.inter.net
www.vocaltec.com
www.yahoo.com.au/Computers/Internet/Internet_Phone.

And there are many others.

Fig. 8.3 shows you how to make an Internet phone call.

There are two keystation models:

- standard keystation (no display)
- executive keystation (1 line × 16 character display)

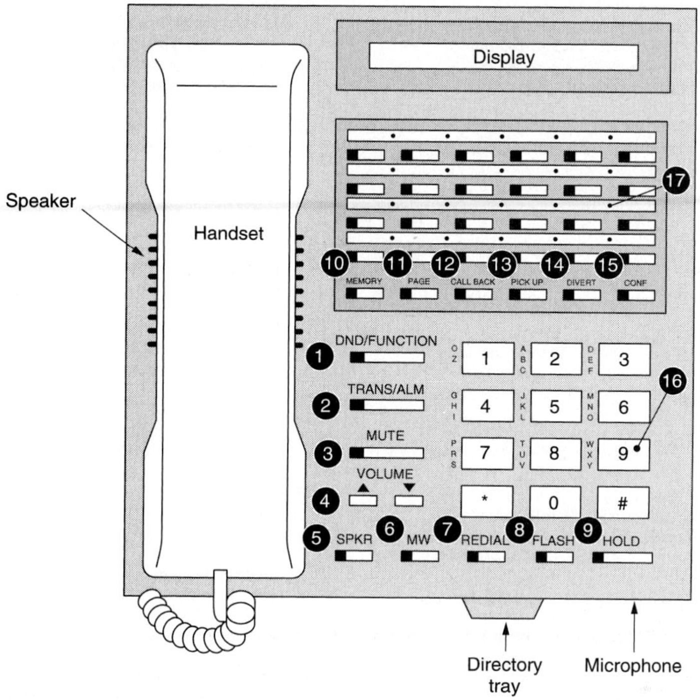

Fig. 8.1(a) Keystation diagram

Activity 8.1

Telephone operations

Refer to the manufacturer's user guide for the system you use at work or the system provided by your tutor/supervisor. In groups, discuss how to:

- answer incoming calls;
- put calls on hold;
- redial the last number called;
- make outside calls;
- transfer calls;
- 'pick up group' (conference).

Each student should complete Activity sheet 8.1 in the workbook.

Keystation key description

Function keys

1. [DND/FUNCTION]
   When the station is in the idle mode this key will initiate 'Do Not Disturb'. In the off-hook mode it is used to prefix function codes.

2. [TRANS/ALARM]
   In the idle mode this key is used to set up an appointment alarm. During conversation it is used to transfer a call to another station.

3. [MUTE]
   Enables and disables the station microphone.

4. [▲] and [▼]
   Increases and decreases the handset, loudspeaker and ringing volume, in increments, each time button is pressed.

5. [SPKR]
   Enables or disables the hands-free mode.

6. [MW]
   Used in conjunction with the Message Wait Facility.

7. [REDIAL]
   the [Redial] key is used to redial the last number called or while listening to ring/busy tone on an outside line call will initiate Automatic Redial.

8. [FLASH]
   Used to recall a parent PABX. Also with network facilities, such as Telstra's Centel™ and EasyCall™.

9. [HOLD]
   Used to place outside line and intercom calls on hold. Also in the idle mode, pressing the [Hold] key will turn the background music on and off.

10. [MEMORY]
    Used to access memory functions.

11. [PAGE]
    Used to access paging facility.

12. [CALL]
    Provides an automatic call back when busy stations or lines become free.

13. [PICK UP]
    Picks up calls ringing at other stations in the same station group.

14. [DIVERT]
    Used to temporarily transfer incoming calls to another station.

15. [CONF]
    Used to set up a conference.

16. Dial keys
    Used to access numbers and functions within the system and via the PSTN.

17. Selection keys
    [DSS] Direct Station Select keys access other stations within the system. Associated LEDs indicate the status of the station, i.e. busy, ringing or idle.
    [SPD] Personal Speed Dial keys give one touch dialling facilities.

Fig 8.1(b) Description of the function keys

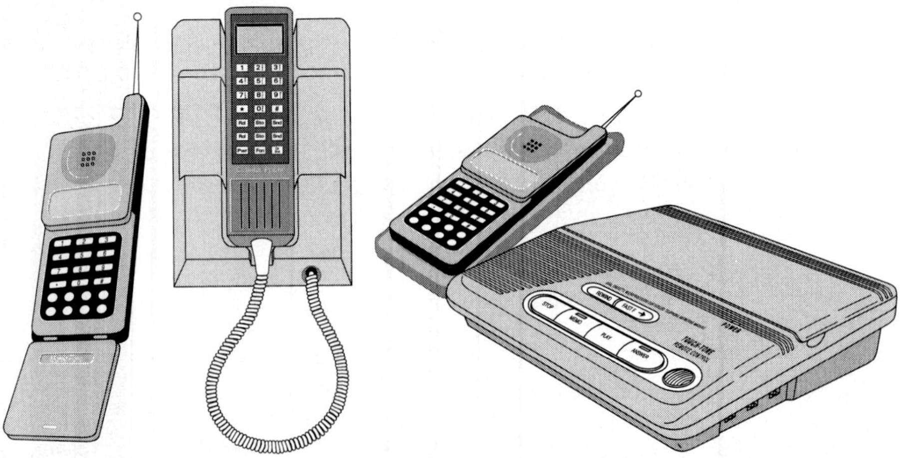

Fig. 8.2 Mobile telephone, car telephone, cordless telephone answering system

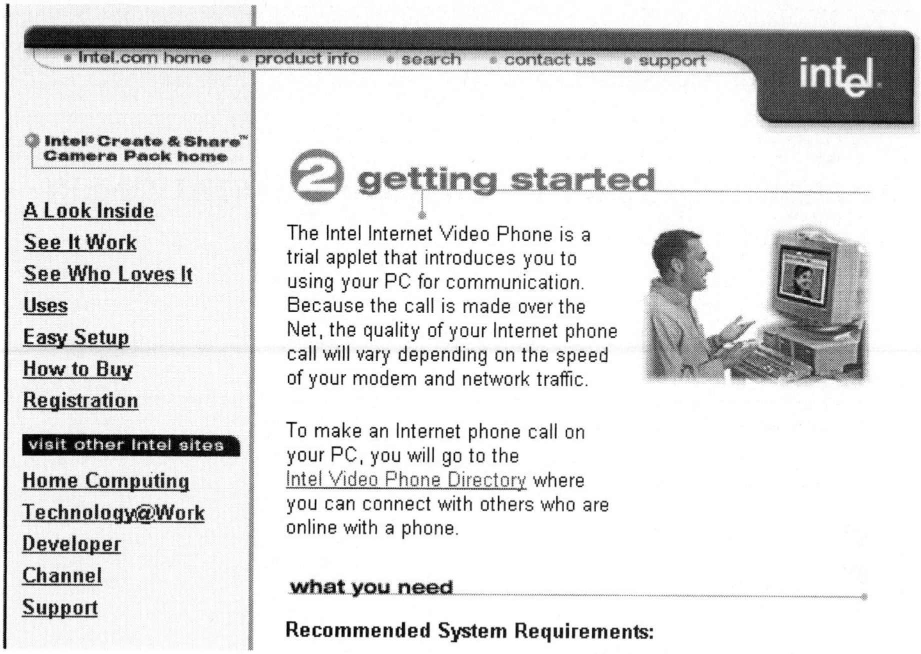

Fig. 8.3 How to make Internet phone calls (www.intel.com.au, with permission)

Incoming calls

Do you really know how to use the telephone?

Most of you have a telephone at home and you probably use it every day. So you think you know how to use it. The telephone rings, you pick up the receiver and say, 'Hallo: how are ya?', 'Great!', and when you finish you say, 'Bye, talk to ya later'. This won't do when you are in a business situation. The correct protocol for answering the telephone is outlined on the next few pages.

For anyone working within an organisation, especially a switchboard operator, there are a number of things you must know to operate the telephone effectively, such as:

- the structural hierarchy of your organisation, with the names and roles of the key personnel in each department or section;
- the products and/or services provided by your organisation;
- the telephone number and location of all personnel.

Business telephone protocol procedures

In most large organisations, the protocol for answering the telephone is entered in the procedures manual so that all personnel answer it in a way that is approved by management. This ensures that everyone deals with customers/clients in a way that will enhance the public relations and corporate image of the business. Each organisation has its own preferences for phrases to be used by the switchboard operator and other personnel when answering the telephone. This chapter makes suggestions that may be useful.

The remainder of the chapter is in the form of a procedures manual for Business Stationery Supplies Pty Ltd, beginning with the contents page. All page numbers for the procedures refer to this chapter.

PROCEDURES MANUAL FOR BUSINESS STATIONERY SUPPLIES PTY LTD

Telephone protocol procedures (to be followed by all staff)

Answering the telephone promptly (Procedure 1)

Try to answer a call on the first ring. This may not be possible if you are on the switchboard, or dealing with someone at the reception desk, but you should try to answer all calls before the third ring. Leaving someone waiting gives an impression of inefficiency, and prospective customers may get tired of waiting and take their business to another organisation.

Identifying your organisation and yourself (Procedure 2)

Your voice is an important component in the image you project to a listener.

- Greet the caller warmly, as you would do if you were speaking to them personally.
- Sound cheerful and enthusiastic (a dull voice makes you sound uninterested).
- Speak clearly and confidently.
- Hold the mouthpiece of the telephone no more than three centimetres from your lips or your voice will not be clear.
- Use good diction, which means pronouncing words correctly and sounding words clearly.
- Speak a little more slowly than you would normally do, but not so slowly that you sound bored or tired.

Example

When you pick up the receiver, say:

'Good morning/afternoon. Business Stationery Supplies. This is Jennifer.'

It is better to say your name last, as callers are more likely to remember the last word they hear. If you are on the switchboard answering an outside call, do not say 'How may I help you?'.

You may answer the telephone hundreds of times a day and you will find there is too much to say. Eventually, what you say will tend to sound 'sing-song' and repetitive, and the words will run into one another. The person on the other end of the phone may then have to ask you to repeat what you said, which wastes their time and yours.

If you are on an extension and a call has been transferred to you, you should identify your department, your name, and say 'How may I help you?':

'Good morning/afternoon. Sales department. This is George. How may I help you?'

If you answer someone else's extension:

'This is George Benson's extension. Sales department. How may I help you?'

Always close telephone conversations with a 'thank you':

'Goodbye, Mr Neill. Thank you for calling.'

Adopting a helpful attitude (Procedure 3)

As a representative of this organisation, you are expected to give the caller every assistance. If you cannot do this, either transfer the caller to someone who can help or take the name and telephone number and call back as soon as possible with the information requested.

Screening calls (Procedure 4)

In some large organisations, where calls are received through a switchboard operator, there is a policy of screening calls. This means putting the caller on hold while you find out whether the person called wants to speak to the caller. There are a number of reasons why someone might not want to speak to the caller, but this is not your concern, and you need to treat this situation diplomatically.

Employing business ethics (Procedure 5)

Although you should always be as helpful as possible when dealing with callers, there are one or two instances where business ethics are involved. Your first loyalty is to the organisation you work for, and there are certain items of information that should not be disclosed to people outside the business. Some examples are:

- personal information about employees, including addresses and telephone numbers, without the permission of that person (this includes 'gossip');
- details about new products before they are released for sale;
- financial details that have not been released to the public officially (which could be construed as 'industrial espionage' and lose you your job);
- an appointment with a client that your boss has forgotten—you should apologise for the inconvenience and indicate how long the visitor will have to wait until your boss is free.

Personal calls

Most organisations do not object to your taking, or making, personal calls—provided you do not waste too much time and it does not interfere with your work.

Transferring calls correctly (Procedure 6)

When transferring calls to another extension, ask the caller politely to hold the line.

Example

'Would you hold the line please, Mr Neill. I'll see if someone else can help you. I won't be a moment.'

Put the call on hold, and speak to the extension. Never leave a caller holding for very long. If you can't get hold of someone quickly, keep returning to the caller:

'I'm sorry to keep you waiting, Mr Neill, I'm still checking for you.'

Put the call on hold again. Then, when you can put the caller through:

'I'm putting you through now, Mr Neill.'

Establishing the requirements of callers (Procedure 7)

You need to establish quickly what the caller exactly requires; to whom they want to speak; and what they want to speak about. This involves effective listening; clarifying by asking questions; knowing key personnel; knowing organisation products and services; and writing everything down.

Effective listening

We have already discussed how important your voice is when answering the telephone. However, if you are to operate the telephone efficiently you must also use your ears effectively. If you don't listen correctly you may make mistakes with names and/or messages, and this could have serious consequences for your organisation. Some important rules for listening effectively are given below.

Stop talking and listen. Listen to what callers say and use verbal cues to encourage them to clarify what they are trying to say, for example: **'I see'**; **'Yes, I understand'**; **'I do agree'**. Don't assume anything—you will probably be wrong. Instead, clarify everything by writing it down. Repeat instructions and messages, using the phonetic alphabet for names and standard grouping of numerals, if necessary, so that there can be no misunderstanding.

Activity 8.2

Telephone voice

Find Activity sheet 8.2 (1) in your workbook. This contains practice scripts. Practise recording these on a cassette recorder, then play them back and make notes of the good and bad points of your telephone voice using Activity sheet 8.2 (2). For example:

- Did you speak too quickly/too slowly?
- Did you speak distinctly?
- Did you sound friendly?
- Did you smile and use other kinds of non-verbal communication?

If you cannot obtain a cassette recorder, practise this with members of your group, at home with your family, or with colleagues at work.

Clarifying (Procedure 8)

Clarifying involves repeating back to the caller what they said. Problems can arise when giving names on the telephone (even if you spell them out) because some letters sound the same. For example 'S' and 'F', 'P' and 'B', 'T' and 'D'. If you are not sure about the spelling of names, repeat the name using the phonetic alphabet recommended by the International Telecommunications Union for use throughout English-speaking countries:

A—Alfred
B—Benjamin
C—Charles
D—David
E—Edward
F—Frederick
G—George
H—Harry
I—Isaac
J—Jack
K—King
L—London
M—Mary
N—Nellie
O—Oliver
P—Peter
Q—Queen
R—Robert
S—Simon
T—Tommy
U—Uncle
V—Victor
W—William
X—X-ray
Y—Yellow
Z—Zebra

Standard grouping of numerals

When giving or taking numbers on the telephone, it is always better to clarify these by repeating them according to the standard grouping:

20 1216	Two zero, one two one six
26 1111	Two six, double one double one
33 6166	Double three, six one double six
489 2666	Four eight nine, two six double six
526 3044	Five two six, three zero double four
9508 2011	Nine five zero eight, two zero double one

It is a good idea to use 'zero' rather than 'nought' or 'oh', as zero is less likely to be misinterpreted over the telephone.

To take clear messages you should also:

- be aware of who is responsible for what in the organisation you work for;
- know the range of products or services provided by the organisation;
- have in mind the kinds of questions that will help you clarify the purpose of the call.

There are two kinds of questioning techniques—open and closed.

Open questions

When you want a customer to explain or discuss an issue, begin your question with:

How, Why, When, Who, What, Where.

For example, 'Who spoke to you about the problem?'.

Closed questions

When all you want is a 'yes' or 'no' answer, begin your question with:

Did, Can, Have, Do, Will, Is, Would.

For example, 'Would this afternoon be convenient?'.

Writing everything down (Procedure 9)

As soon as the caller starts speaking, ask for his or her name and write it down. Also write down the telephone number and address (if relevant). When you speak, use the caller's name often in your conversation: it sounds more friendly. Quickly establish the purpose of the call by using the questioning techniques listed above. Then write down the main points of the conversation. You can then repeat these to the caller so that you both know the information has been communicated correctly.

Using non-verbal communication effectively (Procedure 10)

How can you use non-verbal communication over the telephone? After all, the person you are speaking to cannot see you. Because the person on the other end of the line is not distracted by visual signals, she or he becomes more sensitive to non-verbal communication signals. For example, at some time you have probably phoned a large organisa-

tion with a query about your insurance policy or a complaint about a certain product. When someone in the organisation spoke to you it was quite obvious from their tone of voice and attitude that they were not interested in helping you. You may have heard such phrases as: 'That is not my area' or 'The person who could help you isn't here at the moment'. The speaker's tone of voice and manner told you clearly that they wanted to get rid of you. Nothing annoys customers more than this, and they are likely to take their business somewhere else. These non-verbal cues can create an impression in the listener of lack of interest, which may affect the corporate image of your organisation. Below are some important points to remember.

Do not:

- sound irritable and impatient;
- give customers the feeling that you want to get rid of them;
- give customers the 'run-around' by passing them on from one person or department to another;
- leave the caller sitting on the line any longer than is necessary;
- use slang or offensive language;
- carry on a conversation in the background with someone else while speaking to a caller;
- put your hand over the mouthpiece and say something you think the caller cannot hear (modern telephones are very sensitive and can easily pick up peripheral noise, including conversation, which can prove embarrassing).

Do:

- use the tone and pitch of your voice to sound cheerful and positive;
- smile when speaking to the person on the other end of the phone as this helps you to sound cheerful;
- try to be genuinely interested in the caller's query or problem;
- try to help callers before you transfer them to someone else;
- listen effectively to prevent misunderstandings;
- use empathy to try to appreciate the other person's point of view;
- know who is responsible for the different products or aspects of the business so that you can transfer the caller to the correct person quickly and efficiently.

Activity 8.3

Operating a telephone system

(Role-play this activity in a group of three people.)

Make sure you have completed Activity 8.1 so that you know how to operate the telephone system available to you. **If you don't have access to an operating telephone system, unconnected handsets may be used. (Ask your tutor for equipment.)** There are two scripts for this activity in your workbook. Role-play them in groups of three, observed by your tutor/supervisor. Practise with one script before using the other for official assessment. Remember to use your voice correctly and to use non-verbal communication.

Dealing with difficult customers and solving problems (Procedure 11)

When a customer becomes aggressive over the telephone, it is natural for you to feel aggressive in return. But you must control this feeling, to avoid a more serious situation. For example, the customer could complain about you, leading possibly to a reprimand or even your resignation. Remember that 'the customer is always right'.

- Recognise that their anger is not directed at you personally.
- Remain calm and helpful.
- Ask questions and listen for facts.
  Use the rules of effective listening.
  Make notes of all the facts.
- Use the correct non-verbal signals.
  Sound cheerful and positive, and smile.
- Acknowledge their feelings of anger (empathise).
  Use verbal encouragers.
  Use their name often.
  Show empathy by understanding their point of view.
  Say 'Thank you for letting us know about this problem' to defuse the situation.
- Agree to a course of action.
  Find out what the customer wants you to do.
  Never promise to do something that you cannot do.
  Suggest a solution to the problem and ask the customer if this would be agreeable to him/her.
- Follow that course of action.
  If you are unable to resolve the complaint, refer the customer to someone who can.
- Never become emotionally involved, or the situation will get out of hand.
- Follow up the action.
  Make sure the action has been carried out. If it has not and the customer has to telephone again, the situation will become worse.

Remember to turn complaints into opportunities for customer satisfaction. The satisfied customer will then be an ambassador for you and your organisation.

Terminating calls (Procedure 12)

You should allow the caller to finish the conversation—unless this is taking too long, in which case you should politely but firmly bring it to a close by saying goodbye and thanking him/her for calling.

Activity 8.4

Dealing with angry customers

Form groups of three. There are two scripts for this activity in your workbook. Role-play the scripts, which involve dealing with angry customers. Each person in your group should take a turn at playing each role. This should be observed by your tutor/supervisor. Practise using one script, then use the other for official assessment.

Reconnecting a call that has been disconnected (Procedure 13)

With incoming and outgoing calls, if a call is disconnected the person who made the call should ring again.

Case study 8.1

Dealing with a difficult telephone call

(You may discuss this situation with your group, but write up an individual sheet at the end of the discussion.)

You are the personal assistant to the sales manager of Business Stationery Supplies Pty Ltd, Ian Gardner. A call has been put through to you from the switchboard. The caller, Mr John Moran, wants to speak to your boss, who is out at the moment, and you do not expect him back for at least two hours. From the tone of his voice, Mr Moran is obviously very annoyed. He says that he has been trying to get in touch with Ian Gardner for the past three days. Ian Gardner had promised to send Mr Moran detailed information about products which your organisation hopes to supply to him. They had a meeting last week, and Mr Moran was impressed with the competitive costs of the products but not the service of your company. He is obviously a prospective customer, and he is now threatening to take his business to another supplier. How would you deal with this situation?

Write down your side of the conversation.

Taking and distributing messages (Procedure 14)

A notepad and a message pad (see Figure 8.4) should be kept next to all telephones (as well as the switchboard) so that anyone who answers a call, even if he or she is just walking past, can take down a message. The message should be jotted down quickly and then paraphrased into its relevant details, which should be written down on the message sheet. Every message should include:

- the date and time of the call;
- the caller's name and the name of the organisation;
- the telephone number of the caller (including area code);
- the content of the message;
- the name of the person who took the message.

All messages should be passed on promptly, so that action can be taken immediately.

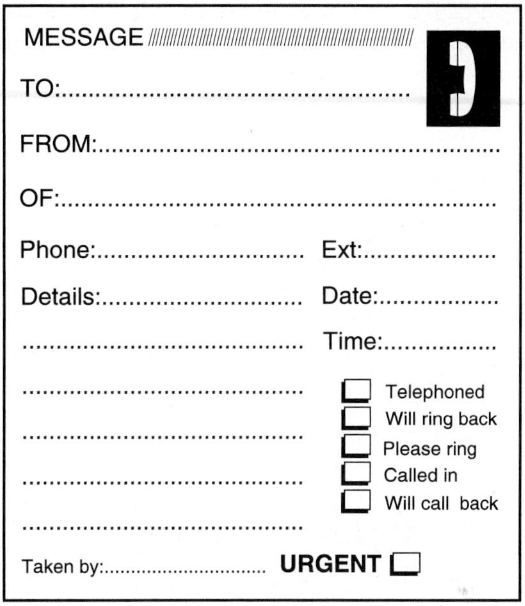

MESSAGE

TO:..

FROM:...

OF:...

Phone:............................... Ext:....................

Details:.............................. Date:..................

.... Time:.................

.... ☐ Telephoned

.... ☐ Will ring back

☐ Please ring

.... ☐ Called in

☐ Will call back

....

Taken by:.............................. **URGENT** ☐

Fig. 8.4 Telephone message pad

Message pads can be bought from stationers and other stores, but you can recycle used paper by cutting it into squares for informal messages within your office. Another method of recording messages within a small business or a small office of three or four people is to make up your own message sheets, which can be placed in a manila folder. Then a permanent record of calls can be kept by photocopying the sheets and filing them away for future reference. If your boss needs the name of someone he/she spoke to weeks or months ago, you can easily find the name and telephone number and details of the message. This is not only useful, but it proves how efficient you are. One disadvantage of small, detachable message pads is that the sheets can become mixed up with other papers and the message lost.

Clarify messages. You should read back a message to the caller to confirm that you have taken it down correctly. If you have any doubts about names or numbers, clarify these using the phonetic alphabet and the standard grouping of numerals (see earlier in this chapter).

Activity 8.5

Recording and distributing messages

Form groups of three. There are two scripts for this activity in your workbook. Use these scripts to role-play answering the telephone and taking down a message (on the message slip provided) using the correct protocol. Each student should take down the message in turn. This will involve changing roles. Practise with one script, then use the other for official assessment.

Electronic message systems

There are also a number of ways in which messages can be communicated, stored and retrieved using digital mobile telephones and computer services. This is a time of rapid technological change, when new electronic methods for communicating messages are constantly being developed and improved. The following are some examples.

Direct-access paging service. This service enables callers to send either tone-alerts and numeric messages direct from their telephone, or numeric and alpha-numeric messages from their visual display terminal via a dial-up modem. See Figure 8.5.

Fig. 8.5 A pager

Personalised answering services. Messages may be left with a centralised, manually operated answering service, which often operates over a 24-hour period. Subscribers phone in to the centralised service to collect their messages from the operator. A charge is made for this service.

E-mail. Electronic messages can be sent from office to office through computers connected by a modem. Recent innovations enable anyone outside the office to send messages using a laptop or notebook computer connected to a data-capable digital mobile telephone using a special interface card. You can call up data from the main computer back at the office, send or receive faxes without a fax machine, check your e-mail, and even surf the Internet. This can all be done from your car or elsewhere when an office is not available. Figure 8.6 gives an example of what e-mail looks like on a computer screen, and Figure 8.7 shows e-mail using a notebook computer and a mobile phone.

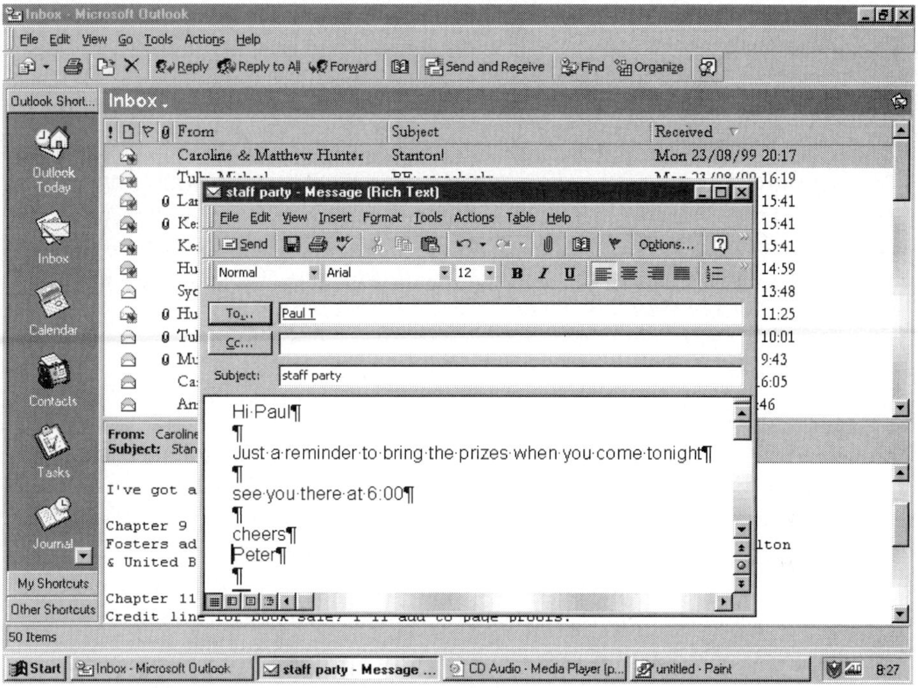

Fig. 8.6 Sending notes using e-mail

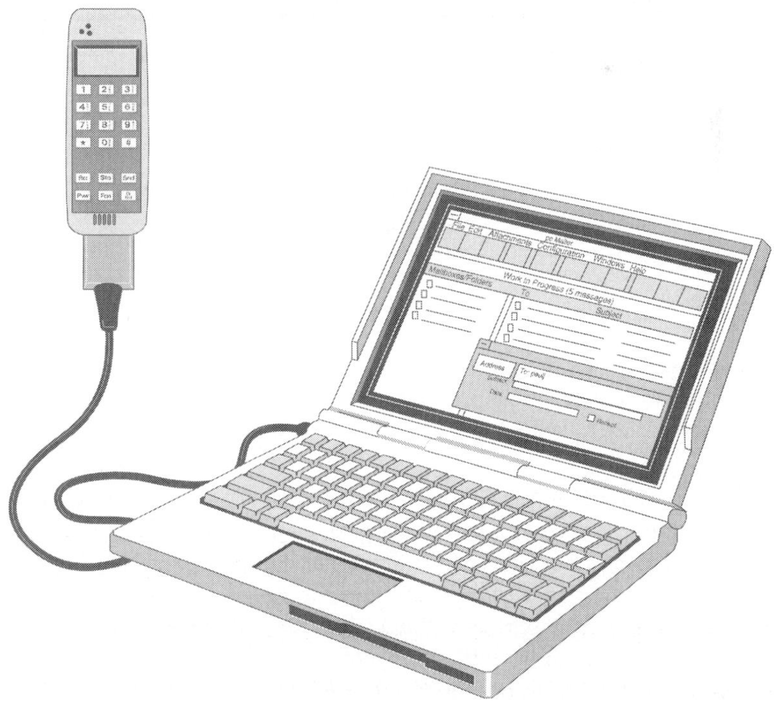

Fig. 8.7 E-mail using notebook computer and mobile phone

Voice mail. Voice mail is a message system linking a telephone or a mobile phone to a voice-activated computer. This is useful for after working hours, allowing a caller to leave a message on the computer. The message is then ready to be attended to the next day.

Voice messaging. With this system, if you are absent from the office, a centralised computer can have your telephone calls rerouted to your own 'voice mailbox'. It will answer your calls, take messages, and even give instructions to callers on how to leave messages. It may be accessed by a special code.

Answering machines. An answering machine is useful for recording messages when the office is closed or you are out of the office. It can also be used to screen incoming calls. Messages can be accessed from another telephone or a mobile phone using a remote device which responds to a security-code number.

You should refer to the manufacturer's manual to learn how to:

- install the tapes;
- record your outgoing announcement;
- screen incoming calls;
- record messages.

One model of answering machine is illustrated in Figure 8.8.

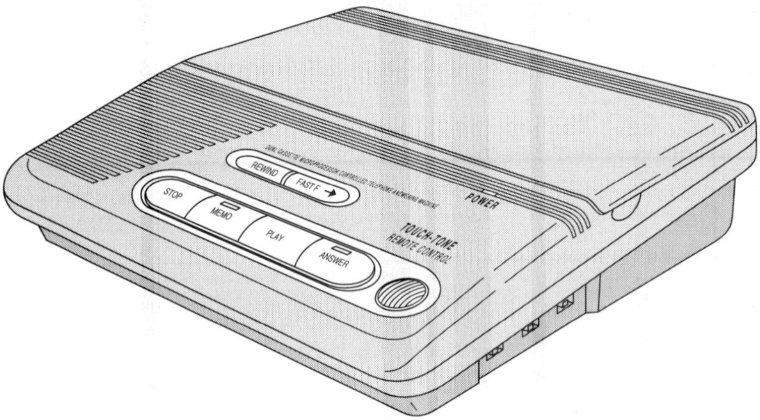

Fig. 8.8 Telephone answering machine

Making outgoing telephone calls (Procedure 15)

Making a telephone call is the easiest, and quickest, method of communicating with clients and customers. The advantages are:

- A telephone call is often less time-consuming and costs less than writing a letter.
- The message is received instantly and communication flows backwards and forwards instantaneously, so a decision can be reached quickly.
- A telephone call is more personal and can be more persuasive.

One disadvantage of a telephone call is that there is no formal record of the conversation, which is often required in business transactions.

Sources of telephone numbers (Procedure 16)

There are many sources of telephone numbers.

Internal telephone directory

An internal telephone directory should be kept next to all telephone extensions and the switchboard, so that callers can be transferred quickly and efficiently. It should be divided into departments, with the names and roles of the key personnel in each department listed in alphabetical order. The directory should be updated as staff join and leave. A list of often used numbers is also useful.

E-mail mailbox. Telephone numbers and mail addresses, as well as e-mail addresses, can be stored in your e-mail mailbox.

Australia's electronic phone book. You can now obtain software which has the complete set of Australian business and residential telephone directories on CD-ROM, and which includes geographical mapping to give users the ability to search for locations. It includes:

- all 55 national telephone directories;
- over 1.28 million business listings;
- over 6.1 million residential listings;
- over 182 000 fax numbers;
- auto-dialling facility.

There is a list-merge facility, which can be used for direct mail, telemarketing or market research.

Telstra telephone directories

Telstra issues telephone directories every year for the use of its customers. They are an invaluable source of information.

White Pages. The *White Pages* provides information on:

- telephone numbers for private and business locations in alphabetical order;
- government services;
- emergency and community services;
- help reference information;
- telephone information services;
- operator-assisted services;
- Australian area codes;
- the international fax network;
- STD and 0011 international calls;
- Telstra FREECALL 1800 (the number is free to the caller except when made from a mobile telephone).

It also provides information on operator-assisted calls for:

- faults and service difficulties;
- repairs and maintenance;

- nuisance calls;
- calls to ships at sea;
- reverse-charge calls;
- conference calls;
- ring-back price;
- wake-up and reminder calls.

Yellow Pages. The *Yellow Pages* provides information on:

- telephone numbers for business enterprises in subject order (then alphabetically within subject);
- emergency numbers;
- A–K fast-find index, L–Z fast-find index.

Activity 8.6

Obtaining telephone numbers

Use the *White Pages* and *Yellow Pages* to obtain the telephone numbers listed on Activity sheet 8.6. Your tutor/supervisor will then ask you to phone one of the following numbers using the correct procedure.

1. Ring the local branch of the Commonwealth Bank and ask them for their hours of opening and closing.
2. Ring international telephone enquiries and ask them for the local time in a country overseas.
3. Ring your local branch of Australia Post and ask them for details of one of the services they offer; for example:
   - express courier service;
   - the cost of insurance for a valuable item;
   - international faxpost.

Leaving messages (Procedure 17)

If you or any other member of staff in your organisation cannot be connected to the person you want to speak to, you may have to leave a message. This could be with the switchboard operator, with someone else in the office, or it could be on a telephone answering system. Most people find it difficult to leave messages, particularly on an answering machine. However, there are one or two rules that may help you to overcome nervousness when making outside calls and leaving messages.

Before you place a call

Make a checklist of the following:

- the name and telephone number of the organisation;
- the name of the person you are calling;

- a brief description of the content of your telephone call or your message;
- any documents you may need for the call (file, letter, invoice etc.).

When leaving the message

- give the name of the person you are calling;
- give your name, your organisation and your telephone number;
- state the message clearly and as briefly as possible (from your checklist);
- indicate whether it is urgent;
- ask the person taking the message to read it back to ensure that it has been taken down correctly (unless you are leaving a message on an answering machine).

Example

Answering machine: This is George Benson's telephone. I am unavailable at the moment. Please leave your name and telephone number and I will ring you back as soon as possible. Leave your message after the beep.

You: (*after the beep*) This is Jennifer Hanson from Business Stationery Supplies. The time is 11 am, Tuesday 3 January. I would like to speak to George Benson regarding delivery of supplies. This is urgent. Could you please telephone me as soon as possible today. The number is XXXX XXXX. Thank you.

You should always give the time and the date of the message left on an answering machine, so that the person receiving it knows when it was recorded. The person receiving the messages may be away and not receive the message until the next day.

Activity 8.7

Operating a telephone answering machine

Check that you know how to operate the answering machine available to your group (or use unconnected handsets). On Activity sheet 8.7 in your workbook there are scripts for practising leaving a message on an answering machine. Use the first one for practice and then choose one of the others for your official assessment. This will be observed by your tutor/supervisor.

PRACTICAL PLACEMENT

Real-life work experience

In the middle of your course, or at the end of it (depending on your specific course), you will be asked to obtain a position in an office to put into practice the skills you have learned. This is an excellent opportunity to experience a real-life office environment. You will observe how people work in group situations and deal with everyday working problems. It also gives management and other members of staff the opportunity to see how you perform tasks such as answering the telephone, dealing with customers, typing letters and other documents, and filing them.

It is your responsibility to telephone a number of businesses to find one that will accept you for one or two weeks (depending on your specific course) so that you may gain practical experience in a real office. You may not be accepted on the first call, and you may have to call many businesses, as some of them don't want to spare the time and effort for someone who is not experienced.

If you are accepted, this can be a good opportunity for you, as it often leads to part-time or full-time employment at the end of your course.

When you obtain work experience, your employer signs one copy of the practical placement form to show that you have been accepted. At the end of the work experience period your employer gives you a report on your performance, which, if good, can be used in your résumé when you look for full-time work.

Summary

- The way in which all personnel in the organisation answer the telephone can have a positive or negative effect on the corporate image of the organisation.
- In the business communication process, the telephone is used more than any other method of communication.
- As many systems are automatic, with no switchboard operator, it is essential for all personnel to answer and transfer calls and deal with clients/customers according to uniform corporate procedures, which can be found in the procedures manual.
- Your voice is an important component of the image you project to a listener on the telephone.
- As a representative of your organisation, you are expected to give a caller every assistance.
- Employing business ethics includes not disclosing personal information about employees; details of new products before they are released for sale; and financial details of the business that are not for public information. Business ethics also includes covering for your boss if she/he has forgotten, or been delayed for, an appointment or a meeting.

- To establish the requirements of callers you will need to listen effectively and clarify by repeating back to callers what they have said. Use the phonetic alphabet and standard grouping of numerals to clarify names and numbers.
- Open questions are used when you want a caller to explain or discuss an issue, and closed questions are used when all you want is a 'yes' or 'no' answer.
- Always write down the main points of a telephone conversation immediately the caller starts speaking. Also write down the telephone number and address (if relevant) so that you will not forget them. Use the caller's name often, as this is more friendly.
- Non-verbal telephone communication refers to the speaker's tone of voice and manner. These non-verbal cues can create in the listener a positive or negative image of your organisation.
- When dealing with angry customers, recognise that their anger is not directed at you; remain calm and helpful; ask questions and listen for facts; use the correct verbal signals and empathise with the caller. Then agree to a course of action and follow up that course of action.
- When taking down a message, always read it back to the caller to confirm that you have taken it down correctly.
- There are a number of electronic message systems, including direct-access paging service, personalised answering service, e-mail, voice mail and voice messaging.
- An answering machine is useful for recording messages when the office is closed, and for screening incoming calls.
- The sources for obtaining telephone numbers include an internal telephone directory (paper or electronic), and the Telstra *White Pages* and *Yellow Pages* directories.

Review questions

(These may be used for group discussion.)

1. There are five aspects of answering the telephone that can affect the public relations and corporate image of your organisation. What are they?
2. Why is it essential that there be uniform corporate procedures for all personnel answering the telephone?
3. As a switchboard operator, there are four things you need to know before you answer incoming calls. What are they?
4. Why is it important to answer the telephone promptly?
5. List the important components of your voice when answering the telephone and identifying your organisation and yourself.
6. List three examples of information you should not disclose to callers outside your organisation. These are part of business ethics.

7. Describe briefly the rules for transferring calls.
8. There are several ways in which you should clarify instructions and messages. What are they?
9. Why is it important to use non-verbal communication in telephone conversations? Give some examples.
10. Describe how you should deal with difficult customers and try to solve their problems.
11. When taking down a message, what are the five important details you should include?
12. Describe briefly two examples of electronic message systems.
13. How can an answering machine be useful for recording messages?
14. Before making an outside call, what should you do?
15. List the five stages of leaving a message.

Competency self-assessment and/or tutor/supervisor assessment

When you have finished reading this chapter, and have completed the activities chosen by your tutor/supervisor (**practical in class, at work or theoretical**), fill in this self-assessment. Then hand it to your tutor/supervisor for official assessment.

Learning outcomes	**Self-assessment**		**Tutor/supervisor assessment**	
I have successfully completed the following learning outcomes	**Yes**	**No** (please tick)	**Yes**	**No** (please tick)
1. Operate a telephone system effectively in the workplace				
1.1 operate equipment to receive and generate calls, using features such as:				
• picking up group	☐	☐	☐	☐
• transferring calls	☐	☐	☐	☐
• putting calls on hold and redialling, according to procedures outlined in the operating manual	☐	☐	☐	☐
• operate a telephone answering system according to procedures outlined in the operating manual	☐	☐	☐	☐

2. Receive and respond to incoming telephone calls in the workplace				
2.1 answer calls promptly and clearly, using designated business protocol procedures	☐	☐	☐	☐
2.2 identify callers correctly and establish their requirements accurately	☐	☐	☐	☐
2.3 give disclosable information only	☐	☐	☐	☐
2.4 transfer calls to appropriate location/person promptly	☐	☐	☐	☐
2.5 record messages effectively using clear and concise language	☐	☐	☐	☐
2.6 distribute messages according to organisational procedures	☐	☐	☐	☐
3. Make telephone calls in the workplace				
3.1 obtain accurate telephone numbers from an appropriate source	☐	☐	☐	☐
3.2 establish contact using designated business protocol procedures	☐	☐	☐	☐
3.3 convey purpose of call clearly and concisely	☐	☐	☐	☐
Date:	Tutor/supervisor signature:			

CHAPTER

9

EFFECTIVE BUSINESS CORRESPONDENCE

COM202A, NOS128, ABD316,
COM501

objectives

The purpose of this chapter is to help you to produce more effective written business communication, in order to complete the following learning outcomes:

1. **Draft simple workplace correspondence from given instructions and information**
    1.1 draft simple correspondence to a standard format consisting of one or two paragraphs and present for approval within designated timelines
    1.2 use clear and concise language
    1.3 use correct spelling, punctuation, grammar and syntax
    1.4 convey intended meaning of correspondence so that it is readily understood by recipient
2. **Draft routine workplace correspondence in response to telephone, oral and written requests for information**
    2.1 clarify and obtain necessary information to complete the task
    2.2 determine and use method of correspondence that suits the purpose of the task
    2.3 use appropriate format for each document produced according to organisational procedures
    2.4 present for approval according to organisational procedures
    2.5 store draft correspondence to enable easy access for future use
3. **Identify appropriate means of responding to requests for information according to organisational procedures**
4. **Compose business memoranda and letters in response to a given need or request**
5. **Compose a report in response to a given need or request**
    5.1 identify and locate information relevant to defined needs
    5.2 access and extract information relevant to the task
    5.3 use report layout as designated by organisational procedures
    5.4 edit, produce and despatch report to nominated person/section on time

INTRODUCTION

When you send out business correspondence to your customers and clients, this reflects the image of your organisation, as do the tasks of answering telephones and receiving visitors. If your letters are carelessly written, badly typed and poorly composed, this will give the people who receive them the impression that the organisation is inefficient and uncaring about standards.

Customer communication

Chapter 4 outlines the major roles of an office, the functions of an office, and the operations carried out in an office, which are primarily for the purpose of communicating with customers.

Whatever the position you hold in an office you will find that 90 per cent of your work is communicating with customers—for example, when you use the telephone; receive and send out mail; receive people at the reception desk; use office equipment to send letters, memos, invoices and photocopy documents, and then file them in the appropriate location. Most chapters in this book are designed to help you understand the ways in which you communicate with customers.

Electronic communication

The most common forms of electronic communication are e-mail and facsimile machine.

E-mail. How to use e-mail, e-mail etiquette and a list of terminology are dealt with in more detail in Chapter 1.

Facsimile machine. The operation of a fax machine in described in Chapter 3.

The electronic office. Producing documents using remote-control software and E-commerce are described in Chapter 4.

Speech recognition. In Chapter 3, we deal with various methods of electronic communication, including producing documents by continuous-speech recognition dictation software.

Help on screen. When composing documents on the screen of a computer, most systems help you with your composition. In Windows a curved red line appears under misspelt words, and a green line appears under badly constructed sentences. There is also a spell-check facility. Many templates and wizards help you with the layout of documents.

Mail merge. One of the most helpful features of electronic document production is the mail merging of a document with names and addresses of customers and clients. This allows you to send out the same document to many customers without having to type them individually, which saves many hours of work.

Guidelines to improve writing business documents

In spite of the above electronic help facilities, most people find writing letters the most difficult and tedious task of any office work. Below are some guidelines to improve business letter-writing. Writing any document involves four stages:

1. thinking and planning;
2. capturing ideas on paper;
3. organising the shape and structure;
4. editing, revising and proofreading.

When writing a letter, you should ask yourself the following questions:

- What is the purpose of the letter?
- What information needs to be included?
- Have I used simple, short words, and sentences that have a clear meaning?
- Is the letter written in a courteous manner?
- Is the letter written with a positive attitude?
- Is the letter clear, concise, accurate and relevant?
- Is the level of language appropriate?
- Is the letter presented tactfully?
- Is the layout appropriate and does it create a good impression?
- Are the ideas presented logically, in the correct paragraph style?
- Have I clearly explained any follow-up requirements?

Finally:

- Have I checked that the spelling is correct?
- Have I proofread the document?

Be careful when using 'spell-check' on your computer word-processing system. The system may indicate that the word is spelt correctly, but you may be using the word in the wrong context. The spell-check will indicate that you have spelt 'their' or 'there' correctly, but you could be using the word incorrectly.

LETTER COMPOSITION—A CHECKLIST

1. *Plan your letter.* Clarify the purpose of the letter. Make a list of the contents of each paragraph. Begin a new paragraph when a new point is being made.
2. *Use the correct layout.* Refer to Figure 9.2 for the correct layout of business letters, and to Figure 9.4 for the correct layout of internal memorandums.
3. *Write from the reader's point of view.* This is called the 'you' attitude. Begin most paragraphs with 'you' or 'your', as this is more friendly and makes the reader feel you are empathising.
4. *Make the first sentence count.* This catches the attention of the reader. For example, you could say 'yes' to a previous request; give an item of good news; or ask a question.
5. *Make your sentences and paragraphs as short as possible.* Using simple words and phrases makes the document easier and quicker to read.

Do not use	Do use
• We wish to acknowledge receipt of your communication	• Thank you for your letter dated . . .
• Receipt is hereby acknowledged	• We received
• We wish to thank you	• Thank you
• It would be appreciated if you would send	• Please send
• I am writing this letter (The recipient knows you are writing a letter.)	

6. *Write courteously.* Writing an abrupt or rude letter will annoy the customer/client, and you may lose their business.
7. *Use a positive approach.* Write in a positive manner and try to eliminate negative attitudes.
8. *Use correct grammar and spelling.* Use spell-check if you are not sure of a word, and keep a dictionary on your desk for extra checking. Incorrect grammar and spelling will reflect poorly on you and your organisation.
9. *Make sure that the information you quote is as accurate as possible.* Do not quote someone unless you have their permission to do so. Do not make a claim without stating 'It is our understanding' at the beginning of the sentence. If you are wrong you could be sued. Do not insult a person's character. This may be libellous.
10. *Use constructive endings to your letters.* This should let readers know exactly what action they are expected to take. For example:
    - Please let me know your progress in this matter.
    - Please phone me if I can help you further.
    - I look forward to receiving your decision in the next few days.
    - A representative will contact you in a few days.
    - I will phone you with the information tomorrow.

LETTER STRUCTURES

Steps in transmitting 'bad news'

1. The first paragraph should start with a neutral comment, indicating some form of agreement.
2. The second paragraph should present an explanation in a positive manner.
3. The third paragraph should state clearly the bad news.
4. The next paragraph should offer an alternative solution.
5. The last paragraph should end the letter on a positive note.

Steps for effective 'persuasive' or sales letters—the 'AIDA' principle

A—The first paragraph should attract the reader's ***attention***.

I—The second paragraph should stimulate the reader's ***interest***.

D—The next paragraph should develop a ***desire*** within the reader to do what you want him/her to do.

A—The last paragraph should encourage the reader to take the specific ***action*** you want him/her to take.

The following are some examples.

Attention. How would you like to cut your water-heating expenses by 30%? Do you worry about repairs you don't have time to do? What would happen to your family if you were injured at work? How would you like to cut your medical bills in half?

Interest. In the second paragraph, show your readers how you can cut their water-heating expenses, or carry out repairs for them at a very competitive rate, or how your insurance product can help to protect their family or reduce their medical bills.

Desire. In the third paragraph, convince your readers that you can satisfy their needs. Make them feel that your product is exactly what they need. Give a few details of the product.

Action. In the last paragraph, you have to make sure that your customer buys your product. Always follow up a persuasive letter as follows:

- Tell your client you will telephone on a specific day.
- Make a special offer of a discount with their first purchase.
- Give the customer a gift with their first purchase.

You can also create attention and interest by using graphics and illustrations and different fonts and colours.

Steps in transmitting 'good news' and neutral letters

1. Begin the first paragraph with the good news.
2. Provide the details of the good news in the second paragraph.
3. Close on a friendly note.

Letter and memo layout

There are three basic styles in common use for letter layout:

- fully blocked style, which is shown in Figure 9.1;
- blocked style, in which the date and complimentary close start at the centre of the page;
- semi-blocked style, which is the same as blocked except that the first lines of paragraphs are indented (shown in Figure 9.2).

Examples of fully blocked and semi-blocked memoranda are given in Figures 9.3 and 9.4.

Facsimile messages

Fax is a very useful method of sending and receiving instantaneous printed communication. One of the disadvantages of telephone communication is that there is no record of the conversation, and this is sometimes needed for legal requirements. This problem is overcome with documents sent by fax. A fax form is used in worksheet 3.4.

Form letters (mail merge)

A form letter is a standard letter which can be written and used many times. A letter can be prepared on a word processor and merged with a list of customers' names and

13 August 199X

(leave 4 blank lines)

Messrs Bonny and Clyde
62 Bright Road
WINMALEE NSW 2777

(leave 1 blank line)

ATTENTION MR J R CLYDE

(leave 1 blank line)

Dear Sirs

(leave 1 blank line)

COMPANY HANDBOOK

(leave 1 blank line)

Our accountant, Mr Scott, has informed me that you are withholding payment of our invoice for printing your handbook because of the price charged. The invoice explains that the reason for this charge was the 8.6 per cent increase in the printing costs since 24 January, and below is Clause 19 of our formal quotation:

(leave 1 blank line)

These prices are subject to rises in wages following the National Wage Tribunal Award or the Graphics Arts Award. Material costs will be negotiated as and when notified.

(leave 1 blank line)

This situation has caused problems that are quite beyond our control.

(leave 1 blank line)

As requested, I enclose a copy of our catalogue.

(leave 1 blank line)

Yours faithfully
DETAILED PRINTING PTY LTD

(leave 4 blank lines)

Debra Pharns
General Manager

(leave 1 blank line)

DP/SL

(leave 1 blank line)

Enc

Fig. 9.1 Fully blocked letter with open punctuation

16 September 199X

(leave 4 blank lines)

Mr M. G. Horeau
Michael Horeau and Company,
Second Avenue,
SYDNEY, NSW 2000

(leave 2 blank lines)

Dear Mr. Bradford,

(leave 1 blank line)

Thank you for your letter dated 12 September, enquiring about electronic typewriters.

(leave 1 blank line)

Our Galaxy electronic typewriters have been designed to produce work to the highest possible standard and, at the same time, they are simple to operate and very cost-effective.

(leave 1 blank line)

The Galaxy has a minimum correction memory of 500 characters and this makes text modifications quite simple.

(leave 1 blank line)

Full details are given on the enclosed pamphlets, and we look forward to hearing from you

(leave 1 blank line)

Yours sincerely,
EASY TYPING COMPANY

(leave 4 blank lines)

Angelo Harvaridis,
Sales Manager.

(leave 1 blank line)

AH/SL

(leave 1 blank line)

Enc.

(leave 1 blank line)

c.c. Sales Department

Fig. 9.2 Semi-blocked letter with closed punctuation

(leave 6 blank lines from the top of the page)

MEMORANDUM

(leave 2 blank lines)

TO:	Simone Seefeld
(leave 1 blank line)	
FROM:	Despina Patouras
(leave 1 blank line)	
DATE:	14 September 199X
(leave 1 blank line)	
SUBJECT:	PROCEDURES MANUAL

(leave 2 blank lines)

The students of ADOA 33 are required to produce a Procedures Manual which will involve work that has been covered in the subjects of Document Production 3 and 4. This manual is part of our assessment for these units.

(leave 1 blank line)

We are both required to compose a report on Business Correspondence for this section in the Procedures Manual.

(leave 4 blank lines)

Despina Patouras

(leave 1 blank line)

DP/SL

Fig. 9.3 Fully blocked memorandum with open punctuation

MEMORANDUM

(leave 1 blank line)

TO: Martha Shaffey

(leave 1 blank line)

FROM: Simone Seefeld

(leave 1 blank line)

DATE: 29 May 199X

(leave 1 blank line)

SUBJECT: TELEPHONE AND ACCESSORIES

(leave 2 blank lines)

As requested in your memo dated 18 May, I have obtained copies of catalogues and enclosed them with this memo.

(leave 1 blank line)

The tone ringer is a portable plug-in unit that works well where there is a lot of noise and distraction. It has a distinctive, high-pitched ring which makes it easier to hear in noisy locations like workshops. It would seem eminently suitable for the Works Manager's office.

(leave 1 blank line)

It would seem wise to remove the permanent phones from a number of points and have the phone sockets fitted, so that we can take a phone to whichever room is in use for a meeting. The sockets can be fitted easily.

(leave 1 blank line)

The extension cord is handy when you have to move about the desk area to find information. There is a plug at one end and a specially designed casing at the other. The standard length is three metres.

(leave 4 blank lines)

Simone Seefeld

(leave 1 blank line)

SM/DP

(leave 1 blank line)

Enc

Fig. 9.4 Semi-blocked memorandum with mixed punctuation

addresses. A persuasive (sales) letter which will be distributed in letterboxes does not require a name and address: it can be addressed to 'Dear householder' or 'Dear customer'.

Internal memoranda

An internal memorandum is used for communication between staff in an organisation. It can be typed or handwritten. Internal memos are now often sent by e-mail. Memos are used to save time, and cost less to produce than a letter. They are less formal than a letter and are not printed on the same quality of paper. The format for memos is illustrated in Figure 9.3.

Correct addressing of letters

The correct method of typing the address on the top of letters, and on envelopes, is illustrated in Chapter 1 and in Figures 9.1 and 9.2.

Activity 9.1

Letter-writing

When replying to the following letters, remember you are an employee of the company we are using throughout this book—Business Stationery Supplies Pty Ltd. Type your letters in the correct business structure and layout (as outlined in the previous pages) using the letterhead provided for you in your workbook (photocopy the ones required for sections 2 and 3).

Section 1
Write a draft letter (or memo, as required) for ***one*** of the following situations. Use the 'bad news' structure.

1. You are the administrative coordinator of your section. Draft an internal memo to all members of staff in your section, informing them that they will be required to work over the weekend in order to complete a very important order for one of your best customers.
2. You have received a telephone call from the local school asking you to say a few words at their annual prize-giving evening. Write a reply politely refusing their request, but suggesting one of your colleagues as a suitable replacement.
3. You have received a letter from the local amateur dramatic society in your area, requesting you to sponsor their group. This is one of the many requests received by your company every year. Write a reply letter refusing to sponsor their group. Write in such a way that none of the members, many of whom are your customers, will be offended.

Section 2
Write a persuasive letter to prospective customers in ***one*** of the following situations (remember to use graphics and different fonts to attract attention).

1. You are starting your own business. Write a 'form' or standard letter to potential customers to introduce your new business. In your letter, describe the product/s or service you are going to provide.
2. You have developed a new product. Write a letter to a manufacturing company describing your product and persuading them to develop and market it.

Section 3

Write a reply to ***both*** of the following situations. Choose the correct letter structure.

1. Your boss has asked you to write a letter to an applicant for a position with your company. This person has been successful. The position is as a clerk in the accounts department.
2. You have received a telephone call today from Ms Jenny Goren, the manager of your local supermarket. She is complaining about the poor quality of the inkjet cartridges you supplied in her last order. Write a reply to this telephone complaint, suggesting what you are going to do about this matter.

When you have completed the draft correspondence, show it to your tutor/supervisor for correction. Using the fully blocked, open-punctuation format (shown in Figure 9.1), type the final versions of the four drafts you have written. When you have finished, hand them to your tutor/supervisor for assessment. Place all your final typed correspondence in a folder.

Case study 9.1

Replying to a letter of complaint

You are the personal assistant to the sales manager of Business Stationery Supplies Pty Ltd. One of your designated duties is to deal with incoming and outgoing correspondence.

Included in the correspondence to which you must reply today is a letter (dated two days ago) from the manager of a fitness and sports centre in your area, who is one of your best customers. She states that she is satisfied with the price and quality of your goods, but she is complaining about the services provided by your organisation. She gives examples of occasions on which staff have been abrupt and unhelpful. She has left messages to which she has not received a reply, and orders have been delivered late on more than one occasion. She is thinking of placing her business with another company.

How would you reply to the letter in order to avoid losing this customer?

1. How would you plan and organise your letter?
2. What questions would you ask yourself before commencing your letter?

(continues)

3. Which letter structure would you use?
4. What would you say in your letter to explain the poor services of your organisation, and what suggestions could you make to prevent this happening in the future?
5. In addition to giving your apologies, what kind of compensation could you offer to this customer for the inconvenience caused?
6. What constructive ending would you use for this type of letter?

Using the letter composition checklist on the previous pages as a guide, write a draft reply. After this has been approved by your tutor/supervisor, type a final version and hand it to your tutor/supervisor for assessment.

REPORTS—RESEARCHING AND WRITING

Report-writing is a vital part of the communication process within an organisation, because:

- Executives need information on which to base decisions.
- Customers expect facts and figures on products and services.
- Employees want information on matters that affect them.
- Government branches require records of business operations.

The purpose of business reports

Reports help solve problems, and in business there are constantly problems to be solved and decisions to be made. For example, the manufacturing department may need to upgrade certain machinery; the office may require to buy, or lease, a certain piece of equipment; or a report may need to be written regarding staff problems.

Another reason for reports is that they prevent problems from occurring. These reports continually update people's knowledge so that they may make day-to-day decisions.

Types of business reports

There are basically two types of reports—long reports and short reports. Both these types may be categorised as follows:

- to analyse a situation;
- to inform as to progress;
- to present a recommendation;
- to suggest a program of action;
- to give an account of conduct;
- to present the results of a survey;
- to report research findings.

To analyse a situation. To study or investigate is to analyse. A report on absenteeism or personnel turnover, or monthly sales figures, would be an analysis. Reports of this nature allow the reader to focus quickly on the main points of the topic.

To inform as to progress. If your company is building a new facility, management will need periodic progress reports in order to judge whether building is on schedule and to plan accordingly.

To present a recommendation. In this type of report a situation is studied and a recommendation made to management on the alternatives—for example, whether to lease or buy new equipment, to hire a new member of staff, or to pay overtime in order to meet organisational goals.

To suggest a program of action. When a problem or situation has been identified, a 'program of action' report will outline how to resolve it. In this technological age it may become obvious that the skills of certain employees need to be updated. Such a report will outline how this should be carried out.

To give an account of conduct. If there has been an incident between employees or between an employee and a customer, and if you were a witness to this incident, you may be asked to write a report on the conduct of the people involved in the incident.

To present the results of a survey. A survey may be carried out to determine customers' attitudes towards your company's products. This type of report will outline the results of the survey, from which management will make a decision on marketing the product.

To report research findings. Many companies have a research (R&D) department for the purpose of monitoring their present products and developing new products. The results of this research are presented in the form of a report to management for policy-making decisions.

Planning and writing a short report

Short reports perform many different functions in a business organisation to keep management in touch with operations. They usually span one page, and the most common are:

- form reports
- memo reports
- letter reports
- committee minutes
- staff circulars
- bulletins.

There are four main categories of short reports:

1. proposal report
2. progress report
3. justification report
4. periodic or routine report.

Many progress reports, and periodic or routine reports, are on preprinted forms. These are filled in with a pen and are often self-carbonated so that copies can be sent to heads of departments on a regular basis. For example, sales representatives fill in weekly progress reports to give to the sales manager at the end of the week. An example of this is given in Figure 9.5.

Customer Contact Report

Enter all visits with customers each week. Indicate your time of arrival and departure as well as product/s discussed. Describe any sale you made or service problem of which you learned. Describe any follow-up you intended to perform.

Date	Time	Name of contact and company	Product discussed	Sales made	Service problems	Follow-up
Mon.	Arr. Dpt.					
	Arr. Dpt.					
	Arr. Dpt.					
Tue.	Arr. Dpt.					
	Arr. Dpt.					
	Arr. Dpt.					
Wed.	Arr. Dpt.					
	Arr. Dpt.					
	Arr. Dpt.					
Thurs.	Arr. Dpt.					
	Arr. Dpt.					
	Arr. Dpt.					
Fri.	Arr. Dpt.					
	Arr. Dpt.					
	Arr. Dpt.					

Signature of sales representative

Note: Complete this form at the end of each work week and mail it immediately to the Sales Manager.

Fig. 9.5 Example of a customer contact report

Steps in planning a short report

The purpose of a short report is to inform, to persuade or to encourage some action by the reader. Short reports present facts, options and alternatives in a way that is easy to read and is presented in concise and logical order.

The following steps should be followed when you are writing any kind of report, whether short or long.

1. Identify the purpose of the report. Clarify your task precisely.
2. Consider the reader/s of the report. Take into account their knowledge and familiarity with the topic, as these will affect both the content and language of your report.
3. Identify your information needs. List these as well as the sources that will provide the information.
4. Research your information from both primary and secondary sources. (This is covered in greater detail in long reports in the next section.)
5. Sort your information, discarding any irrelevant material and organising what is important into numbered headings.
6. Arrange your sections in suitable sequence, as shown in Figure 9.6.

	MEMORANDUM REPORT
Headings Who the report is written to → By whom and date → Subject of report →	TO: Martha Patouras, Manager, Administration Department FROM: Debra Smith DATE: 3 May 199X Office administrative coordinator SUBJECT: Conservation of resources within the office
Introduction **Purpose statement** • State the problem →	At the last last general meeting it was pointed out by the Chief Accountant that he had noticed many instances of wastage of resources around the office. It is now company policy to conserve as many natural resources as possible—not only to reduce costs but also to help preserve the environment. I was asked to make all staff aware of these conservation issues.
Body **Information** • Discuss and explain the facts → • Summarise and show how to carry out the recommendations →	There are many ways in which natural resources can be conserved in all departments: • switching off light and power when not required • recycling paper in all departments —using readdressable internal envelopes —using china cups rather than polystyrene cups —many other ways
Conclusion • Present the recommended solution → • Outline the conclusion →	I have asked managers and supervisors of each department (with the cooperation of staff) to let me have a list of other ways in which everyone can conserve resources in their department. When I receive the lists from all departments, I will incorporate them into a long report which I will hand to the Chief Accountant, with a copy to you, at the next general meeting.

Fig. 9.6 Plan for writing a short report and an example of a short memorandum report

Activity 9.2

Writing a short report

(You may discuss this with the members of your group, or you may complete it on your own.)

Draft a short memorandum report to the manager of your department on one of the topics listed below:

1. health and safety issues in the office;
2. training of new staff in the use of computers and other electronic machines;
3. how to conserve resources when dealing with incoming and outgoing mail;
4. how to conserve resources when using machines in the office;
5. any other topic approved by your tutor/supervisor.

When your draft has been approved by your tutor/supervisor, type a final short report for assessment.

Planning and writing a long report

Short reports are usually only one page long. Formal or long reports can span hundreds of pages. They can be written to summarise research into and present recommendations on future possible action; evaluate action already taken; and provide solutions to problems. What differentiates reports from other business documents is that all reports have their content organised under main headings and subheadings. The presentation of a formal report can influence its outcome as much as the findings contained in it, so careful planning is required.

Steps in planning a long report

There are seven steps in planning a long report:

1. clarifying the purpose of the report;
2. considering the reader;
3. researching the subject;
4. interpreting the data and deciding how to organise and present them;
5. preparing the outline;
6. writing the report;
7. packaging the report.

Clarifying the purpose of the report. Determine what it is that the reader wants and needs to know. This has to be established before any more work is done. It may be helpful to think of the title of your report before you begin planning it. Titles should not begin with 'Why', because this indicates a preference.

The purpose of the report might be to:

- provide information;
- report on a project;
- recommend solutions to a problem;
- analyse a situation;
- explain a situation;
- report research findings.

The report provides information needed to make a decision or solve a problem. Thinking about what information will be needed to make that decision is a good point for starting your planning.

Considering the reader. Thinking about who will read or use the report will help you to make other decisions. Consider the reader's needs, not your own. Also, write to the reader, not for yourself:

- Is the report for your boss, the manager of your department or even your tutor/supervisor?
- How much does the reader of your report know about the topic of the report? (The aims and objectives of the report should be at the level of the reader's knowledge.)
- Will the reader appreciate a light touch, or is it a formal report?
- How will the reader be affected by the language you use? Will you need to explain some of the terms?
- What is the reader's primary interest? (Conservation of the environment, cost-cutting, human resources matters?)

Researching the topic. You will need to decide what information should be included in the report. You should collect as much data about the topic as you can from primary and secondary sources (see Table 9.1) and then sort these into major headings and subheadings (as shown in Figure 9.8).

Primary sources	Secondary sources
Personal observation (going out and seeing what is taking place)	Internet
Interviews (talking to people involved in this area)	World Wide Web
Surveys (people filling in questionnaires)	Pamphlets
Consultants and experts in the specific area (talking to them)	Books
Experiment (making changes and seeing what impact those changes have)	Journals
Organisation reports, files, minutes of meetings etc.	Magazines
	Newspapers

Table 9.1 Sorting information into primary and secondary sources

Interpreting data. When you have collected as much information as possible (see Figure 9.7), sort it into major topic areas. These will form the major headings of the body of the report. You should interpret these facts in a logical sequence so that they support your argument or statement. Write the headings down on a sheet of paper. Then sort the

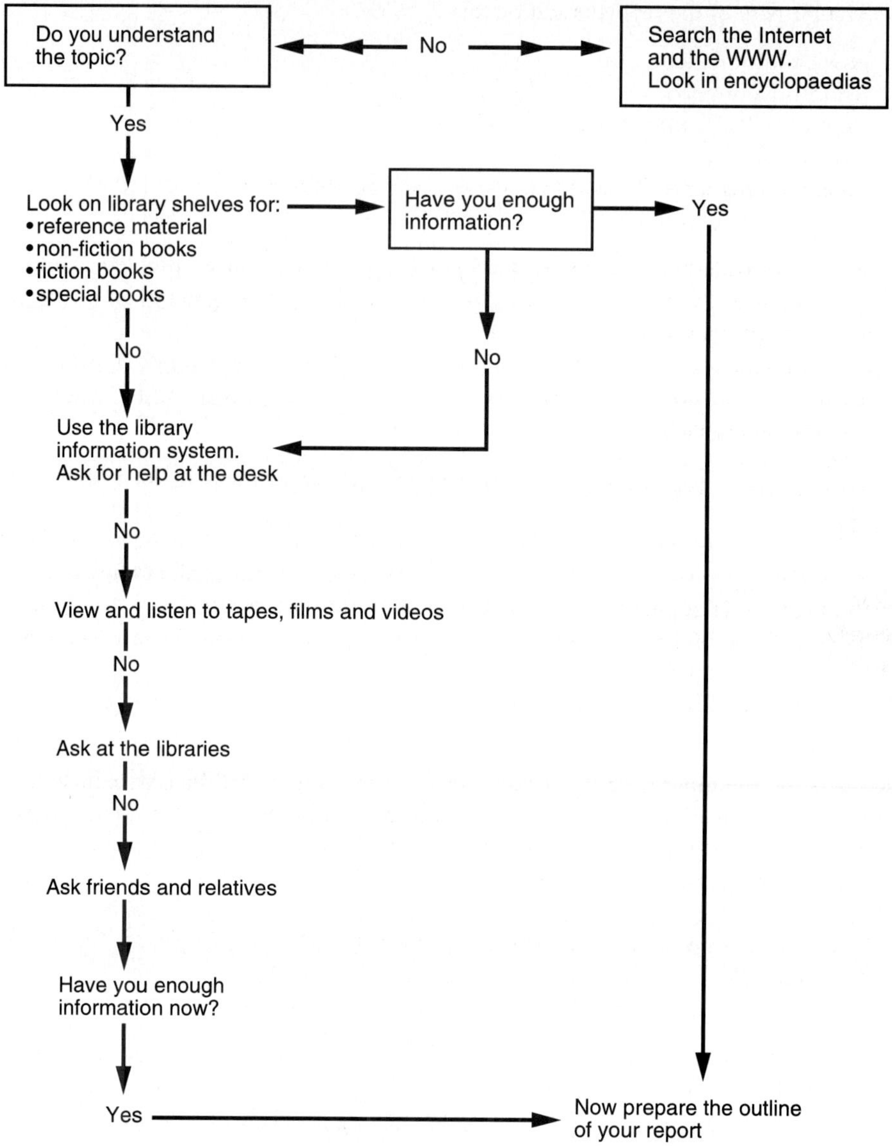

Fig. 9.7 The steps necessary to search for information

information in the major areas into subsections, which will form the subheadings under the major headings, as shown in Figure 9.8.

Organising data. You have now collected your facts and selected those that meet your objective. You use logic to interpret these facts, draw conclusions and organise them for presentation in the form of an argument. You use a structure that will:

- compare similar facts;
- choose those facts needed for the argument;

- evaluate and select material suited to the purpose;
- keep to the subject;
- sift significant points from irrelevant ones;
- combine all parts of the topic sequentially;
- achieve a balanced argument.

You should use **effective reasoning**, and this will happen only if you have:

- avoided generalising;
- avoided drawing hasty conclusions;
- used sufficient data to make your argument valid;
- used specific examples;
- isolated cause and effect correctly;
- defined the question adequately;
- been explicit in concluding from evidence;
- used results that can be duplicated;
- explained any contradictory evidence.

You should also use **deductive reasoning**, which begins with a general statement, moves to particular details or applications in specific instances, and reaches a conclusion.

Facts can be organised in many different ways—by historical sequence, problem by problem, geographically, or in order of importance. The most common way of organising a long report is chronologically, which is shown in Figure 9.9.

Preparing the outline. The principles of outlining are:

- ***division.*** Here the facts are divided into parts, which are then further subdivided. This will give main topic headings, which can be further subdivided in subheadings.
- ***consistency.*** Once you have decided on your main headings and their order, list and number them. Keep to a consistent format, as follows:
  —If your first main topic is expressed as a sentence, then express all main topics as a sentence.
  —If you use a phrase for one, then use a phrase for all.
  —Express all names, dates, abbreviations consistently.
  —If you use a capital letter for 'Manager', keep it consistent throughout.

To present the report in an objective manner, it is general practice to write business reports using the **third person** and the **past tense.** You have completed your research and any experiments, and you are now writing a report on something that happened in the past.

Incorrect sentences	**Correct sentences**
I did this study	The purpose of this study was to . . .
I found	The findings indicate that . . .
I recommend	It is recommended that . . .

Parallelism means that the relationship of ideas can be reflected by giving headings or points of equal importance the same treatment—by, for example:

- using initial capitals for all major headings;
- using the same font style;
- indenting, capitalising or enhancing text consistently throughout the report.

An example of the above points when outlining your report is given in Figure 9.8. This shows a table of contents, with major headings and subheadings, and numbering system for headings and pages.

TABLE OF CONTENTS

Fig. 9.8 Major headings, subheadings and numbering system

Writing your report. Now write your report using the rules outlined at the beginning of this chapter. Isolate the key concepts under each major heading and its subheadings, and use the key words in your topic sentence in each paragraph. Pay attention to the following so that you achieve clarity and effectiveness.

- When making a point, give an example taken from a primary or secondary source.
- Compare or contrast findings or facts with others to provide a frame of reference.
- Give reasons for statements, assumptions and proposals.
- Support your statements with statistics, bar or pie graphs, illustrations, and other materials such as quotes from publications.
- In the body of your report state facts; do not draw conclusions, solve problems or make recommendations. These come at the end of the report.

- At the end of each paragraph, or below each chart, statistic or graph, you must source your material (state where you have obtained it). This is very important, because if you do not state where you have obtained your information you could be accused of plagiarism (which is copying someone else's work). This is an infringement of the *Copyright Act 1968* (Cwlth), and could result in fines (see Chapter 3).

Packaging your report. When packaging your report, you should again consider the reader. Is this a report to a prospective partner in a small business—a florist, a boutique, a garage, a children's clothing shop, music or computer shop? You could make your report illustrate the type of business with pictures, pie graphs, cartoons, in the more persuasive style outlined at the beginning of this chapter, and your folder could be a much brighter colour. Or is this a formal report to the managing director, or board of directors, of your organisation? If so, the style of writing and formatting your report should be very businesslike; the colour and presentation of the folder, and all other sections of the document, should reflect this. Formal, long reports are packaged in the form outlined in Figure 9.8, and as follows.

1. **The cover.** The cover of your report communicates to the reader the title of the report, to whom it is written, by whom it is written and the date, and the name or logo of the company to make it as attractive as possible. If your report is a business report the cover should be in business colours—blue, grey, green, burgundy, black or white. If the report is not formal, the cover could be in colours to reflect the business—pink, red, light blue, yellow or gold.

2. **Letter (or memo) of authorisation.** This is a letter, or memo, written by the person who is requesting the report. This gives you the authority to write the report.

3. **Letter (or memo) of transmittal.** This is a letter, or memo, written by you (the writer of the report), which you include with the report when you send it to the person it is written for.

4. **Title page.** This is the first page of the report, and sets out:

- the full title of the report;
- the name of the person to whom it is written;
- the name of the person who prepared the report;
- the name of the organisation (if applicable);
- the date the report was written;
- the place the report was written (if applicable).

All these should be centred horizontally and vertically, as shown in Figure 9.9.

5. **List of figures, tables, graphs.** List these items on a separate sheet of paper at the beginning of the report.

6. **Executive summary (synopsis, or abstract).** This is a one-page, or shorter, summary of the entire report. It contains, in about three paragraphs, the purpose of the report and the most important findings and recommendations, stated in the briefest possible terms.

The purpose of the executive summary is to enable the reader to get to the essence of the report without having to read it in full. If a report is many pages long, the executive summary may well be the only part of the report a busy executive has time to read.

STRESS AT WORK

prepared for

Debra Smith
Occupational Health & Safety Officer,
Business Stationery Supplies Pty Ltd

by

(Your name)

3 May 200X

Fig. 9.9 Example of a title page

7. **Introduction.** The introduction is an important part of the report. It should indicate the scope of the report and present an overview of what is to follow. The introduction outlines the aims, objectives and method of the report, and follows the content in the same order as the body of the report. It is always the first heading of the report, is numbered '1', as shown in Figures 9.8 and 9.10, and is then broken into subheadings. The headings of the body of the report are then numbered in sequence after the introduction.

8. **Numbering.** Scholarly or scientific reports often use the author/date (Harvard) system, but most business reports use the decimal system of numbering. This is shown in, Figure 9.10.

9. **The body of the report.** The body of the report is the section that contains the text to be written, already identified under headings and subheadings in the outline. Some parts of the body require special attention:

- In the section on findings, present data or facts only, with explanations or examples but no discussion or conclusions. These are under separate headings at the end of the report.

1. **INTRODUCTION**

1.1 **Authorisation**

This report was requested by Debra Smith, Ocupational Health and Safety Officer, Businesss Staionery Supplies Pty Ltd. The request was made as a result of several cases being reported of stress-related illness within the organisation.

1.2 **Aim**

To identify the nature and causes of stress-related problems in organisations
To determine the sources of stress in a business environment
To link signs of excess stress to poor health habits
To point out the significance of these to managers
To recommend action to overcome the stated stress problems

1.3 **Method**

1.3.1 researched various texts on stress management
1.3.2 interviewed managers
1.3.3 put out a survey for employees to complete
1.3.4 analysed data on staff absenteeism

Fig. 9.10 Example of an introduction using the decimal numbering system

- When you are presenting your discussion or analysis, use only evidence or expert opinions consistent with the findings from other sources. Evaluate findings logically, compare, point out advantages or disadvantages, or measure against standards.

10. Conclusions. Conclusions are reached from your analysis or discussion in each section, and these are outlined under a separate heading at the end of the report.

11. Recommendations. Recommendations are made based on your conclusions, and are listed under this heading at the end of the report.

12. Endnotes or footnotes. Material taken from other sources and used in the report must be referenced. A number is placed at the end of the paragraph where the material is inserted and the source is then quoted at the foot of the relevant page, or at the end of the report as 'endnotes', which are put in numeric order. This is shown in Figure 9.11.

13. Bibliography. The bibliography sources primary and secondary material used in your report. This is positioned after the endnotes and often contains the same references as sourced in the endnotes. But endnotes are listed in numeric order and bibliography references are listed in alphabetical order of author/s, as follows:

ENDNOTES

1.	Stoner, J.A.F., Collins, R.R. & Yetton, P.W., *Management in Australia* Sydney: Prentice-Hall of Australia Pty Ltd, 1995, p. 505	three authors
2.	Drucker, P.F., *Innovation and Entrepreneurship: Practice and Principles* London: William Heinemann Ltd, 1996, p. 156	one author
3.	Stoner *et al.*, op. cit., pp 505-507, 509-511	2nd entry (*et al.* means all the authors, op. cit. means the same publication. (This saves you having to repeat all the reference details.)
3.	Drucker, P.F., op. cit. p. 49	
4.	Dullforce, W., 'Innovation: Why the Answer Lies in a Range of Approaches', *The Financial Times*, 6 February 1997	journal reference
6.	Rance, C., *Management of New Ideas,* www.mang.net.au	Internet reference

Fig. 9.11 Endnote list

- the author's surname, followed by initials or given names (if more than one author, all the names are entered);
- the name of a government department, if this is used;
- the title of the book, journal or magazine, underlined or in italics;
- if quoting an article in a book or journal, the title of the article in quotation marks;
- for books, the place of publication, publisher and year of publication;
- for journals, the name of the journal underlined or in italics, and volume number.

14. Quotations. When using quotations, the following rules apply.

- Quotations smaller than five typed lines should have quotation marks around them and be incorporated in the text.
- Quotations longer than five typed lines should be indented and single-spaced. It is not necessary to use quotation marks.
- Quotations may be abbreviated by omitting sections of the original passage. This is indicated by inserting three spaced full stops in the passage (e.g. 'It is ... important to have one day free ... to prepare for the following week').

15. Appendixes. Supplementary or detailed material which is used in the report but is too detailed or too long to include in the body of the report is added to the end and listed at the end of the contents.

Activity 9.3

Writing a long report

(This is an individual activity.)

Each student will produce (out of class time) a long report on any topic chosen by him/her (with the approval of the tutor/supervisor). The report should be 1500 words long (the body of the report), and should include the following. (If you have difficulty thinking of a subject for your report, any one of the chapters in this book could be used as a topic.)

Requirements for the report

- a cover
- title page
- memo of authorisation
- memo of transmittal
- table of contents
- executive summary (synopsis)
- introduction
- analysis, discussion, under major headings
- conclusions
- recommendations
- endnotes or footnotes
- bibliography
- appendixes.

Note: You should use graphs, statistics, graphics and/or illustrations to enhance your report.

Further:

- The report must be typed.
- The body of the report should be in double-line spacing.
- The main headings should be numbered and typed in bold capitals.
- The subheadings should be numbered in decimal form, typed in lower case and underlined.

Refer to your tutor/supervisor for the time you will be allowed to produce this long report.

Summary — Business correspondence

- Business correspondence sent to customers and clients reflect the 'image' of your organisation, as do the ways in which you answer telephones and receive visitors. If your letters are badly typed and poorly composed, this will give the people who receive them the impression that the organisation is inefficient and uncaring about standards.
- Writing any document involves four different stages: thinking and planning; capturing ideas on paper; organising the shape and structure; and editing, revising and proofreading.
- You should plan how you write your letters: use the correct layout; write from the reader's point of view (the 'you' attitude); make your sentences and paragraphs as short as possible; use correct grammar and spelling; and make sure the information you quote is as accurate as possible.
- Some standard letter structures include steps in transmitting bad news, and steps for effective 'persuasive' letters (the AIDA principle). There are three basic styles for letter layout.
- E-mail and fax messages are useful and quick methods of sending and receiving instantaneous printed communication.
- One of the most helpful ways of producing large numbers of form letters is using electronic 'mail merge', which saves hours of time.
- Internal memos are used for written communication between staff within an organisation.

Review questions

(These may be used for group discussion or formal written assessment.)

1. Writing any document involves four different stages. List these on a separate sheet of paper.
2. Before you commence writing any document, you should ask yourself a range of questions. List these.
3. List the stages in the letter composition checklist.
4. Write down the steps in transmitting bad news.
5. Write down the steps for writing effective persuasive (sales) letters using the AIDA principle.
6. Outline the steps used in a 'good news' letter.
7. What is the 'you' attitude when writing letters, and why should you use it?

Summary — Researching and writing reports

- Report-writing is a vital part of the communication process within an organisation, because reports help to solve problems and help decisions to be made. They also prevent problems from occurring. Short reports perform many functions in a business organisation, especially keeping management in touch with operations.
- There are seven steps in planning a long report: clarifying the purpose; considering the reader; researching the subject; interpreting the data and deciding how to organise and present them; preparing the outline; writing the report; and packaging the report.
- There are two sources of information—primary and secondary.
- You should organise data into a structure that will: compare similar facts; choose those that are needed for the argument; evaluate and select material suited to the purpose; keep to the subject; sift significant points from irrelevant ones; combine all parts of the topic sequentially; and achieve a balanced argument.
- When preparing the outline, use division, consistency and parallelism.
- When writing your report, isolate key concepts to be placed under each major heading and subheadings.
- Your report packaging should include the following sections: a cover, title page, memo of authorisation, memo of transmittal, table of contents, executive summary (synopsis), introduction, analysis, discussion in the body of the report, conclusions, recommendations, endnotes, bibliography, and appendixes.

Review questions

(These may be used for group discussion.)

1. Why are reports used in business and other organisations?
2. Why are reports important to the people who write them?
3. There are basically two types of reports. What are these?
4. Both these types of report may be categorised in seven ways. List these.
5. List the six different types of short reports.
6. There are four main categories of short reports. What are these?
7. List the six steps in planning a short report.
8. List the seven steps in planning a long report.
9. There are two sources of information—primary and secondary. List under each heading the different sources of information.
10. Outline the steps necessary to search for information.

11. When you have collected your facts you use a structure to organise them. Explain the points of this structure.
12. You also use effective reasoning. How do you do this?
13. What is deductive reasoning?
14. What are the principles of outlining your report?
15. Write down examples of the sentences you should use and the sentences you should not use.
16. Explain what parallelism is.
17. Write down the different sections of a long report, and explain what these are.

Competency self-assessment and/or tutor/supervisor assessment

When you have finished reading this chapter, and have completed the activities chosen by your tutor/supervisor (**practical in class, at work or theoretical**), fill in this self-assessment. Then hand it to your tutor/supervisor for official assessment.

Learning outcomes	**Self-assessment**		**Tutor/supervisor assessment**	
I have successfully completed the following learning outcomes	**Yes** (please tick)	**No**	**Yes** (please tick)	**No**
1. Draft simple workplace correspondence from given instructions and information				
1.1 draft simple correspondence to a standard format consisting of one or two paragraphs and present for approval within designated timelines	☐	☐	☐	☐
1.2 use clear and concise language	☐	☐	☐	☐
1.3 use correct spelling, punctuation, grammar and syntax	☐	☐	☐	☐
1.4 convey intended meaning of correspondence so that it is readily understood by recipient	☐	☐	☐	☐
2. Draft routine workplace correspondence in response to telephone, oral and written requests for information				
2.1 clarify and obtain necessary information to complete the task	☐	☐	☐	☐
2.2 determine and use method of correspondence that suits the purpose of the task	☐	☐	☐	☐

2.3 use appropriate format for each document produced according to organisational procedures	☐	☐	☐	☐
2.4 present for approval according to organisational procedures	☐	☐	☐	☐
2.5 store draft correspondence to enable easy access for future use	☐	☐	☐	☐
3. Identify appropriate means of responding to requests for information according to organisational procedures	☐	☐	☐	☐
4. Compose business memoranda and letters in response to a given need or request	☐	☐	☐	☐
5. Compose a report in response to a given need or request				
5.1 identify and locate information relevant to defined needs	☐	☐	☐	☐
5.2 access and extract information relevant to the task	☐	☐	☐	☐
5.3 use report layout as designated by organisational procedures:	☐	☐	☐	☐
• cover	☐	☐	☐	☐
• title page	☐	☐	☐	☐
• memo of authorisation	☐	☐	☐	☐
• memo of transmittal	☐	☐	☐	☐
• table of contents	☐	☐	☐	☐
• executive summary (synopsis)	☐	☐	☐	☐
• introduction	☐	☐	☐	☐
• analysis, discussion under major headings and subheadings, numbered correctly	☐	☐	☐	☐
• conclusions	☐	☐	☐	☐
• recommendations	☐	☐	☐	☐
• endnotes or footnotes	☐	☐	☐	☐
• bibliography	☐	☐	☐	☐
• appendixes	☐	☐	☐	☐
5.4 edit, produce and despatch report to nominated person/section on time	☐	☐	☐	☐
Date:		Tutor/supervisor signature:		

CHAPTER 10

JOB-SEEKING

COM202, NCS003

objectives

The purpose of this chapter is to teach you how to look for a job, and how to plan and prepare your application for a job, including the steps involved and the documents you will use in preparing for your interview, in order to complete the following learning outcomes:

1. **The steps involved in obtaining a position**
   - 1.1 research employment opportunities
   - 1.2 carry out preparation and planning
   - 1.3 make telephone contact
2. **Prepare job application documents**
   - letter of application
   - your résumé
   - your job folder
   - application for employment form
3. **Make preparations for a job interview**
   - appearance
   - checklist for interview
   - prepare and anticipate questions
   - after the interview

INTRODUCTION

In previous chapters you have read about the role of an office, and the titles and duties of personnel working in an office. You have learnt how to operate equipment, and how you should deal with people on the telephone and at the reception desk. You have learnt how to write business correspondence, and what is involved in a work environment. Now is a good time to look at job-seeking, because a lot of preparation is required before you actually start looking for a job.

Applying for a job is like running a race. You have to:

- prepare well in advance;
- follow the rules and regulations;
- dress appropriately;
- adopt the correct mental attitude;
- possess the relevant qualifications.

What are employers looking for?

You have seen people holding positions in a travel agency, in an insurance office or many other offices. How did they obtain these positions, and why were *they* chosen rather than someone else? When an employer puts an advertisement in a newspaper for a vacant position, what kind of person are they looking for to fill that position? The structure of all office positions consists of four parts: job analysis, job description, job specification, and job evaluation. (This is described in more detail in Table 11.1 and Figure 11.1 in Chapter 11.)

Employers are looking for people with:

- a willing attitude
- reliability
- flexibility
- productivity
- skills and qualifications.

Recruitment and selection process

Once a position has been advertised, the procedure for selecting the best person for the job usually follows the stages shown in Figure 10.1. This process is very expensive. The organisation's costs include the time involved in assessing applications and interviewing candidates, and the cost of advertising.

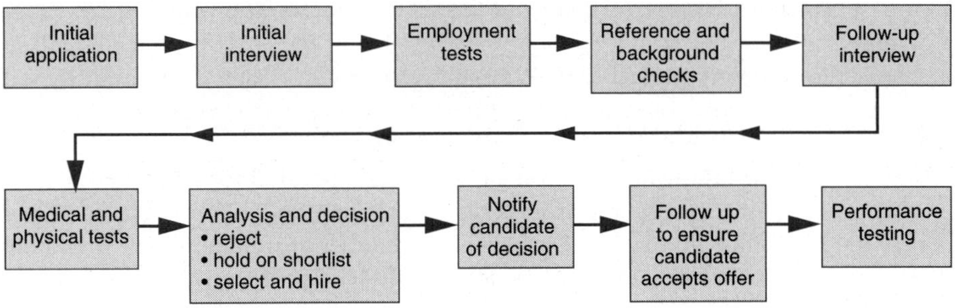

Fig. 10.1 The recruitment and selection process

THE STEPS INVOLVED IN OBTAINING A POSITION

The steps involved in obtaining a position (winning the race) are:

- discovering employment opportunities;
- deciding what type of job you are interested in;
- deciding what type of job you are looking for;
- applying for the position, including preparing for the interview.

Employment opportunities

You are applying for your first position, or you would like to change your present position. How do you go about this? Where do you look? What should you do? There are many places and ways in which you could look for job opportunities.

- The **classifed sections of local and state newspapers** list vacant positions in subject order and then in alphabetical order.
- **Public sector** positions are advertised by departments or statutory authorities in major newspapers and public service journals.
- **Private employment agencies** will ask for a copy of your résumé (CV) and give you a series of tests to determine your skills and qualifications. They will then try to help you obtain a position linked to those skills and qualifications. Always register with more than one employment agency, as they often have an exclusive agreement with one or two large organisations to supply candidates for vacant positions. This applies particularly in the city, and if you register with more than one agency you are covering a wider selection of job vacancies.

Not all vacancies are advertised. Other opportunities for finding a position are as follows.

- You can make a **direct approach to employers**. Advertise yourself by sending an application letter, with brief résumé, to a number of organisations. Managers often like this approach, as it shows personal initiative.

- You can approach **friends, relatives and former workmates** who work in large organisations.
- Check the **company noticeboard**. A large organisation, a college or university will often post an internal vacancy on a noticeboard. If no suitable person can be found internally the position is then advertised externally.
- You may be **'headhunted'**. Once you have been in a position for a number of years, and have established a reputation for excellence in your particular area of business, you may be approached by another department, or another organisation, that would like to employ you. It is more common for people in top executive positions to be approached from outside the organisation, but personnel at a lower level are often approached within an organisation.
- **Work experience** may lead to employment. One way organisations recruit is to accept students for one or two weeks' work experience. This gives the organisation the opportunity to determine how you perform the duties allocated to you. If you seem to be performing quite well, they may offer you a temporary or part-time position with a view to future full-time employment. This benefits both parties: the organisation has no advertising and interviewing costs, and it gives you a chance to view the people you will work with, and the organisation, before you commit yourself to a full-time position.

The more positions you apply for, the more chance you have of getting a job.

Deciding what type of job you are interested in

The ideal job is working in a position you love, and being paid to do it. You should try to achieve this by analysing your aptitudes, your likes and dislikes, your abilities and your values. Are you artistic? Are you good at figures? Or do you possess very good word-processing or desktop-publishing skills and enjoy using them? You should use your Activity sheets 11.1 and 11.2 from Chapter 11.

Link all the above abilities and skills, and decide:

- what kind of work you would like to do;
- whether any work (other jobs, or work experience) you have had in the past was the kind you would like to continue doing;
- what your limitations are—there is no point in applying for a position you have no chance of getting because you lack the basic skills or aptitudes; but do not narrow your chances by thinking you are not good enough because you lack only one requirement of the position, as training may be given once you have been employed.

Look at the duties and qualifications outlined in the job advertisement in Figure 10.2, and ask yourself the questions shown in the figure.

Deciding what type of job you are looking for

Once you have analysed your preferences, skills, work experience and training, you should look for positions for which you could apply. Look through the job advertisements in the newspapers and cut out any positions that seem suitable. Consider the options available to you, as shown in Figure 10.3.

Administrative Assistant

Accounting Department

A vacancy exists within the Accounting Department for an Administrative Assistant. ← **Is this the kind of position you would like?**

Duties and responsibilities:

- Assist with secretarial and administration support in Group Accounting Services ← **Do you have the necessary experience?**
- Updating of Accounting Policy and Procedures Manual
- **Other duties include:**
  - sending/receiving and delivering facsimiles ← **Would you like to perform these duties?**
  - delivery of accounting reports and memos
  - assisting in data input accounts payable ledger
  - verifying system-generated reports from accounts payable system
  - assisting in preparation of invoice batches from accounts payable for filing
  - liaising with Travel Department invoice documentation for processing
  - maintaining photocopier routine maintenance and paper levels

Qualifications and skills required: ← **Do you have the required qualifications?**

- Year 12 education
- Office Administration course preferred
- Good verbal and written communication skills ← **Do you have good verbal and written communication skills?**
- Good telephone manner
- Self-motivated ← **Are you self-motivated?**
- In-depth knowledge of: MS Word for Windows
  Excel (Windows)
- Good word-processing and typing skills ← **Do you have in-depth knowledge of Word for Windows, Excel, and good typing skills?**

Responsible to Accounting Manager
Hours of Work: 9 am to 5 pm Monday to Friday

Fig. 10.2 Duties and qualifications for an advertised position

Applying for the position

Before you apply for a job, you should research it and the organisation, and prepare and plan as follows.

Preparation and planning

If the position has been organised for you by an employment agency, ask your consultant for more details.

If the position was advertised in the newspaper, check the details carefully; if you feel you do not have enough information, phone the company and ask whether a job

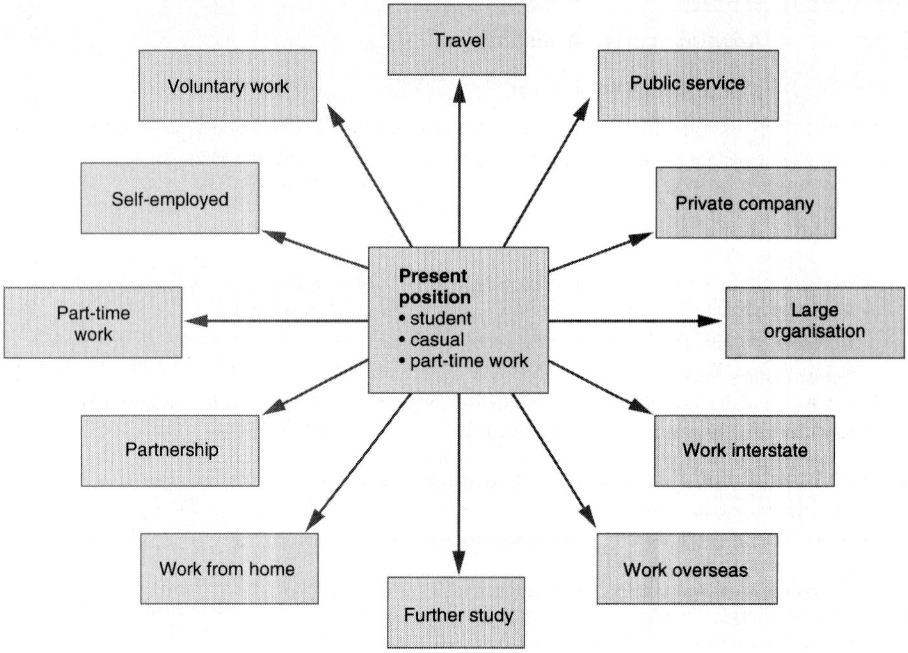

Fig. 10.3 Options available to you

description could be sent to you. If a job description is not available, ask for more details over the telephone.

Telephone contact with employer

When phoning an employer to ask for more details about a vacant position, you should remember to:

- make a list of your own questions;
- make a list of the answers to questions that could be asked by the employer (this will help you to sound confident and will be a test of your telephone techniques: remember to speak clearly and a little slowly);
- address the person by name;
- have your documents with you so that you can give answers to any questions asked;
- check the name and position of the person to whom you should send your details;
- ask questions clearly, concisely and courteously;
- keep the telephone conversation as short as possible;
- thank the person you are speaking to for their help and taking the time to speak to you;
- indicate your willingness to go for an interview.

JOB APPLICATION DOCUMENTS

The purpose of job application documents is to obtain an interview, and ultimately to make you the preferred person for the position.

Letter of application

Your letter of application is important, as this is the first document the person who is reading your application will see. It should be well presented and easy to read, and it should show at a glance why you are the best person to fulfil the requirements of the position. It should also demonstrate your excellent skills in typing business letters, which may be a requirement of the job.

- Address the letter to the person indicated in the advertisement.
- State, in the subject heading of the letter, the position advertised. There may be more than one position advertised in a large organisation, and this makes it easier for the reader to see which position you are applying for.
- State in the first paragraph where you obtained details of the position. If this is from a newspaper, include the title and date.
- The second paragraph should indicate why you are interested in this particular position with this specific company.
- The third paragraph should give a brief description of your qualifications and skills linked to the job specification, to show that you are qualified to apply for the position. It is a good idea to put this in bold point form so that it catches the eye of the person reading it.
- State in the fourth paragraph that you are enclosing a copy of your résumé.
- The last paragraph should indicate that you are willing to go for an interview and where you can be contacted.

This is illustrated in Figure 10.4.

5 Gowie Avenue
Bankston, VIC 3199

3 January 199X

Miss D Georgiou
Recruitment Officer
Business Stationery Supplies Pty Ltd
163 Collins Street
MELBOURNE VIC 3100

Dear Miss Georgiou,

APPLICATION FOR POSITION OF RECEPTIONIST (Position 123)

I wish to apply for the position of receptionist (Position 123) advertised in *The Australian* dated 3 January.

This position appeals to me, as I have an outgoing personality and I like meeting new people.

I have trained in the duties required for a receptionist position, in a one-year Advanced Certificate in Office Administration at Hometown College of TAFE, which I have just completed. This includes:

- reception protocol
- telephone techniques
- file management
- office equipment
- word processing, including: MS Word for Windows
Excel (Windows)

A copy of my résumé, including certificates, work experience and references, is enclosed for your information.

I am available for interview at any time during working hours, and can be contacted on telephone number XXXX XXXX.

Yours sincerely

Your name

Enc:

Fig. 10.4 Letter of application

Activity 10.1

Applying for a job

Form into groups of three, and choose an advertisement from a newspaper for a job that you would like to apply for at the end of your course. (Choose one for which you will be suitably qualified at that time. Complete the following activities:

1. With another member of the group, role-play making telephone enquiries about this position.
2. Write an application letter for this position.
3. Fill in the 'application for employment' form provided in your workbook (Activity sheet 10.1). Your tutor/supervisor will observe the telephone role-play and assess your performance. You will produce this letter and application form, together with the advertisement and your job folder, for a role-play interview in Activity 10.3.

Your résumé (curriculum vitae)

A résumé is a concise and orderly listing of information about a person. It contains details required by potential employers so that they can evaluate the contents in relation to the requirements of the position being offered. A résumé is sent with a job application letter, or produced at an interview. It contains:

- an information section with personal details, such as name, address, date of birth and telephone number;
- education details and qualifications;
- work experience listed in chronological order, starting with the most recent;
- the names, addresses and telephone numbers of referees (usually three).

An example of a traditional résumé is given in Figure 10.5.

Computer CV templates

There are many computer systems incorporating templates and wizards that will set up the framework of a curriculum vitae (résumé). This is very helpful for people who do not know how to set out a résumé. But one of the advantages of setting out your own is that you can make yours appear different and stand out from the other dozens of applicants' CVs. This is outlined in the following paragraphs. Figure 10.6 shows how to create a résumé.

The layout of the traditional résumé is perfectly acceptable, but this is where you can gain two steps in your race to obtain the position. The advice given by a consultant from a well-known large employment agency is never to pay for someone else to write your résumé. Résumés that follow the same pattern look the same. One that stands out from the rest is more likely to attract attention.

RESUME

NAME:	Margaret Mary Jones
ADDRESS:	45 Gowrie Avenue, FRANKSTON, VIC 3199
TELEPHONE NO.:	XXXX XXXX
DATE OF BIRTH:	XX/XX/XXXX

TERTIARY EDUCATION Jan 199X-Dec 199X	Monash University Bachelor of Business Major studies: (list subjects) Minor studies: (list subjects) Diploma in Business Studies Hometown College of TAFE Major studies: (list subjects)
SECONDARY EDUCATION:	Year 12 Higher School Certificate Frankston High School Major studies: (list subjects) School Captain
WORK HISTORY: Apr 199X–Nov 199X	Administrative Coordinator Business Stationery Supplies Pty Ltd, Melbourne Major duties: (List all other positions you have held.)
REFEREES:	Names, addresses and telephone numbers of three referees

Fig. 10.5 Traditional résumé

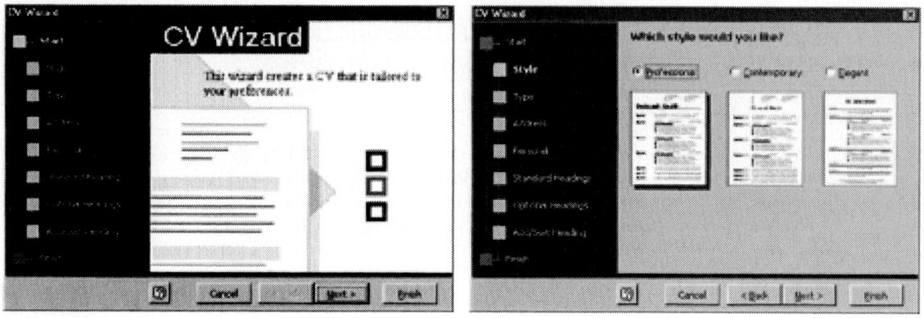

Fig. 10.6 Creating a résumé

You have probably been told that a résumé should be only one or two pages long, but a résumé should be as long as your experience. If you are straight from college and have not had a permanent position, you will need to include your work experience while at school or college; your résumé will not be long. If you have been working for 15 years and have had a varied career, you will want to point this out, and your résumé will be much longer.

If you want your résumé to stand out, and if you want to get those two steps ahead, the following instructions may help you.

- Every person is different, with a different personality, skills, hobbies, work pattern, goals, beliefs and values. You should make your résumé reflect your own personality and make the readers feel that you are the person they are looking for.
- Follow the traditional résumé format, but make it draw attention to the particular accomplishments that make you more suitable to fill the position than other applicants. For example, if you are applying for a position that requires excellent word-processing or desktop-publishing skills (and this is usually the case today), your résumé will not stand out if you type it in a traditional format. Show what desktop-publishing skills you have in the presentation of your résumé.
- The first page should be a presentation of the essential 'you'. List your personal details; your best skills and anything you are outstanding in can be shown here, linked to the specific requirements of the position.
- A résumé should always be easy to read. If it takes effort on the reader's part to decipher what your duties were in each position, he or she may not bother to read it. List your duties in point form so that they can be read at a glance.

An example of this kind of résumé is given in Figure 10.7.

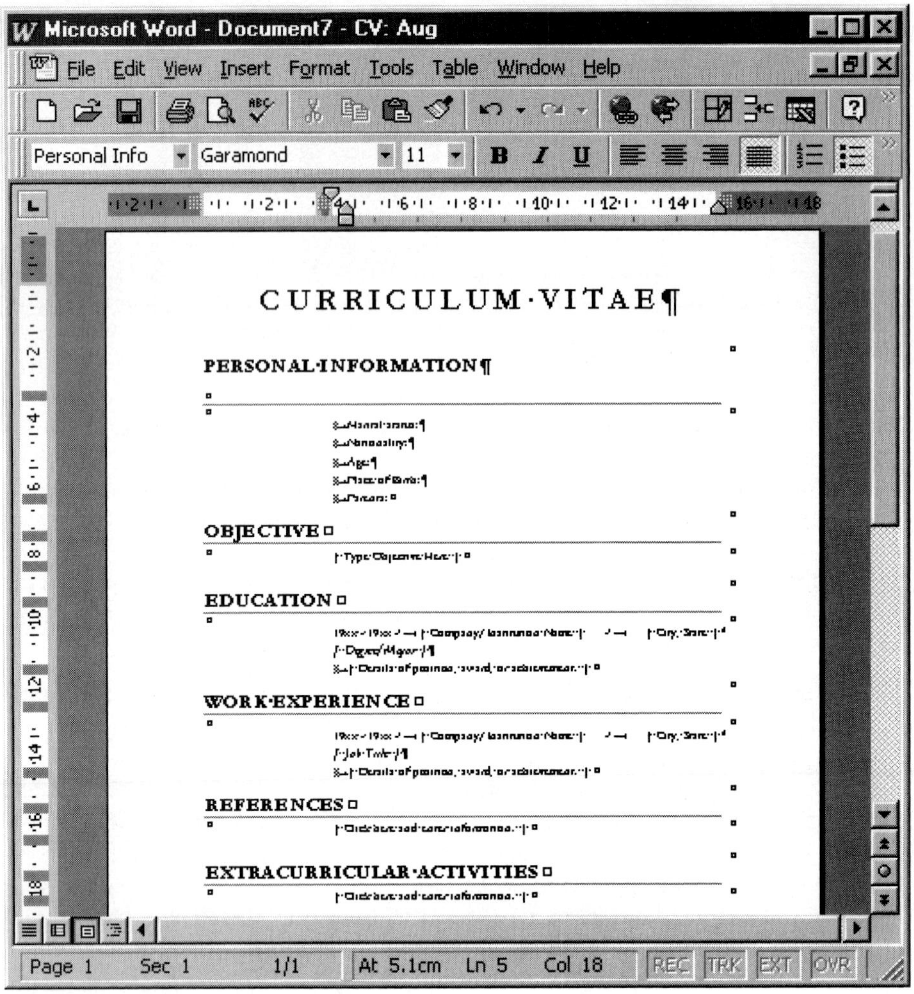

Fig. 10.7 Personalised résumé

Activity 10.2

Résumé

With the other members of your group, discuss how you can personalise your résumé to reflect your personality and particular skills and abilities. List the information, personal qualities, and skills and aptitudes that should be included.

1. Produce a draft personalised résumé, and show it to your tutor/supervisor to make sure you have included all relevant information.
2. Produce a final, well-presented, personalised résumé for inclusion in your folder.

Your job folder

A job folder holds all the documents you are taking to an interview for a position. It should include:

- your up-to-date résumé;
- your school and college certificates;
- references from previous positions of employment and from people who have known you for many years.

References can be either verbal or written. These are reports from your present or previous employer to your prospective employer. They may be confidential (you might not see them). Names, addresses and telephone numbers of referees are listed on the last page of your résumé.

You can also enclose examples of your work, such as business letters. Your résumé should be an example of your work.

You should keep your job folder for the rest of your working life and update the contents as you change jobs. Never send original documents with an application letter in the post, or leave them with a prospective employer, as these could be misplaced. Always use copies. Originals can be shown if necessary.

Presentation of your folder

The presentation of your folder is important, as this can make it stand out from the rest. There are some attractive folders that can be bought, with plastic inserts and paper the same colour as the folder. The colours are light grey, green or blue (do not use bright colours), and present a businesslike, professional appearance.

Application for employment form

Just before an interview you may be asked to complete an 'application for employment' form. It is wise to take with you information you will require to complete the form. All this information should be in your folder. An example of an application for employment form is given in Figure 10.8.

APPLICATION FOR EMPLOYMENT

PERSONAL DETAILS

Surname Mrs/Ms Miss/Mr ______ First Names ______

Address ______ Postcode ______

Age ______ Date of Birth ______ Place of Birth ______

Phone no. Home ______ Business ______

Driver's Licence YES ☐ NO ☐ Height ______ Weight ______

EDUCATION

	YEARS	INSTITUTION	LEVEL REACHED
Secondary			
Tertiary			
Other			

Interests and Hobbies ______

Business College, Courses and Studies since school ______

EMPLOYMENT EXPERIENCE Starting with most recent employer

NAME AND CITY	FROM/TO	POSITION HELD	RESPONSIBLE TO
1.			Name
			Title
			Phone No.
2.			Name
			Title
			Phone No.
3.			
4.			
5.			

SPECIAL REMARKS

(Any information that could be useful about your experience and abilities or duties and responsibilities.)

______

______

SUMMARY OF EXPERIENCE Tick for work experience

CLERICAL	BOOKKEEPING	TYPING	SECRETARIAL	OFFICE MACHINES	P/R SERVICE
☐ Filing	☐ General Books	☐ Electric-Copy	☐ Shorthand	☐ Photocopier	☐ Receptionist
☐ Statistics	☐ Bank Reconciliation	☐ Executive IBM	☐ Dictaphone	☐ Calculator	☐ Switchboard
☐ Wage Records	☐ Posting Ledger	☐ Word Processing	☐ Own Letters	☐ Accounting Machine	Type:....................
☐ Stock Records	☐ Trial Balance	☐ Mag Card IBM	☐ Private Secretary	☐ Mini Computer	☐ Sales
☐ Costing	☐ Profit and Loss	☐ Correspondence	☐ Technical	☐ V.D.U.	☐ Customer Enquiries
☐ Data Processing	☐ Computer Records	☐ Figure	☐ Legal	☐ Key Punch	☐ Insurance
☐ Purchasing	☐ Kalamazoo System	☐ Technical	☐ Medical	☐ Telex	☐ ....
☐ Cash Handling	☐ Debtors/Creditors	☐ Balance Sheet	☐	☐	☐ ....

OTHER EXPERIENCE ______

I consent to your checking references with previous employers. You may give relevant information to prospective employers.

NOTE: Discrimination because of race, colour, religion, sex, national origin or age is prohibited by various laws. Completion of related questions above is optional.

Signature ....

Fig. 10.8 Application for employment form

YOUR INTERVIEW

The interviewer's purpose is to:

- find out what kind of person the applicant is;
- check the factual details supplied on the application form or résumé;
- fill in any gaps in the information supplied;
- supply the applicant with more details about the position;
- enable the applicant to ask questions about the position;
- compare the applicant with the requirements for the position and assess his/her suitability;
- select the best person for the job.

Your interview is the last stage in your race to the winning post. Below are a few hints to help you with your interview.

Your appearance

First impressions always count, so dress neatly and to suit the occasion. Dress to suit your personality so that you feel good, as this gives you confidence. But remember that how you dress is a form of non-verbal communication, so dress in a businesslike manner. (Dress is discussed in more detail in Chapter 7.)

Checklist for the interview

- Research the company before the interview, so that you can answer questions on why you want to join that particular organisation.
- Find out as much as you can about the position, so that you can answer questions on why you would like this particular one.
- Try to arrive a little early.
- If you are not familiar with the area, and the location of the office for the interview, have a trial run so that there will be no delays on the day. This will help to give you confidence.
- Give a firm handshake.
- Do not sit down until you are invited to do so.
- Sit or stand without slumping, slouching or giving the wrong signals with your body language.
- Do not smoke—even if you feel you need to.
- Listen effectively. If you are not sure about a question, ask for clarification.
- Answer all questions directly, without evasion.
- Give examples of your experience, but do not exaggerate.
- Let the interviewer bring up the matter of salary. You should give the impression that you are more interested in the position than in the salary you are going to receive.
- Thank the interviewer/s for their time, and shake hands on leaving.

Preparing and anticipating questions

Questions you could be asked

- Why did you apply for this particular job?
- Why have you chosen this particular type of work?
- What experience have you had in this particular type of work?
- What do you know about our organisation?
- What are your goals for the future?
- Do you like accepting responsibility?
- What do you think you could contribute to our company?
- Do you like working as part of a team?
- What do you feel are your strengths and what are your weaknesses?
- Are you willing to work occasional weekends or flexitime?

Questions you could ask

- Is this a permanent position?
- Could you give me more information about the position?
- Would I be given any training?
- Are there any prospects for advancement?
- Will I need to do further studies?
- What would be the hours of work?
- Would I need any special equipment or clothing?
- When can I expect to hear the outcome of this interview?

Answers you should not give

- I never thought about it.
- I just want any position.
- I don't know anything about the company or the industry.
- I don't know.

These answers will lose you the job.

Answers you should give

- I have talked to people who have done this kind of work.
- I know you are one of the top insurance companies, and my parents have one of your policies.
- This is the kind of position I have always wanted, and I hope to rise in the company.
- My strengths are punctuality, reliability and loyalty, and I am a willing worker.

Be open, honest and try to relax (difficult in this situation, but tell yourself that you can only do your best—if you don't get this particular position there are others). This interview will have been good practice for others in the future. Make a note of the things you think you did wrong and try to rectify these next time.

After the interview

If you were chosen for the position, congratulations. You won the race! If you did not: go through the above checklist and try to work out where you went wrong. You may not have done anything wrong: remember, only one person can be chosen for each position.

Activity 10.3

The interview

With the other members of your group, plan and arrange an interview for the position you applied for in Activity 10.1.

1. Place all your documentation in your job folder ready to bring to the interview. You can keep this folder for real interviews in the future. The standard of presentation of the folder should be as high as possible. It will be kept in your possession at all times. (Your tutor/supervisor will check it for assessment and hand it back to you.) For this interview, include the letter and application form from Activity 10.1.
2. Make a list of the questions you would ask about the position you are being interviewed for.
3. Make a list of questions the interviewers could ask you.
4. Two members of your group will act as interviewers. Show them your job advertisement so that they can plan the questions they will ask you linked to the vacant position.
5. Arrange a place and time for your interview (organised by your tutor/supervisor).
6. Arrive on the day dressed appropriately for an interview, and bring your job folder.

Your tutor/supervisor will assess your interview.

Case study 10.1

Selecting the right person for the job

You are the personnel officer for Business Stationery Supplies Pty Ltd. There is a position vacant in the accounting department for an administrative assistant. You have interviewed three people for this position and now have to decide who would be the best person to fill the vacancy.

The job description/specification is outlined in Figure 10.2 and a brief description of each interview is listed below.

(continues)

Applicant 1
Ruby Ayers is 35 years of age and has held many positions over the past 15 years. She has no formal qualifications other than year 12. Some of her former positions were as an accounts clerk for five years; a receptionist for six years; and her most recent position was as a personal assistant for four years. She had to leave her last position because of rationalisation and downsizing in the organisation.

She has had experience in performing all the duties listed in the job description/specification.

At the interview, she was very confident and was smartly and appropriately dressed. She is planning her future career and is willing to study to improve her position. She has a pleasant manner.

Her hobbies are reading, playing squash and golf.

Applicant 2
Leanne Ginis is 19 years of age and has just completed an Advanced Certificate in Office Administration at the local TAFE college, majoring in accounting subjects, and has gained good grades in the skills required for this position. This course had been her first choice on completing year 12. If she is chosen to fill this position, she is willing to study part-time to obtain a more senior position. She has had some experience in a part-time weekend clerical position in a medical clinic.

Her hobbies include playing in the local netball team.

At the interview she was not very confident; but she appeared very willing to learn and appeared to be a methodical and hard worker from the results of her studies. She has a pleasant personality and is a little shy.

Applicant 3
Will Heinz is 20 years of age and has also just completed an Advanced Certificate in Office Administration at the local TAFE college, majoring in accounting subjects. He has gained excellent results in all subjects and has qualified in the skills required for this position. This course was his second choice, as he would have preferred to attend university in the accounting stream. But he has enjoyed the course and thinks it will be useful in his future career. He seems very ambitious.

His hobbies are playing squash (he reached the junior championship) and rock-climbing.

At the interview he was very confident and was smartly and appropriately dressed. He also has a friendly and pleasant manner.

1. Choose which applicant you think would be most suitable to fill the vacancy.
2. State in detail the reasons why you have selected this particular applicant.
3. State in detail why you rejected the other two applicants.

Activity 10.4

Writing to the applicants

In Case study 10.1 you received three applications for a vacant position for an administrative assistant in the accounting department. You now have to complete the following tasks:

1. Write a letter to the three applicants, inviting each of them to attend an interview next week. Select the date and time and allow each applicant an hour for the interview. Use the names in the case study and make up addresses.
2. Write a letter to the person you have decided you will appoint, offering him/her the job.
3. Write a letter to the other two people informing them that their application has been unsuccessful in this instance.

Note: Refer to Chapter 11 for letter styles and structures and the correct method of writing business letters. Choose the correct style and structure for each of the letters you write in the activities above.

Summary

- Applying for a job is like running a race: you have to be trained, follow rules and regulations, dress appropriately, adopt the correct mental attitudes, and possess the relevant qualifications.
- Employers are looking for people with a willing attitude, reliability, flexibility, productivity, and skills and qualifications.
- The steps involved in obtaining a position include: discovering employment opportunities; deciding which type of job you are interested in and looking for; applying for the position; and preparing for an interview.
- Applying for a position includes: preparation and planning; telephone contact with the prospective employer; completing documentation (letter of application, résumé, job folder).
- Your letter of application is very important, as this is the first document the prospective employer will see. It should be well presented and easy to read, and should show that you are the best person to fulfil the requirements of the position.
- A résumé (curriculum vitae) is a concise and orderly listing of information about an applicant, including personal details, education and qualifications, work history, and the names and addresses of three referees.

- A job folder holds all the documents you are taking to an interview, and should include an up-to-date résumé; your school and college certificates; and written references from previous employers or people who have known you for many years.
- A successful interview requires preparation, including dressing in an appropriate way; planning answers to questions you could be asked and questions you could ask; getting information about the company; and having a trial run to discover how to get to the interview.

Review questions

(These may be used for group discussion.)

1. Explain why applying for a job is like running a race.
2. In general, what qualities are employers looking for in an applicant for a position?
3. List the steps involved in obtaining a position.
4. How would you go about looking for a position, and where would you look?
5. Describe briefly the type of job you are looking for.
6. List the steps involved in preparation and planning before you apply for a position.
7. List the documents you should prepare that are necessary when applying for a job.
8. Why is your letter of application important, and what should it include?
9. How can you make your résumé stand out from those of the rest of the applicants?
10. How can you make your job folder stand out from the rest of the applicants' folders?

Competency self-assessment and/or tutor/supervisor assessment

When you have finished reading this chapter, and have completed the activities chosen by your tutor/supervisor (**practical in class, at work or theoretical**), fill in this self-assessment. Then hand it to your tutor/supervisor for official assessment.

Learning outcomes	**Self-assessment**		**Tutor/supervisor assessment**	
I have successfully completed the following learning outcomes	**Yes** (please tick)	**No**	**Yes** (please tick)	**No**
1. The steps involved in obtaining a position				
1.1 research employment opportunities	☐	☐	☐	☐
1.2 carry out preparation and planning	☐	☐	☐	☐
1.3 make telephone contact	☐	☐	☐	☐
2. Prepare job application documents				
• letter of application	☐	☐	☐	☐
• my résumé	☐	☐	☐	☐
• my job folder	☐	☐	☐	☐
• application for employment form	☐	☐	☐	☐
3. Make preparations for a job interview				
• the correct appearance	☐	☐	☐	☐
• use checklist for interview	☐	☐	☐	☐
• prepare and anticipate questions	☐	☐	☐	☐
• attend an interview	☐	☐	☐	☐
— on the day and time specified by my tutor/supervisor	☐	☐	☐	☐
— with an advertisement for a position	☐	☐	☐	☐
— with my job folder	☐	☐	☐	☐
— dressed correctly	☐	☐	☐	☐
— answered questions in the correct manner	☐	☐	☐	☐
— asked the correct questions	☐	☐	☐	☐
• my interview was successful	☐	☐	☐	☐
• after the interview	☐	☐	☐	☐
Date:	Tutor/supervisor signature:			

section four

ORGANISATION

CHAPTER 11

PERSONAL EFFECTIVENESS AT WORK

ORG101A, ORG201A, NOS121, 7365T

objectives

The purpose of this chapter is to help you to improve your work effectiveness through planning and organising your personal daily routine, in order to complete the following learning outcomes:

1. Plan and organise a personal daily work routine based on given tasks in a workplace-related context

1.1 identify and prioritise nominated tasks for a given day
1.2 separate tasks into achievable components
1.3 select appropriate resources to complete tasks
1.4 reprioritise tasks in response to reviewed instruction
1.5 complete tasks within designated timelines

INTRODUCTION

Many years ago, as a secretary, you would probably have worked for only one person; if that person went away from the office for a few days, you would use this quiet time to catch up with your filing or reorganise your work area, or finish some routine jobs. You did not receive a high salary, but it was a pleasant job with little stress.

Today, as a secretary, personal assistant or administrator, your position is likely to be very different. Because businesses are cutting costs through 'rationalisation' and 'downsizing' there are fewer members of staff to complete the tasks that need to be done, and you will probably be working for many people.

You find you are going to work earlier in the morning and leaving later in the evening, in order to complete your tasks. You are missing tea breaks and working through your lunch hour. You may even have to work over the weekend to catch up on jobs that you did not complete during the week. You leave work feeling exhausted and depressed when you realise it is going to be just the same tomorrow. Eventually, the standard of your work falls and you begin to have symptoms of work-related stress (see Chapter 6). You have frequent headaches, you sleep badly at night. To avoid 'burnout', which is the result of long-term stress, you will need to reorganise your working day so that you can work more effectively.

Efficiency versus effectiveness

People often confuse the two terms 'efficiency' and 'effectiveness'. Working efficiently is assuming that your work objectives/goals are correct and achieving them with the minimum of effort and resources. Working effectively is choosing the right work objectives/goals from a number of alternatives and reaching those objectives/goals in the time allocated.

PERSONAL FACTORS THAT HELP YOU WORK EFFECTIVELY

Know your job

It is difficult to say no to someone who asks you to do something at work. It may be only a very small task, but if you keep accepting these small tasks (when they are not specified as part of your job) you will not have enough time in which to complete your specified tasks. The first objective of any new job (or any job, if you find yourself with too much to do) is to find out what your specified duties are.

Most large organisations issue a 'job specification' when advertising a vacant position. The duties have been worked out through 'job analysis' by the organisation's management. This is done by establishing the tasks that need to be carried out in order to meet the requirements of the position. Then the necessary qualifications and skills are decided on. The structure of all office positions in an organisation consists of four parts: job analysis, job description, job specification, and job evaluation (see Table 11.1). These details are specified in both internal and external advertisements. A job advertisement is shown in Figure 11.1.

Job analysis	Job description	Job specification	Job evaluation
The process of breaking a job down into its component parts through interviews and questionnaires	The written statement of the duties of the job	The list of what is needed to perform the job: • qualifications • experience • education • practical skills • knowledge • mental skills • social skills • amount of responsibility • physical demands	Includes the complexity and degree of difficulty of the duties, the skills and experience required, the level of authority and responsibility, to determine the payment given for the position

Table 11.1 Structure of office positions

Once you have established your duties and you know your job, there are a few points that need to be looked at before you commence. Without the correct attitudes and motivation you will never achieve effectiveness (see Figure 11.2).

Aptitude/ability/skills

Your abilities and skills should meet the requirements of your present position, but if you want to rise in the organisation you will need to improve your qualifications and skills. This could be why you are studying in your present course. Many large organisations have the facilities to enable you to train while in your present position, or will pay your fees and the cost of books so that you can become multi-skilled, which will enrich your job or help you obtain a better job.

Personality/interests

Most of us cannot change our personality. This is something we were born with, and is linked to our individual culture and the way in which we were brought up. But you should certainly think about your personality when you choose the kind of job you would like to work in. For example, it would not be wise to pick a position as receptionist in a doctor's or dentist's surgery if you are very shy when meeting people and cannot change this attitude. You can be taught how to answer the telephone and how to greet people at reception (see Chapters 7 and 8), but if you feel that this really is not what you like doing, then you should choose something else. You may be good with figures and

Administrative Assistant

Accounting Department

A vacancy exists within the Accounting Department for an Administrative Assistant.

Duties and responsibilities:

- Assist with secretarial and administration support in Group Accounting Services
- Updating of Accounting Policy and Procedures Manual

Other duties include:

- sending/receiving and delivering facsimiles
- delivery of accounting reports and memos
- assisting in data input accounts payable ledger
- verifying system-generated reports from accounts payable system
- assisting in preparation of invoice batches from accounts payable for filing
- liaising with Travel Department invoice documentation for processing
- maintaining photocopier routine maintenance and paper levels

Qualifications and skills required:

- Year 12 education
- Office Administration course preferred
- Good verbal and written communication skills
- Good telephone manner
- Self-motivated
- In-depth knowledge of: MS Word for Windows
  Excel (Windows)
- Good word-processing and typing skills

Responsible to Accounting Manager
Hours of Work: 9 am to 5 pm Monday to Friday

Fig. 11.1 Job advertisement

Individual attributes	+	Work effort	+	Organisational support	=	Improved personal effectiveness
Aptitude/ability Competency Skills Personality Values/attitudes Interests (linked to job specification) Task requirements		Motivation to work Job satisfaction		Resources Equipment Interpersonal relationships Leadership Corporate culture		

Fig 11.2 Improved effectiveness equation

would be quite happy working in an accounting firm, or as an administrator, where you meet people occasionally but not all day and every day. The ideal situation is when:

you love what you are doing and you are paid to do it at the same time.

Activity 11.1

Personal analysis

Form into a group of four students (or a group of people at work) and choose four job vacancies (with job descriptions) from one of the newspapers. The group should discuss the vacancies and decide which person is the most suitable for which position. Each advertisement should then be given to the person chosen. Each of you should now complete Activity sheet 11.1 provided in your workbook and decide whether you think you are suitable to fill this position. (Complete it as though you have finished your present studies.)

Values/attitudes

Most of us are somewhere in the middle range of the ideal situation. You are ecstatic when you first get a new position but slowly your attitude changes. In all jobs you will find you are enthusiastic and motivated at first, but you can become disillusioned once the job becomes routine. Training to enrich your present job could improve the situation, or you could find another job. What else could be done?

There are many theories on how people remain motivated at work. One of the most common is Herzberg's two-factor theory, which is based on identifying different working conditions as sources of satisfaction and dissatisfaction in a specific job. These are called 'hygiene' factors and 'motivator' factors, as shown in Table 11.2.

Dissatisfaction factors (hygiene)	Satisfaction factors (motivators)
Company policy and administration	Achievement
Relationship with supervisor	Recognition
Work conditions	The work itself
Salary	Responsibility
Relationship with peers	Advancement
Status	Growth within the firm
Security	

Table 11.2 Hygiene and motivator factors

According to Herzberg, changing any of the factors on the left will not make you any happier in your job. For example, raising your salary and providing you with luxurious surroundings and the latest technology will not make you any happier at work if you do not get on well with your workmates, or you hate your supervisor, and you think some of the policies put in force by management are unfair. But if you are recognised for some achievement, such as working late to complete an urgent order or successfully completing a project, this gives you a feeling of satisfaction (an intrinsic reward). If you are then promoted or given a raise in salary (an extrinsic reward) this also makes you feel appreciated. You feel involved and part of the organisation. This motivates you to do even better.[1]

Activity 11.2

Job satisfaction

Form into a group and discuss what factors make you satisfied in your present position (or the factors that you expect to find satisfying in a position you would like to have) and what factors make you dissatisfied. List these on Activity sheet 11.2 provided in your workbook.

Positive attitudes

Your attitudes towards work and life in general will depend to a large extent on your cultural background, which will influence the ideas, beliefs, thoughts and values that you hold about yourself and other people. These determine how you think, feel and act in any situation. There are two types of attitude:

- *Positive attitudes.* These make you feel in control and confident, and will help you reach your personal and work goals.
- *Negative attitudes.* These make you feel out of control, give you low confidence, and will prevent you from reaching your goals.

You should never use **negative self-talk**, such as the following:

> 'I will never get this done.'
> 'I'm not competent to do this project. I will give it to someone else.'
> 'I hate my job.'
> 'I hate Mondays.'
> 'Oh no! It's work tomorrow.'
> 'I don't like my supervisor.'

Any of these may become a 'self-fulfilling prophesy' and you *will* hate your job and you will *not* be competent or effective.

Positive thinking involves being cheerful and smiling. This is contagious, creates better interpersonal relationships, and makes everybody feel better. Think:

> 'I'm lucky to have this job.'

---

1. See Bailey, Schermerhorn, Hunt and Osborn, *Managing Organisational Behaviour in Australia*, Jacaranda Wiley, Brisbane, 1986.

'What are the good points about this job?'
'If I really try I'm sure I can do this project.'

You will then complete the project and possibly be promoted.

Always use self-motivation and self-discipline.
The more you do, the more you will be able to do.

PROCEDURES TO IMPROVE WORK ROUTINE

The following procedures will help you improve the planning and organisation of your work routine and help you work more effectively.

- Personalise your work station.
- Organise your work space.
- Improve your time management by organising your daily work routine, through:
  —goal-planning
  —time schedules
  —completing 'to do today' lists
  —time logs
  —prioritisation.

Personalise your work station

As mentioned in Chapter 4, many offices in large organisations have an open-plan layout, where you will occupy a work station divided from other people by only screens and plants. You may be in the centre of the room with no natural light. This can be depressing with grey walls, desk and chairs, also damaging to your eyes if the only light is from overhead fluorescent tubes. Personalising your work station makes it your own territory (however temporarily). Providing your own pot plants, table lamp, photographs, pictures, calendar and desk diary makes the area particularly your own. Everything will look brighter and more cheerful, giving you a feeling of belonging and boosting your morale. A desk lamp will provide better lighting (refer to Chapter 6) and a bright spot in which you are happy to work.

Organise your work space

Arrange your work tools around you for ease of use. If you use files often, place the filing cabinet or lever-arch files near to your desk so that you do not have to get up every time you need to retrieve a document. Divide the surface of your desk into three work areas (see Figure 11.3).

The **primary work area** is the distance from elbow to hand when sitting at your desk. Use this area for the items you use most often, so that they are within easy reach. Examples are the computer keyboard, your 'to do today' list, diary, pen and pencil holder, telephone, teledex and notepad.

The **secondary work area** is the distance within the reach of your arm. Use this area for those items which you use fairly often but not all the time: for example, customer files, ruler, stapler, hole-punch, telephone directories.

The **reference work area** is out of reach of your arm at the back of your desk. Use this area for those items which you use the least, such as a dictionary, reference manuals, calculator and sticky-tape.

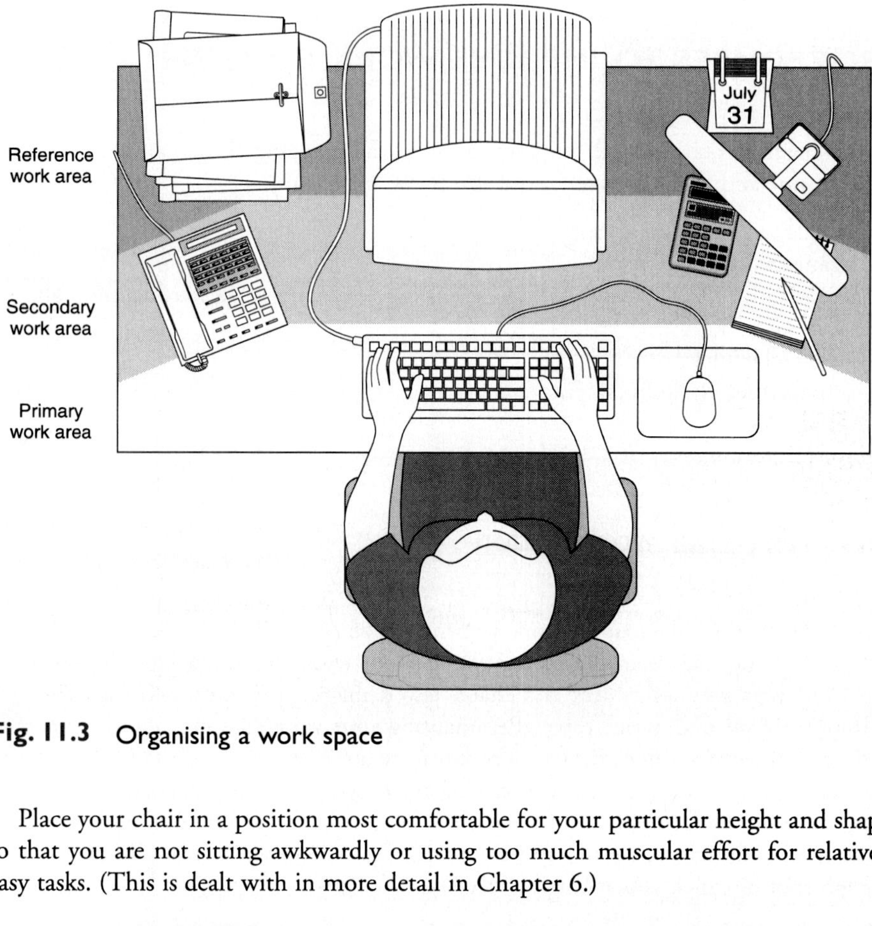

Fig. 11.3 Organising a work space

Place your chair in a position most comfortable for your particular height and shape, so that you are not sitting awkwardly or using too much muscular effort for relatively easy tasks. (This is dealt with in more detail in Chapter 6.)

Activity 11.3

Work areas and conditions

Draw the work area you use at the moment, including your equipment and any other articles you keep on your desk or table. This may be your desk in the office, or a desk in your bedroom, or a table set aside somewhere at home. Make a list of the equipment and articles. Think how these could be reorganised into primary, secondary and reference work areas. Now use Activity sheet 11.3 in

your workbook to draw your reorganised work area. What kind of chair do you use? If it is not ergonomically designed, how might you rearrange it so that you can work more comfortably (refer to Chapter 6)? Also, look at the lighting of this area. Is there any natural light? If not, what could you do to improve the lighting to protect your eyes?

When you have organised your work station, it is time to prioritise your duties into order of urgency and organise the time in which you have to complete these duties. This is called time management.

Improve your time management

Time is a valuable resource, and using it effectively requires organisation and discipline. The basis for using your time effectively is planning your day—how many tasks you have to do, and how long it takes you to do each task. First of all, try to cut down as much as possible on time-wasters, such as those listed in Table 11.3.

Time-wasters	Solutions
Unclear objectives	Better planning of time and projects
Personal disorganisation	Use better time management
Being unable to say 'no'	Be decisive—accept extra tasks only when you have finished priority tasks
Leaving work unfinished	Finish one job before you start another
Too-long telephone calls	Follow telephone techniques in Chapter 8
Lack of self-discipline: • not following up • carelessness	Make sure instructions have been carried out. Do things right the first time, or you will have to do them again
Postponing tasks you dislike doing	Do the unpleasant jobs first to get them out of the way
Not listening effectively	Ask for clarification if you are not sure what someone has said
Badly planning meetings	Refer to Chapter 13

Table 11.3 Time-wasters

Your time management will be much more effective if you establish your goals, linked to the organisation's goals, and write everything down.

Electronic aids to better time management

There are many types of planners that will help you plan your time, such as computer packages incorporating daily planning, appointments and reminder systems. Figure 11.4 is an example of a completed template file being used to create a new appointment book. Figure 11.5 shows a palm computer organiser, which stores approximately 6000 addresses, five years of appointments, 1500 to-do items, 1500 memos, 200 e-mail messages and other applications.

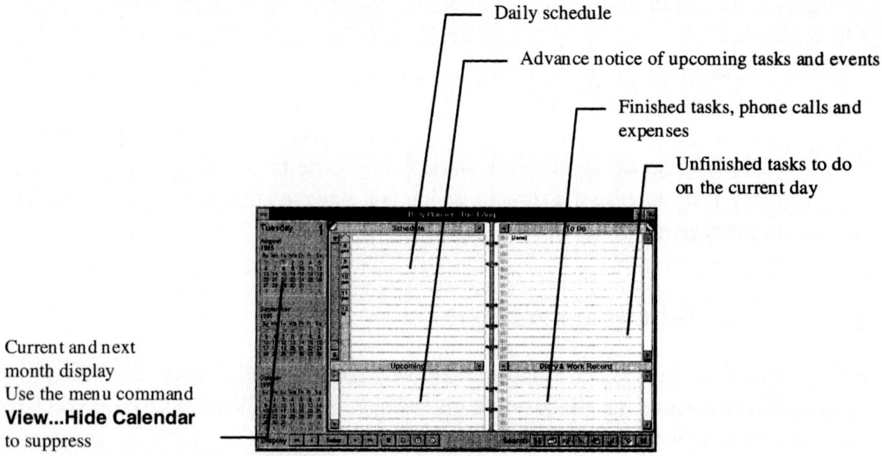

Fig. 11.4 Microsoft electronic daily, weekly and monthly planner

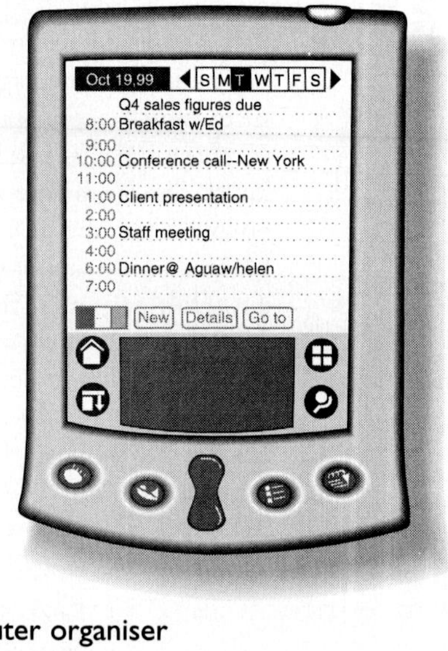

Fig. 11.5 Palm computer organiser

There are a number of practices you can adopt in a paper-based system to organise your working day and become more effective. These include using wall planners, desk diaries, 'to do today' lists and desk calendars.

Activity 11.4

Time-wasters

Form into groups of three or four. Refer to the list of 'time-wasters' in Table 11.3 and discuss ways in which you think you waste time (either as a student or at work). Through brainstorming with other members of the group you may discover other ways in which you are wasting time. List these other time-wasters on Activity sheet 11.4 in your workbook. Discuss with your group ways in which you could change these work habits. There are some solutions given in Table 11.3, but as a group you may be able to suggest more. Fill in possible solutions on the activity sheet.

Organising your daily work routine

The greatest time-waster is personal disorganisation—you have no objectives, no logically prioritised tasks, you don't finish one task before starting another one, and you work on tasks in no particular order. The practices below will help you to improve your personal organisation. The first thing you should do is plan your goals and the time you have available each day, each week, and even months ahead, and write this down. It is a well-known fact that writing down your short-term and long-term goals will help you to complete them. Remember: **write them down**.

Goal-planning. Work out your work goals from your job specification, linked to the organisation's goals:

- **short-term goals**—days, weeks, months;
- **medium-term goals**—one to two years;
- **long-term goals**—three to five years.

If one of your tasks is to type out the financial report, this is obviously linked to the financial goals of the company, and your report must be finished before the end of the financial year. If typing the sales report is one of your responsibilities and the monthly sales meeting is to be held next week, your report must be finished in good time for the meeting. These deadlines should be written down on some kind of time schedule.

Time schedules. There are many systems of paper-based time schedules that will help you to plan your work goals and manage time. The golden rule is to write everything down and cross it out when it is completed (tasks, telephone messages, reminders). You think you won't forget, but you do! Remember: **write it down**.

Activity 11.5

Goals

On Activity sheet 11.5 in your workbook, list your short-term, medium-term and long-term goals. If you are a student these will include your course goals; if you are working these will include your work goals, linked to the firm's goals. Also include your personal goals.

'To do today' lists. You should list the tasks you have to do today. Work out which are the most important. You may have to type a report and have it finished by 2 pm (this is urgent). There may be items of stationery that you need to order (this is fairly important) and copies of letters that you need to file away (this is not very important). Think of all the tasks you have to do today and write them down. This will ensure that you do not forget to do something that has to be done today. You could keep a note in the diary/tickler file (refer to Chapter 2), but it is also a good idea to have a preprinted form that you keep for this specific purpose on your desk. You can buy 'to do' pads from the stationers or you can make one up yourself quite easily (see Figure 11.6). As you think of tasks you have to do tomorrow, jot them down on your 'to do' list ready for when you come in to work the next day.

Diary/tickler file. Place any documents that require urgent attention in the diary/tickler file (as described in Chapter 2). You will take them out first thing in the morning, as you should do every day.

Time log. You then work out how long each of today's tasks should take to complete, what time you should commence the task, and when you hope to complete it.

Prioritisation. Put your tasks in order of priority.

- **Urgent tasks.** These need to be done immediately. Complete them at the beginning of the day, while you are still fresh.
- **Important tasks.** These need to be done before the end of the day.
- **Low-priority tasks.** These may be completed when you have finished the first two priorities.

To do today

Date: 14 January 199X

1. Pick up mail from postbox
2. Order stationery
3. Report photocopier fault
4. Telephone Sue Windas to cancel appointment for manager at 11 am tomorrow
5. Types sales report for meeting next week
6. Filing
7. Photocopy agenda for meeting tomorrow
8. Reorganise stationery store ready for deliveries next week
9. Check diary/ticker file
10. Telephone Mr Bond to make an appointment at 11 am tomorrow
11. Balance petty cash ready to give to accountant the day after tomorrow
12. Type a new internal telephone directory including new staff
13. Enter appointments for manager into her diary
14. Make up files for 6 new customers to enter into system
15. Type letter for manager to go out by 4 pm today

Fig. 11.6 'To do today' list

Activity 11.6

Prioritising tasks

In groups, sort the tasks given in the 'to do today' list in Figure 11.6 into 'urgent', 'important' and 'low priority'. Discuss this with your group members first and decide why you think one task is more urgent than another. List these roughly on the draft 'to do today' list provided in Activity sheet 11.6 (1) in your workbook, with the urgent tasks at the beginning of the day, then the important tasks, and the low-priority ones last. Then write your final decisions on Activity sheet 11.6 (2).

Energy cycles

Try to perform difficult tasks when you are fresh and energetic. Most people are fresh first thing in the morning after a good night's rest. It is a well-known fact that energy flags at 4 in the afternoon, and production in factories and offices is at its lowest at this time. But it is also a fact that managers often want tasks completed at the last minute, so you should try to conserve some energy for this period. Complete as much work as you can in the morning and early afternoon. Take frequent breaks between tasks and, if you have been sitting at the computer for a long period, stand up and walk about a little. Make a visit to the bathroom or have a drink. Try not to drink too much tea and coffee, as this can give you too much nervous energy and could cause more stress. (This is covered in more detail in Chapter 6.)

DECISION-MAKING AND PROBLEM-SOLVING

However well you have organised your time and your tasks, you will find that problems do occur and you will have to make a decision on how to solve them. You are not working alone but with other people, and this can cause difficulties. In all work situations, getting the job done takes group effort. We have discussed in Chapters 8 and 9 how you should deal with difficult customers on the telephone and at reception. The following are some suggestions for dealing with problems and making decisions.

Identify and analyse the problem

To clarify a problem, good listening skills and effective communication are required. There are very few 'good listeners', as most people are too busy putting forward their own opinions to listen to the ideas and opinions of other people. Think of any social or business group situation—everybody is talking and nobody is listening.

Effective listening

In the business communication process, listening is used more often than any other skill.

Business skills		Total communication
Writing	(letters, memos, faxes etc.)	9%
Reading	(" " " ")	16%
Speaking	(telephone, face-to-face)	35%
Listening	(" ")	40%

Effective listening is important, because it:

- prevents misinformation;
- improves morale, rapport and trust levels with other people;
- improves teamwork.

The communication process

To work more effectively both personally and in a team situation, you should understand the communication process. Lack of communication between management and staff, and poor communication between members of staff and between customers and staff, cause many problems in business.

Problem-solving and conflict resolution

(Refer to Chapter 7 for more details.)

- *Establish the criteria for the solution of the problem.* Brainstorm several ideas for solving the problem, or discuss suggestions with several other people.
- *List and explore alternative solutions.* From the brainstorming session or the suggestions, decide which ones are the most suitable.
- *Select the best alternative.*
- *Implement the decision.*
- *Evaluate the decision over time.*

This applies in both personal and team situations.

Case study 11.1

Planning and organising a daily work routine

(You may discuss this case study in small groups before completing the 'to do today' list individually.)

You are the receptionist at Business Stationery Supplies Pty Ltd. Below is a list of the routine tasks you perform every day in addition to your other reception and switchboard duties.

8 am	Pick up the mail from the post office. Sort and distribute the mail back at the office.
9 am	Check the diary/tickler file and retrieve documents that have to be dealt with today.
12 noon to 1 pm	Take a lunch break.
4 pm	Take cash to the bank and outgoing mail to the post office.

Each day you also have to file all the documents for your section. This usually takes you about 30 minutes.

Today you had planned to do quite a few other small, routine tasks which usually take you about 30 minutes altogether, and you have been asked to do the following additional tasks.

1. At 9.30 am your manager asks you to send a fax to the Sydney office. She has to receive a reply by 10.30 am so that she can telephone an important client with the information she receives.

(continues)

2. The general manager's secretary contacts you to say that her stock of letterhead paper is low and she must have a further supply by this afternoon in order to type the letters she has been given today. You discover that stock has run out in the store. You will need to telephone the printer to deliver six reams as early as possible this morning.
3. Your supervisor has asked you to photocopy a number of documents, which she requires first thing tomorrow morning. This will take you about 30 minutes.
4. At 10 am an important customer telephones with a complaint. He is very annoyed and threatens to take his business to another company. The stationery supplies he ordered have not been delivered. You will need to check with the warehouse why the order has not been delivered, and ring the customer back with an explanation before 2 pm.
5. Today is the day you have set aside to check the stationery store and reorder supplies. (This is part of your specified duties.) This will take you about one hour.

Requirements

You are to organise all the routine and non-routine tasks listed above on your 'to do today' sheet (Case study sheet 11.1 in your workbook). Prioritise them against the times listed according to their urgency.

List on a separate sheet of paper, against each task, the resources you will require to perform the task.

Summary

- Because businesses are cutting costs through 'rationalisation' and 'down-sizing', there are fewer members of staff to complete the tasks that need to be done, and you may find that you are working for many people.
- To avoid 'burnout', which is the result of long-term stress, you will need to reorganise you working day so that you are working more effectively.
- Working effectively is assuming that your work objectives/goals are correct and achieving these with the minimum of effort and resources. Working effectively is choosing the right work objectives/goals from a number of alternatives and reaching those objectives/goals in the time allocated.
- Personal factors such as knowing your job and having positive attitudes and values will help you to work more effectively.
- The structure of office positions is determined by job analysis, and outlined in a specific job description or job specification.
- Procedures to improve work routine include personalising your work station, organising your work space, improving time management, organising your daily work routine and writing things down.
- Decision-making and problem-solving may be improved through effective listening; exploring alternative solutions and selecting the best alternative; then implementing and evaluating the decision over time.

Review questions

(These can be used for group discussion or official assessment.)

1. Why should you 'know your job'?
2. What is a job specification?
3. List the factors that should be taken into account if you want to improve your personal effectiveness.
4. List the sources of dissatisfaction in a job and the sources of satisfaction, according to Herzberg.
5. Describe the procedures that will help you to improve your competency and effectiveness.
6. What are the three work areas of the surface of your desk? List briefly the work tools you would place in each work area.
7. What are the factors that can waste time in your working day?
8. Next to the above factors, state solutions that you can put in place to prevent wasting time.
9. Why is it important to prioritise the tasks you have to complete each day?
10. Why is it important to plan your goals?
11. Name as many different methods of time schedules and organisers as you can think of.
12. List the reasons why it is important to listen effectively.
13. List briefly the stages in the problem-solving and decision-making process.

Competency self-assessment and/or tutor/supervisor assessment

When you have finished reading this chapter, and have completed the activities chosen by your tutor/supervisor (**practical in class, at work or theoretical**), fill in this self-assessment. Then hand it to your tutor/supervisor for official assessment.

Learning outcomes	Self-assessment		Tutor/supervisor assessment	
I have successfully completed the following learning outcomes	**Yes** (please tick)	**No**	**Yes** (please tick)	**No**
1. Plan and organise a personal daily work routine based on given tasks in a workplace-related context				
1.1 identify and prioritise nominated tasks for a given day	☐	☐	☐	☐
1.2 separate tasks into achievable components	☐	☐	☐	☐
1.3 select appropriate resources to complete tasks	☐	☐	☐	☐
1.4 reprioritise tasks in response to reviewed instruction	☐	☐	☐	☐
1.5 complete tasks within designated timelines	☐	☐	☐	☐
Date:	Tutor/supervisor signature:			

CHAPTER

12

TEAMS

TEM101A, 7365Y,
TEM201A, NOS142, 7365S,
ORG301A, NOS146

objectives

The purpose of this chapter is to help you to understand the importance of group/teamwork in an organisation, and to explain how team climate and environment can influence team performance, in order to complete the following learning outcomes:

1. **Organise own schedule in a workplace-related context to meet designated group/section goals**
   1.1 identify and prioritise nominated tasks
   1.2 select appropriate resources to complete tasks
   1.3 review and reprioritise tasks in response to changing circumstances
   1.4 complete tasks to designated timelines
2. **Participate as a team member to complete a series of work-related tasks within designated timelines**
   2.1 participate in identifying team goals and determining tasks necessary to achieve them
   2.2 participate in allocation of responsibilities to team members to ensure that designated team goals are met within agreed timelines
   2.3 complete agreed tasks to meet team goals and objectives
   2.4 participate in review and reprioritise tasks in response to changing circumstances
3. **Quality teams**

INTRODUCTION

Many of the tasks performed in a work situation are in groups or teams. In today's rapidly changing work environment it is unlikely that any one person will have all the skills needed to complete every task. Therefore a workgroup is formed of people with a range of skills, which are pooled; once these people are able to relate to one another and link their knowledge and skills to the knowledge and skills of the other members of the group, the group is able to perform effectively.

WORKING IN A TEAM

A group in an organisation is a collection of people who interact with each other, and work together to complete a particular task, achieve a goal, or solve a problem. The group's role is to make decisions, plan, achieve goals, reflect on and assess its own performance and, where necessary, plan again.

Types of workgroups

Workgroups can be divided into two main types: formal and informal.

Formal group. A formal group in an organisation is established by management to perform specific tasks in order to meet the goals of the organisation. It may consist of a committee, a department, a section, or some other unit in the workplace. In these groups, relationships and procedures are regulated through formal processes, as in formal meetings. (**Formal meetings** are described in Chapter 13.)

Informal group. Informal groups form naturally around the workplace as, and when, the people involved feel like it. These could be 'interest' or 'friendship' groups.

Types of teams

The terms 'team' and 'group' are often used in the same context, but, while a team can be classified as a group, a group may not conform to the specific criteria for a team. The key elements of a team are that members are operating within a charter. Members of a team see themselves as having specified roles, and they acknowledge that the team is accountable for achieving a specific organisational goal.

- A **project work team** is created to complete a particular task or project. Members are chosen because their qualifications and experience are useful to the team's purpose. The members are chosen by management and the team disbanded when the project is finished.
- A **participative work team** is able to take some initiative and to participate in decision-making within the group. It is created to complete a range of duties.
- A **self-managed team** creates new ideas, makes decisions and achieves results. This type of team is suited to a dynamic and flexible organisation, in which teams need to respond quickly to changing circumstances.

Group dynamics

Any group has a 'personality' of its own, which is composed of the individual personalities of the people in it. A good example of this is student classes. Each group of students is different from any other group of students, and one person can change the whole dynamics of a particular class. This is caused by the 'synergy' of the group. **Synergy** is described as 'the whole being greater than the sum of the individual parts'. Group synergy can be either positive or negative. In an organisation, good team effectiveness can act as a powerful socialising agent. Examples of this are helping new employees with organisational policies and procedures, and working together to get jobs done more quickly and to a higher standard, using cooperation and reciprocation.

Group behaviour

In addition to completing tasks and achieving goals, members of a group or team relate to one another. Each person brings interpersonal concerns to the team, and these concerns become part of the group process. There are three interpersonal concerns within groups:

- inclusion,
- control, and
- acceptance.

In a team with inclusion, control and acceptance, group leaders and members are accessible to one another. No one person monopolises the conversation or the decision-making, or interrupts while others are talking. As a member you are given space to put forward your views.

Activity 12.1

Group dynamics

Discuss with the other members of your class, or each individual group (or the group of people you work with), the following questions. Before you start your discussion, list the discussion topics (you may be able to think of more than the ones listed below). Each member should write answers underneath each question at the end of the discussion.

- What type of group are you working in? College, school or work?
- What are the goals and objectives of your group/team?
- How many members are there in your group/team?
- What are the dynamics of your particular group? Are members friendly and cooperative?
- Describe the personalities of one or two members of your group/team. Do these people contribute to the end goals of your team, or are they disruptive?
- Is your group working effectively together towards your end goal, and is the synergy positive or negative?
- If the synergy is negative, what do you think should be done to make it positive?

Advantages of teams. The advantage that working as a team has over people working alone will depend on the qualifications and competencies, and effectiveness, of the people in the team. As mentioned above, the synergy of a team can be either positive or negative. Usually, the more people involved in making a decision, the greater the likelihood of a successful outcome, but if the synergy is negative the meeting can become bogged down with irrelevancies and no decisions are made.

Working in a team is also better for the morale and motivation of the workforce. People usually experience greater job satisfaction working in groups than working alone. Although success can depend on the specific problem, difficult problems have been found to be more effectively handled by a group, whereas relatively simple problems may be handled more effectively and quickly by one person working alone.

Disadvantages of teams. While many groups or teams are successful, they can also have disadvantages. Groups take a lot longer than one person to make a decision and to take action. Teams take time to develop a structure and are more effective over a longer period of time, and in some situations speed will be essential.

Groupthink

In 'groupthink' no one person in the team challenges the ideas of the group, or wants to be seen as being apart from the rest of the group. This impairs the group's performance.

Teams suffering from groupthink fail to consider problems objectively or to research adequately. They do not consider alternatives and may take unnecessary risks. Groupthink may even lead members to look after their own interests rather than those of the organisation or government department employing them. Teams that suffer from groupthink:

- fail to express doubt because of social pressure from other members;
- appear to be invulnerable;
- stereotype outsiders unfavourably;
- are ready to ridicule critics;
- have a false impression of internal harmony by avoiding disagreements.

Team climate

The climate of a team is created by its cohesiveness. When a team has a common commitment to a task and is considering a problem objectively, the climate is positive. If there is a lack of unity, dissatisfaction and low morale in a team, the climate is negative and the team will be less effective.

Group cohesiveness is the degree to which members are attracted to and motivated to remain part of a group or team.

Team environment

A poor physical environment can influence team performance, creating a feeling of low status, poor concentration, and lack of motivation. Make sure any group or team you work in has the following:

- *Quiet environment.* Any group of people meeting to create new ideas, make decisions and achieve results needs a quiet location with few distractions and interruptions in order to perform effectively.

- *Good seating arrangement.* In any group situation, whether informal or formal, a good seating arrangement can have a positive outcome, as people can then communicate more effectively. They can see one another without having to lean forwards or backwards and are not separated. Make sure that the view of the speaker in a formal team situation is not blocked by other people. A round table or 'U'-shaped table for formal team meetings are considered to be more appropriate.
- *Small size.* The interaction between members of any group is influenced by size. Small groups usually work more effectively. In large teams, the skill of the leader, the climate and environment, and the individual personalities affect the structure, cohesiveness and outcome of decisions made within that group or team.

Team-building techniques

To build an effective team out of a group of people, there are several preconditions that exist.

Group size. An effective group needs to be small. For the activities in this book, groups performing practical tasks should comprise no more than three people, unless four are specified.

Group matched to task. Group members who are similar in age, education and experience (homogeneous groups) work better together for simple tasks, where goals and objectives require cooperation, good working relationships and speed of completion. Group members who differ in age, education and experience (heterogeneous groups) work more effectively with creative and more complex projects, if speed is not a consideration.

Group cohesion. Cohesion occurs in a group that works well together. Group members recognise each other's contribution to the group goal, and operate well by using the following stages of formation.

Group contribution. In any team situation, it is essential that each member be allowed to contribute his or her special skills and attributes. Every one of us is good at something: one person is artistic, another good at figures, someone else has good organisational abilities. If all members contribute these special skills to the group goal, this will produce good synergy.

Group objectives. The group identifies and establishes the common objectives required to complete the goal. All members of the group must come to an understanding and agree on how this will be achieved.

Group structure. Coordination is very important, and the appointment of a leader is essential to achieving the group objectives. The leader is responsible for making sure that the group objectives are understood and are completed in the designated time.

Leadership. There are some people who seem to be natural leaders. No-one is a born leader, but it is widely recognised that leaders have some common characteristics:

- intelligence
- social maturity
- motivation and achievement orientation
- self-confidence and enthusiasm
- good communication skills
- commitment to the job.

Team decision-making

The purpose of any group/team in a business is to achieve a corporate goal or goals. How successful your goal or team is depends on the factors discussed above. It will also depend on how decisions are made and communicated within the team. Work groups/teams are an integral part of business, and depend on their members to reach objectives and the end goals of the organisation. The way in which decisions are made can affect the relationship of members within the group and the completion of the tasks allocated to reach those goals.

Authoritative decisions. Authority, after group discussion, is one of the most common ways of arriving at decisions. The group discusses the issues, then the leader or even an outside supervisor or manager makes the decision. This may be the quickest way of reaching a decision, but over a long period it can foster resentment, low morale and lack of purpose within the group.

Consensus decisions. Decisions made by consensus are made by agreement among the majority of team members—often by voting with a 'show of hands' or by vocal 'yes' or 'no'. The disadvantages of decisions arrived at by consensus are that they take time and require good communication skills.

Majority decisions. Decisions made by majority are when agreement is reached among the majority of team members. These decisions are often made when there is insufficient time for any other method, or when it is clear that not all members will agree. This can leave a minority that is dissatisfied with the decisions.

Compromise decisions. Decisions reached by compromise take place when there are some members of the team who clearly will not agree on anything. The group leader will make the decisions or decide on a compromise that will appeal to both sides.

Whatever the decision taken, you should try to end with a win/win situation, where each person or group is satisfied with the outcome of the decision. This is outlined below.

Activity 12.2

Team decision-making

Most of the activities in this book have been performed in a group. Discuss with the other members of your group the following questions:

1. Was your group a cohesive group?
2. Was your group an effective one? If not, why was it not effective?
3. How did your group arrive at a decision? By authority, consensus, majority or compromise?
4. Did your group have good communication skills? If not, how could these have been improved?

(When you have finished your discussion, each member should write a short report on the effectiveness of their group.)

TEAM EFFECTIVENESS AT WORK

We have discussed the different types of workgroups or teams and the different types of group dynamics and group behaviour. We have outlined team climate and environment, team-building techniques and how teams make decisions.

An effective workgroup or team is one that achieves high levels of both task performance and human resources maintenance over time. Every manager should promote workgroup effectiveness.

The structure of a group or team may develop slowly or quickly. Tuckman[1] has identified four stages in the development of groups or teams:

- forming
- storming
- norming
- performing.

Team forming. The forming stage occurs when a team is established and people begin to interact within the team. The group's membership, leadership and goals may be uncertain at this stage. Except in formal teams, where the procedures are regulated, members will spend this early stage finding out about each other, looking for inclusion in the group, establishing their own status and the status of others. The real purpose of the team will be established, with differences becoming apparent.

Team storming. When conflict emerges the team has reached the storming stage. This conflict may be about leadership; about goals, or the way they are (or are not) being achieved; or about how individuals perform or feel in the group. Personal agendas, or lack of commitment to the group's goals, also arise at the storming stage.

Team norming. As members interact, similarities and differences emerge. A pattern of relationships develops. The norming stage is when members of a team start to feel part of the group and a common approach and shared goals emerge. The group norms or acceptable behaviour, attitudes, work patterns and related behaviour emerge in this stage. The group becomes cohesive and the accomplishment of work goals begins.

Team performing. Once it has established its goals, and its structure and identity are formed, the team concentrates on solving problems and making decisions. Team members' activities in each stage are listed in Table 12.1.

In all work settings you will be required to work with other people in a group situation. Whether it is a small task that has to be completed by one or two people, or a decision that has to be made by a number of people, it will be affected by interpersonal relationships.

Good interpersonal relations is entering into and maintaining effective peer relationships in the work setting. This is vital if work is to be performed effectively and completed according to the goals set by the organisation. There are three interpersonal concerns within groups. As the group moves through each stage listed in Table 12.1,

1. Tuckman, B.W. 'Development Sequence in Small Groups', *Psychological Bulletin*, May 1965.

Stage	Activities
Forming	**Clarification of goals** commitment to group's purpose establishment of relationships making tentative contributions communication between members
Storming	**Critique of group's performance** presentation of alternative points of view emergence of conflict over power or leadership provision of negative and positive feedback discussion of problems or concerns resolution of conflict
Norming	**Offering support to other members** problem-solving decision-making improvement of plans verification of goals development or extension of skills
Performing	**Improvement in ways of doing things** development of interpersonal relationships acceptance of delegated tasks involvement in decision-making achievement of performance

Table 12.1 Team members' activities in each stage

these concerns develop in the order given earlier in the chapter, of inclusion, control, and acceptance. Good communication between all members and with the leader is one of the most important skills in dealing with interpersonal concerns.

Communication skills in groups or teams

The communication process is dynamic and interactive. People place meaning and structure on the variety of messages received from others. This interaction may be direct or indirect. **Effective communication** occurs when the intended meaning of the sender and the perceived meaning of the receiver are one and the same.

The communication process takes place in various situations for different reasons, with the potential for many interpretations. This has seven main elements (as shown in Figure 12.1):

1. sender
2. message
3. receiver
4. feedback

5. channel
6. context or setting
7. noise or interference.

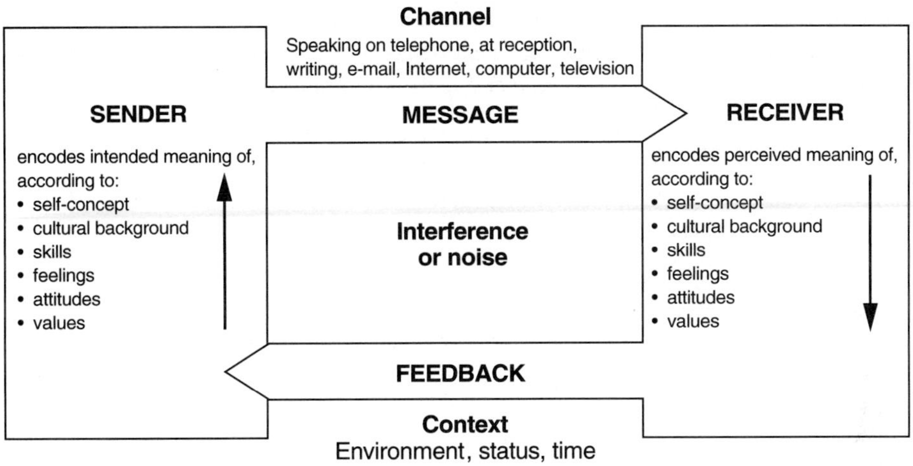

Fig. 12.1 The communication process

The sender. One of the most important skills any person in an organisation can have is the ability to communicate. How a person communicates depends on their background of accumulated attitudes, experience, skills, cultural conditioning and individual differences.

The message. This is the idea or feeling transmitted from the sender to the receiver to achieve understanding. It makes a connection between the sender and the receiver.

The receiver. The receiver decodes or interprets the message to achieve understanding. He or she is also a unique individual, with a different background of cultural experience and conditioning and a particular personal perception of the message.

Perception is the way people understand or give meaning to their environment. This means that the way a message is perceived by the sender may be quite different from the way the receiver perceives the message. All people are limited by their background experiences, and this affects how they communicate with one another.

Feedback. Feedback is essential in interpersonal communication. It is the other person's response to a sender's message (whether verbal or non-verbal). It indicates understanding by both sender and receiver and how the message is being received.

- *Verbal communication.* This comes in the form of words. It can be spoken words between two or more people, or written words in letters, memos and messages.
- *Non-verbal communication.* This is a message sent by any means other than words or graphics—by facial expressions, body movement, posture or dress (even the way you speak—implying something—on the telephone or in reception).

The communication channel. This is the means or technique used to send a message—a letter, a telephone call, a conversation, a radio or television program.

The context. This is the environment or setting within which communication takes place—a formal meeting, a workgroup or a business lunch. The time of the message or the status of sender and receiver all present a different environment for communication.

Noise or interference. Something can interfere with the way any message in any environment is received. For example, physical distractions (noisy office; the mood of the sender; lack of empathy between sender and receiver; a hot, humid day; how tired or frustrated the sender or the receiver is) can cause communication barriers.

Communication barriers

Common causes of stress in a business situation are communication barriers between individual workers, workers and managers, and among management. These are caused by:

- lack of courtesy by the sender or the receiver;
- inappropriate choice of words;
- inadequate feedback;
- inappropriate message;
- inappropriate channel;
- receiver inattention;
- non-verbal communication that does not support the words used;
- poor layout and presentation in letters, memos and other documents;
- poor timing.

At all stages of development of a group or team, leaders and members should communicate. Effective interpersonal communication within a group or team has two components: spoken and unspoken.

Spoken communication skills. Good spoken communication skills allow all members actively to participate in the interaction process. Members share ideas and their knowledge. In formal situations (e.g. meetings) members should be given the opportunity to speak and to share their ideas, qualifications and experience.

Most of the time we can work amicably with other people, and sometimes we feel a natural friendliness towards one person or a group of people. But, because of personality differences or cultural differences, we do not always 'get on' with everybody. Some individuals seem to have a natural antipathy towards other individuals. This can lead to problems between people thrown together in a work situation.

Group problem-solving

Solving problems in a group has four major advantages. It leads to:

1. greater knowledge and information (synergy);
2. more approaches to the problem;
3. increased acceptance of solutions;
4. better comprehension of the decision.

This is why the activities in this book are set for students in small groups of three or four people. The more experienced students will contribute many ideas to the brainstorm sessions, which will benefit all members of the group, and inexperienced students will gain knowledge from the more experienced.

Brainstorming

Brainstorming is a process where all members of the group express any ideas or solutions that occur to them without logic or deep thought—quickly and at random. Often an illogical, random thought from one person will lead to a logical thought from another, merely through the association of ideas. Brainstorming works effectively if all members of the group are allowed a chance to put forward ideas through effective communication and effective listening.

Activity 12.3

Team effectiveness

When you form into groups for the activities in this book, use the team-building techniques given earlier in this chapter.

Form into a group of four members and perform the following activities. Enter the required information on Activity sheet 12.2 in your workbook.

1. Choose a leader of the group by mutual agreement.
2. Brainstorm ideas for a group project. Some suggestions are:
   (a) a survey on a service area in your college/school/work (e.g. reception area, cafeteria, carpark facilities);
   (b) a team project, such as:
   - organising an outing,
   - completing a task given to your group by your tutor/supervisor, or
   - visiting Telstra, a print works, or a conference centre.

   Once you have chosen your team activity, it must be approved by your tutor/supervisor.
3. All the group members should now discuss how this project is going to be planned and organised, setting deadlines for each stage.
4. The leader should allocate tasks to each member of the group, recognising the skills and abilities of each individual.
5. Each member prioritises his or her tasks and provides the resources to complete the tasks.
6. Organise and produce a short report for your tutor/supervisor, including the following:
   - why you decided on this particular activity;
   - the aims and objectives of the activity;
   - how you organised the activity;
   - the results of the activity.

Interpersonal conflict

Most people in an office environment work with other individuals fairly agreeably most of the time. Accommodation or adjustment can be reached to achieve and maintain group harmony. But conflict can arise between workers, and this has to be solved before it affects the effectiveness of a working environment. Conflict can arise in three different ways:

1. **Interpersonal conflict** occurs when two or more people disagree over an issue or when they experience emotional opposition to one another.
2. **Substantive conflict** involves disagreements over resources, rewards, policies and procedures, and the allocation of roles.
3. **Emotional conflict** occurs when there are feelings of anger, dislike, resentment and distrust, resulting in personality clashes. We all experience the feeling of liking some people better than others. You often meet someone for the first time and either like them a lot or instinctively dislike them.

In a work situation there is always conflict between people at some point, and this has to be solved through conflict resolution.

Conflict resolution

Conflict resolution solves interpersonal conflict in three ways—through lose/lose strategy, win/lose strategy, or win/win strategy. How this operates is shown in Table 12.2.

Conflict type	Possible solutions	Results
Lose/lose	Avoidance of conflict 'Smoothing'—recognising both sides of conflict Compromise—each party gives up something of value	Conflict may appear to have been solved; disagreements are smoothed over; future conflict over the same issue will arise again
Win/lose	Competition—conflict resolved through force by someone with superior skills Authoritative command—someone with authority specifies the outcome	Only one person has gained anything and the problem is not resolved; future conflict over the same issue may arise again or lead to related conflict
Win/win	Confrontation of the issues Problem-solving to resolve the conflict issues on both sides	Conflict has been resolved on both sides; each person, or side, has won something and the conflict is unlikely to arise again

Table 12.2 Conflict resolution

Case study 12.1

Conflict resolution

You are the office supervisor of our company, Business Stationery Supplies Pty Ltd. A conflict situation has arisen.

The receptionist, Gina, has complained to you that she is being given tasks to complete by Soula, personal assistant to the manager. Gina says these are not her specified tasks and she just does not have time to do them. She says the reception area is very busy and because she is doing these tasks for Soula she is not completing some of her own specified tasks. Gina is a popular member of staff and has a friendly personality, which is suitable for reception duties, but she does spend a lot of time chatting to visitors and other members of staff.

Soula is an efficient and hard-working person but has been under stress for a number of months. She seems to have too much work, as she is coming in to work early in the morning and leaving late most evenings. This is why she has given tasks to Gina to complete. The situation at present is that Gina and Soula have had arguments. Both of them are dissatisfied with the situation and one or both of them might resign.

How would you deal with this conflict situation? Which conflict resolution strategy would you use to solve this problem?

QUALITY TEAMS

Workgroups/teams enable organisations to accomplish tasks that individuals cannot do alone. An effective group/team has positive synergy and can become a key human resource of the organisation. An effective group or team that is working towards the organisation's goals can have a positive effect on individuals in the team, which will improve their performance. It is therefore important for managers to recognise that:

- group/team activity is important in organisations;
- groups/teams can have a positive or negative effect on the business and the people who work in it;
- an understanding of group/team behaviour can have a desirable effect on workers in the organisation.

The use of joint labour–management committees to discuss a wide range of company conditions has been growing for a number of years—conditions such as occupational health and safety, training, or production policy.

Total quality control

TQC uses the experience, expertise and creativity of people in teams or groups (sometimes called *quality circles* or *quality improvement teams*) to analyse problems and plan to solve them.

Quality circles

A quality circle is a group of workers who voluntarily meet on a regular basis to identify, analyse and solve job-related quality problems and to develop employee potential. The proposed solutions, designed to lead to cost reduction and higher productivity, are submitted to management for adoption or rejection. This originated in Japanese organisations, and worked well in the Japanese culture.

Solving problems using quality circles operates effectively in a manufacturing work environment. Working with each group is a **facilitator** or coordinator, who is a company employee and serves as consultant to a number of circles. When a solution to a problem has been developed by the circle, the facilitator arranges to have the solution considered by the appropriate managers, and helps to install the acceptable improvements. The facilitator may report to a **steering committee** or a senior manager with top-level responsibility for company programs. How a quality circle operates is shown in Figure 12.2.

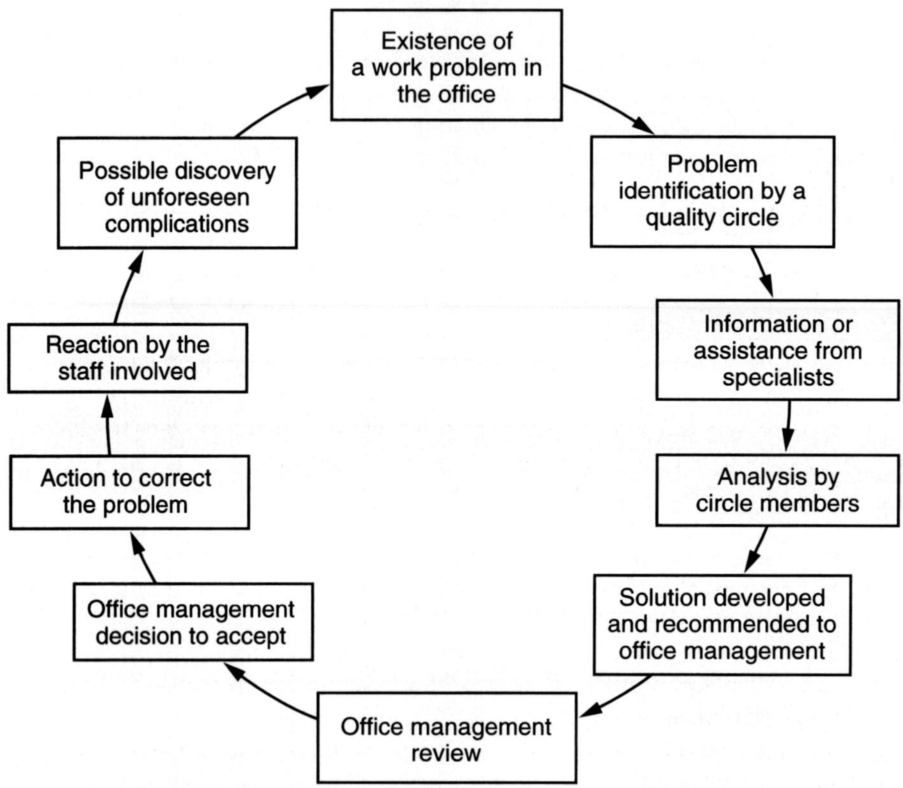

Fig. 12.2 Problem-solving in a quality circle

Nominal group technique (NGT)

NGT is a variation of the quality circle and brainstorming techniques, and is particularly useful for solving problems related to the improvement of office productivity. The nominal groups meet regularly to decide what aspects of their work should be measured and how the measurement should be accomplished.

The NGT is applicable to clerical workers as well as middle management, and is very successful in determining the effectiveness of a new administrative office system.

This technique is different from other group processes, in that individuals work alone in small, non-interacting groups, where non-verbal generation of ideas takes place. Each group consists of five to nine workers of equal status, who are guided in their exchange of ideas by a facilitator. Each of the members of the group is polled in writing individually, and asked to state his/her ideas about indexes that may be used to measure productivity. After discussion of the indexes, the group votes to select the most important items, which are weighted and listed in sequential order.

Summary

- There are two types of groups in an organisation, formal and informal.
- Any group has a 'personality' of its own, which is composed of the individual personalities of the people in it. This is called 'group dynamics'.
- Team-building techniques include keeping the group small, matching the group to the task, group cohesion, group contribution, group objectives, group structure and leadership.
- Group problem-solving has many advantages, but 'groupthink', where members do not think critically about an issue, can be a disadvantage.
- Brainstorming is a process where all members of a group express any ideas, or solutions, that occur without logic or deep thought, quickly and at random. This can lead to a logical thought from another member of the group merely through the association of ideas.
- Team climate is created by its cohesiveness. Group cohesiveness is the degree to which members are attracted to and motivated to remain part of a group or team.
- Team environment can affect team performance adversely unless there is a quiet environment and good seating arrangement. Group size also can have an influence on team cohesiveness.
- Team effectiveness means achieving high levels of both task performance and human resource maintenance over time.
- The stages in the development of groups or teams are: forming, storming, norming, and performing.
- A group or team cannot work effectively unless there are good interpersonal relations, which include: inclusion, control, and acceptance.
- Effective communication in a group occurs when the intended meaning of the sender and the perceived meaning of the receiver are one and the same.

- Interpersonal conflict occurs when two or more people disagree over an issue. This may involve substantive conflict (disagreement over resources or rewards) and emotional conflict (when there are feelings of anger and dislike resulting in personality clashes).
- Conflict resolution solves interpersonal conflict in three ways: through lose/lose, win/lose, or win/win strategies.
- The use of joint labour–management committees to discuss company business and solve problems includes quality circles and total quality control (TQC).

Review questions

(These can be used for group discussion.)

1. What are the two types of group in an organisation?
2. Describe synergy briefly.
3. Write down briefly the stages of team-building techniques.
4. List the advantages of group problem-solving.
5. What is groupthink, and what are its disadvantages in a team situation?
6. What is brainstorming, and how does it work?
7. What are the three different ways in which conflict arises between people?
8. Describe the structure of group or team development (e.g. forming, storming, norming, performing).
9. Explain what good interpersonal relations are, and why they are important in a group/team situation.
10. What are communication barriers? Make a list of how these are caused.
11. Explain good communication skills.
12. Conflict resolution solves interpersonal conflict in three ways. What are these? Which one is the most successful?
13. Describe a 'quality circle' and how it operates.
14. Describe nominal group technique (NGT) and how this operates.

Competency self-assessment and/or tutor/supervisor assessment

When you have finished reading this chapter, and have completed the activities chosen by your tutor/supervisor (**practical in class, at work or theoretical**), fill in this self-assessment. Then hand it to your tutor/supervisor for official assessment.

Learning outcomes	Self-assessment		Tutor/supervisor assessment	
I have successfully completed the following learning outcomes	**Yes** (please tick)	**No**	**Yes** (please tick)	**No**
1. Organise your own schedule in a workplace-related context to meet designated group/section goals				
1.1 identify and prioritise nominated tasks	☐	☐	☐	☐
1.2 select appropriate resources to complete tasks	☐	☐	☐	☐
1.3 review and reprioritise tasks in response to changing circumstances	☐	☐	☐	☐
1.4 complete tasks to designated timelines	☐	☐	☐	☐
2. Participate as a team member to complete a series of work-related tasks within designated timelines				
2.1 participate in identifying team goals and determining tasks necessary to achieve them	☐	☐	☐	☐
2.2 participate in allocation of responsibilities to team members to ensure that designated team goals are met within agreed timelines	☐	☐	☐	☐
2.3 complete agreed tasks to meet team goals and objectives	☐	☐	☐	☐
2.4 participate in review and reprioritise tasks in response to changing circumstances	☐	☐	☐	☐
3. Quality teams				
• group cohesiveness	☐	☐	☐	☐
• quality circles	☐	☐	☐	☐
Date:	Tutor/supervisor signature:			

CHAPTER

13

MEETINGS

ORG402, NOS139

objectives

In this chapter you will learn how to organise and hold formal business meetings, using practical methods of practice and assessment, in order to complete the following learning outcomes:

1. Procedures for different types of business meetings
- 1.1 different types of business meetings and procedures
- 1.2 meeting terminology
- 1.3 documentation used for business meetings

2. Plan and manage the diaries of another/others
- 2.1 establish priorities of the person/group
- 2.2 identify contact/time
- 2.3 make and record appointment times
- 2.4 advise executives of appointment times in advance

3. Prepare documentation for meetings
- 3.1 prepare a notice of meeting
- 3.2 prepare agenda and identify any other relevant material
- 3.3 check documentation for accuracy, ensuring that all errors are corrected
- 3.4 despatch documentation to participants within designated timelines
- 3.5 prepare a spare master set of documents

4. Make meeting arrangements
- 4.1 identify requirements of meeting and participants
- 4.2 prepare meeting room with supplies and documents
- 4.3 advise participants of any changes to original meeting details

5. Record minutes of meeting
- 5.1 take concise and appropriate notes to ensure an accurate record of meeting
- 5.2 prepare minutes so that they reflect a true and accurate account of meeting
- 5.3 format minutes in accordance with organisational procedures
- 5.4 check minutes for accuracy and submit for approval by nominated person
- 5.5 despatch minutes to the nominated persons within designated timelines

INTRODUCTION

A meeting is when two or more people get together to discuss anything. When you meet your friend to decide where you are going at the weekend, you discuss various options and make suggestions. You listen to one another's views, discuss them, then reach a decision and organise the outcome. This is basically the same process followed in business meetings, which can also involve only two people who are making a decision on an administrative matter, or 1000 people who are attending the annual general meeting of a public company.

Upper-level management of an organisation needs to make decisions on important matters concerning the business; for example, to set monthly and yearly goals in order to achieve the aims and objectives of the business, and to decide what is required to achieve these goals. This involves setting deadlines and planning administrative procedures to run the business effectively. Managers and all key personnel in each department meet regularly to provide management with the information necessary to make these decisions.

Why are meetings necessary?

Meetings are a valuable group communication process, and can be more effective than communicating individually—if the meetings are held using the correct procedures. The synergy and brainstorming process of a meeting can also provide many more effective solutions to problems than one person can provide alone. A number of people put forward different points of view and provide different perspectives. But the procedures for a meeting need to be handled and organised effectively, otherwise they can be time-consuming and unproductive, which can be costly to the organisation. Sometimes a meeting can be unproductive—for example, when members are influenced by one aggressive member (see 'groupthink' in Chapter 12).

When should a meeting not be held?

A meeting is not necessary if:

- the problem can be solved by one or two people;
- the number of people who need to be consulted is small;
- the objectives of the meeting could be achieved by a number of telephone calls;
- the information could be distributed in another way;
- the authority to make a decision is not held by any participant of the meeting.

Other names for 'meeting'

There are many terms given to numbers of people coming together.

- A **committee** or **subcommittee** is a group with skills and knowledge formed to complete a specific task, which may involve presenting information at a large, formal meeting.
- A **convention** is a meeting of delegates representing various groups, and is usually the largest gathering of people.

- A **conference** is a meeting for consultation, discussion and information. It usually has a theme, a purpose or problem to solve, and may incorporate an exhibition. It is a gathering of people interested in the same theme.
- A **seminar** is usually a meeting of a small number of people, organised to discuss a particular topic. This often takes place in a college or university.

Computer videoconferencing

Organising meetings and conferences is expensive and time-consuming. People have to travel to the location of the conference and often stay overnight for one or two days. All these people have to be catered for. The use of videoconferencing is growing, as it is becoming easy and relatively cheap to send images and sound over the Web.

E-mail and the Web

Web cameras can enhance e-mail and other communications. People meet between many locations using their PC screen and telecommunications, a PC video camera (with a built-in microphone) to send images over the Internet via CD-ROM. The meeting or conference can be recorded on videotape.

You should check with your ISP before setting up a Webcam site. Because it projects continuous pictures, you need a semi-permanent connection using USB connectivity. Videoconferencing is expensive for individual people to set up, but is worthwhile for organisations thereby saving the cost of arranging a conference in an expensive location.

Internet relay chat (IRC)

IRC allows many users on different systems at different locations to converge on one 'room' and have a discussion, similar to a conference call or party line. IRC is used for both entertainment and serious discussion purposes.

FORMAL AND INFORMAL MEETINGS

In business, there are basically two types of meetings—informal and formal.

Informal meetings

Informal meetings have no rules or regulations and may be called on the spur of the moment, or occur when two or more people meet casually in another person's office. They can be held by any number of people for any number of reasons in any available location. An informal meeting:

- is held in any convenient location;
- has no agenda;
- has no chairperson;

- has no rules and regulations;
- has no formal records;
- is held on an ad-hoc basis.

Formal meetings

These are meetings in which there are predetermined rules and regulations and official records are made of the proceedings; examples are:

- the annual general meeting of a public company;
- local council meetings;
- meetings of charity organisations such as Rotary, Lions, the Salvation Army and the Red Cross.

Meetings for these organisations are required because finances have been injected into the businesses by outside people, such as: shareholders in a public company; taxpayers to the local council; and donations from the public to charity organisations. These people have a right to know how their money has been used, and whether it has been used in an honest and ethical way.

Formal meetings have:

- a notice;
- an agenda;
- a prearranged time;
- a prearranged location;
- invited members;
- predetermined rules and regulations;
- official minutes;
- an appointed chairperson.

Rules and protocol for formal meetings

The rules and protocol for a formal meeting are as follows.

- Questions must be addressed through the chairperson.
- Participants must remain seated while the chairperson speaks.
- Participants must rise to address the chairperson.
- Participants must rise to propose a motion or move an amendment.
- Only one person may speak at a time.
- Only one motion or amendment may be put forward for discussion at a time.

Styles of meetings

Decision-making. Examples of decision-making meetings are when management needs to make a decision on a policy, a procedure, goals, manufacturing deadlines or sales targets; or when a group requires a decision on a proposed outing, building or action. Representatives from each department, or area concerned, are present at the meeting to contribute information and help make a decision.

Brainstorming. Brainstorming involves getting a number of people together to produce ideas. Each member of the group suggests ideas quickly and without deep thought. This leads to other members of the group contributing more ideas, which can lead to specific suggestions and subsequent action.

Information. Management may need information for a specific project and will call a meeting of key personnel to obtain that information. A certain direction may be taken through people giving and sharing information.

Opinion-seeking. Your supervisor may need to put in place certain policies and procedures, and will call a meeting of all personnel involved to obtain their opinions on the best way of putting these policies into operation.

Motivational or team-building. All sales organisations hold regular weekly, monthly and yearly meetings to provide guidance, suggestions and psychological support in order to help salespeople improve motivation and enhance productivity.

Sales. Management holds regular sales meetings with key personnel from the sales department, manufacturing and production departments to make sure that sales targets are being met. If targets are not being met, suggestions are made at the meetings to improve productivity and sales.

Mutual interest. Meetings are held by groups with the same interests, such as philatelic societies, sports clubs, entertainment and music societies, charity groups and parent/teacher organisations.

Case study 13.1

Meetings

You are the office supervisor of Business Stationery Supplies Pty Ltd. There are a number of situations that have arisen in your business where decisions have to be made, and you have been asked to organise meetings to determine how to resolve these situations. On Case study sheet 13.1 in your workbook, write against each situation:

(a) whether the meeting is an informal or formal meeting;
(b) which members of staff in the organisation would be asked to attend the meeting;
(c) the type and style of meeting you would use;
(d) which location you would choose for the meeting;
(e) the resources you would need for each meeting.

Situation 1
Your business has been very successful over the past year, and your managing director is thinking of opening up a new branch in New Zealand. He has asked you to plan and organise a meeting to discuss this expansion.

(continues)

Situation 2
You have noticed that quite a few people in the office are using electrical equipment incorrectly, and there are other hazardous situations you think should be attended to.

Situation 3
You have been asked to organise the monthly sales meeting.

PLANNING AND ORGANISING A MEETING

Planning and organising a formal meeting consists of five stages:

1. planning
2. informing
3. preparing
4. controlling
5. summarising and recording.

Planning

Location for the meeting

The location in which the meeting takes place will depend on the type and style of the meeting.

- A formal meeting will be held in a board room or small conference room, or a large conference hall or public building if it is an annual general meeting of a public organisation with thousands of shareholders.
- An office will be suitable for a meeting of two or three people.
- A meeting room is used for numbers not exceeding 10 people.
- A school hall will be used for a parent/teacher meeting.

The location chosen for some meetings will partly depend on the facilities available. A sales meeting will require equipment such as a video player, a whiteboard, and facilities for refreshments. If there are many people in a large hall, a microphone will be needed.

Timing

The timing of a meeting is very important. Business meetings should:

- be held on a regular basis on the same day and at the same time, so that staff can fit these times into their daily, weekly and monthly planners—if staff are not informed of a meeting in plenty of time there can be a clash of interests, and some people will not be able to attend;

- be held during working hours when possible, as too many meetings held after hours can cause personal problems;
- be held just before lunch or before the end of the day so that time will be kept to a minimum;
- not be held at busy times, such as the end of a financial year or close to production deadlines, when staff are under increased pressure of work.

Setting the date

Preparing includes setting the date for the meeting. The date for a formal meeting, such as an annual general meeting, will have been made at the last meeting, but if a date has not already been agreed one of your duties will be to contact participants to establish a time that is the most convenient for everybody. It is useful to make a 'preferred meeting date list' containing the proposed dates for the meeting and a list of the participants' names (see Figure 13.1). Phone each person on the list (or their personal assistant) and note the preferred date. The meeting can then be set for the convenience of the majority of the participants.

Proposed dates for next sales meeting, June 199X

Participants	Dates proposed 2	9	16	Date telephoned
Chairperson (Gerrard)			X	6/5
G. Anton		X		6/5
J. Ginis	X			6/5
T. Leenstria			X	7/5
S. Tran			X	7/5
S. Vaughan			X	7/5

Fig. 13.1 Preferred meeting date list

Informing

Once the date, time and location of the meeting have been decided, all participants should be informed. In the case of informal/internal business meetings or sales meetings, a telephone call or an internal memorandum is all that is required (see Figure 13.3). The date, time and location can then be recorded in your diary and in the diary of your boss. You should phone the other participants (or their personal assistants) and inform them also of the date, time and location, so that these can be entered in their diaries. For formal meetings a written notice is required (see Figure 13.2).

NOTICE OF ANNUAL MEETING

Notice is hereby given that the annual meeting of Business Stationery Supplies Pty Ltd will be held on 3 January 199X in the board room, 163 Collins Street, Melbourne, at 2.00 pm.

All members of the company are invited to be present.

Agenda

1. Welcome and apologies
2. Minutes of last annual meeting
3. Business arising
4. Chairperson's report
5. Financial statement and balance
6. Correspondence
7. Any other business
8. Date of next meeting

Secretary

Fig. 13.2 Notice of meeting and agenda

INTERNAL MEMORANDUM

TO: *John Ginis, Sue Tran, George Anton, Tony Leenstra* **DATE:** *19/6/9X*

FROM: *Jan Pryor, office supervisor*

SUBJECT: *Notice of Health and Safety Committee meeting*

Would you please attend a meeting in the board room on Tuesday 26 June at 9.00 am to discuss the new health and safety regulations with the H&S Committee.

This should take about an hour and a half

Fig. 13.3 Internal notice of meeting

Participants

The people who attend a meeting are chosen according to the type and style of meeting. The costs of meetings are high, as the hourly rate for a top executive is high, so only the following should be invited:

- the minimum number of people necessary to achieve the objectives (at official meetings there is often a quorum required, which is the minimum number of people laid down by the company rules or legal regulations);
- those people who have the necessary information or expertise for the decision-making process;
- those people who are personally involved in a situation that needs to be resolved.

Preparing

Documentation for formal meetings

The preparation for formal meetings must be carried out according to the procedures outlined in the Memorandum and Articles of Association of each individual organisation. These procedures must meet the legal requirements of the Corporations Law, as defined in the Corporations Act, and would normally include the following documentation:

1. ***Minutes of the previous meeting.*** These and any other relevant papers and documents should be kept in a safe place ready to be taken to the meeting on the stated day.
2. ***Notice of the meeting.*** All participants of a formal meeting are informed of the date and time by a written notice, which is sent well in advance of the meeting so that members have time to prepare their contribution. There is often a legal requirement for the notice to be sent out a week or two weeks before a large public organisation meeting. The notice should include:
   - the name of the organisation or meeting group;
   - the type of meeting (e.g. second annual general);
   - the date, time and location;
   - the order of agenda items with brief details.
3. ***Agenda.*** The order in which topics are discussed in a meeting should follow the items on the agenda. The chairperson should make sure that all members keep to the order of the agenda, and a time limit should be placed on each item. The chairperson's copy of the agenda usually has a column down the right-hand side so that she/he can make notes during the meeting.

Activity 13.1

Preparing papers for a meeting

In Case study 13.1 there are three situations in which you have to organise meetings. Using the names of people in your class (or at work), carry out the following tasks:

1. Type (or write very neatly) a correct notice for each meeting.
2. Work out the topics you would need to discuss at each meeting and type an agenda listing the items for discussion.

On the day of the meeting

The **receptionist** should be advised that a meeting is being held, and where, so that participants can be directed to the meeting room on arrival.

The **switchboard operator** should be informed of the members of staff attending the meeting, and should redirect calls or take messages.

A 'meeting in progress' notice should be placed on the door of the meeting room to ensure that there are no interruptions.

Preparation of meeting room

From your list of office-bearers and members, name cards should be prepared and placed on the meeting table in front of the participants' seats. Names can be checked off your list as people arrive.

Supplies that may be required are:

- a clock for timing the agenda items;
- pens, pencils and notepaper for all participants to take notes;
- a jug of iced water and glasses on the meeting table;
- spare copies of the agenda;
- a tape-recorder (if this is allowed);
- chairperson's agenda and brief;
- refreshments on a side table (tea and coffee, iced water, cups and saucers, biscuits; a light lunch of sandwiches and fruit may be needed).

Documents may also be required, such as:

- copies of documents previously circulated;
- the minute book;
- a calendar or diary;
- Memorandum and Articles of Association, or rules governing the meeting;
- proxies and apologies;
- ballot papers.

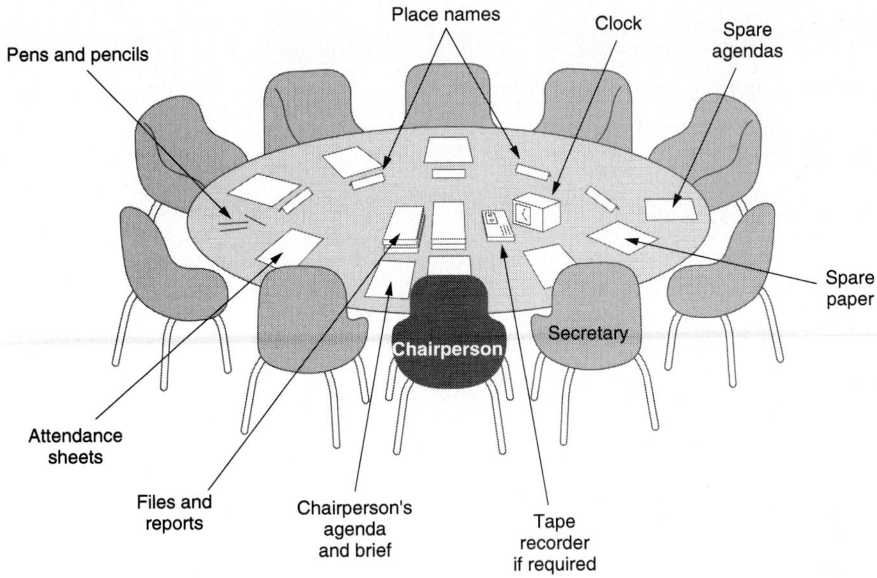

Fig. 13.4 Meeting preparation

Controlling

A successful meeting is achieved when all members participate in the meeting process. Each member should be allocated a role with clearly defined duties. Like all group activities a meeting needs a leader, and this is the role of the chairperson, who controls and leads other members of the group.

Duties of the chairperson	Duties of meeting members
• formally open meeting • check quorum is present (if necessary) • preside over meeting • control and lead the members • make sure only one person speaks at a time • encourage group participation • ensure all members contribute • encourage questions • remain impartial and make no contribution to discussion • adjourn meeting (if necessary) • put motions and amendments • check and sign the minutes	• plan and prepare for meeting • contribute information • allow other members to put forward their views • use effective listening • evaluate information • be objective • use empathy • use critical thinking • accept leadership of chairperson

Duties of minute-taker
• prepare notice of meeting and agenda • prepare the location of meeting • provide notepaper, drinking water, refreshments • assist the chairperson • record attendance and any apologies • take the minutes • record topics discussed on the agenda, comments made, decisions reached, and the name of any person who has to carry out any action • prepare draft minutes and check with the chairperson for any errors • type final minutes • circulate minutes • follow up any action required

Table 13.1 Duties of chairperson, meeting members and minute-taker

Meeting process

After checking that a quorum is present (if this is a requirement), the chairperson will welcome members to the meeting. At a formal meeting the chairperson will open the meeting with, for example:

> **I have pleasure in declaring the annual general meeting of Business Stationery Supplies Pty Ltd open.**

He/she will then read the agenda items, and explain the inclusion of any special items. An opportunity will be given for members to request a change in the order of the items on the agenda. The minutes of the previous meeting are read and, if everyone agrees, the chairperson signs them as a 'true record'. Discussion commences in the agreed order of the agenda. Anyone wanting to speak should wait until the previous person has finished, then raise a right hand and ask permission of the chairperson.

After discussion, a vote may have to be taken on how many members are in favour, or against, the motion. Voting includes a **show of hands**, in which the chairperson asks those in favour to raise their hands, then those not in favour to raise their hands, and a count is made; and a **poll**, which is a secret vote where members post a slip of paper with their vote written on it into a locked box. A count is made later.

Summarising and recording the minutes

Minutes are a formal record of all decisions and events that have taken place at a meeting. The purpose of minutes is to:

- record the meeting for legal requirements;
- record decisions formally, so that everyone knows what action is to be taken and who is responsible for taking it;
- give authority in writing to the person taking the action;

- act as a historical record of past proceedings that can be referred to if necessary;
- act as a reporting and reminder system;
- give the meeting a structure, which is followed at every meeting.

Minutes are written in the past tense, as they are a recording of an event that has already taken place (see Figure 13.5). The following details are included in the minutes:

- *members present:* a list of members present, stating who is in the chair and who is the secretary;
- *apologies:* a list of members who were invited but have notified the secretary that they cannot attend;
- *minutes of last meeting:* minutes of the previous meeting, to be agreed to as a correct record of the proceedings or amended if necessary;
- *matters arising from last meeting:* any questions that arise from the reading of the last minutes;
- *correspondence:* letters or other documents that have been sent out regarding matters discussed in the last meeting, and any replies received to be shown to the members;
- *reports:* reports on any actions that have been taken; financial reports and reports on any other matters;
- *agenda items:* items listed on the present agenda requiring attention;
- *any other business:* items not on the agenda which members want to mention at the end of the meeting;
- *closure:* a note made of the time the meeting ended;
- *date and place of next meeting:* so that members can make a note in their diaries.

After the meeting

The following activities will need to be performed after the meeting:

- Clear the room of unwanted articles.
- Advise the receptionist and switchboard operator that the meeting is over.
- Ensure that all documents are returned to files.
- Prepare draft minutes as soon as possible for approval by the chairperson.
- Request agenda items for next meeting.
- Distribute final minutes.
- Follow through on any tasks assigned at the meeting.

Business Stationery Supplies Pty Ltd

MINUTES of the annual meeting of Business Stationery Supplies Pty Ltd, held in the Board Room, 163 Collins Street, Melbourne, on 3 January 199X at 2 pm

		ACTION
PRESENT:	E. Gerrard (chair) G. Anton (company secretary) J. Ginis T. Leenstra S. Tran S. Vaughan	
APOLOGIES	S. Krantz	

MINUTES OF LAST MEETING:
It was RESOLVED that the Minutes of the last meeting held on 3 January 199X be accepted as a true and accurate record. Moved by S. Tran, seconded by S. Vaughan. ACCEPTED

BUSINESS ARISING
G. Anton reported that the investigation into opening a branch in New Zealand was being held at the present time, and a report would be presented on the results at the next general meeting. GA

CHAIRPERSON'S REPORT
The chairperson reported that the year had been a very successful one, and he congratulated all managers and staff of all departments for their magnificent effort over the previous year. A bonus would be issued to all personnel. EG

FINANCIAL STATEMENT AND BALANCE SHEET
The company secretary tabled her report on the yearly trading figures, which showed an improvement over the previous year's figures of 6.1%.

CORRESPONDENCE
A letter was read by G. Anton received from Building Enterprises Pty Ltd tendering for the proposed building of the new branch in New Zealand. It was moved by J. Ginis and seconded by T. Leenstra that plans should be inspected and a report presented at the next general meeting. RESOLVED

DATE OF NEXT MEETING
The next general meeting will be held on 30 September 199X.

CLOSE OF MEETING
There being no further business, the meeting closed at 3.10 pm.

Chairperson .... Date ....

Fig. 13.5 Minutes of meeting

MEETING TERMINOLOGY

Refer to Chapter 12 for definitions of the following: **group dynamics; brainstorming; team-building techniques; conflict resolution.**

Articles of Association, Memorandum of Association: the rules and regulations of the organisation
Addressing the chair: a member wanting to speak should seek the consent of the chairperson, and all remarks should be directed to the chairperson
Adjournment: when the chairperson stops the meeting because of shortage of time
Ad-hoc meeting: an unplanned meeting, arranged for a specific purpose
AOB: any other business
Casting vote: a vote (usually a second vote) granted to the chairperson for use in the event of equal votes for, or against, a motion
Chairperson: the person responsible for conducting and controlling the meeting according to rules and regulations
Closure: the chairperson ends discussion on a matter before the meeting
Committee: a small group of office-bearers formed to run a larger group, or formed to complete a specific task
Constitution: covers the rules and regulations under which the meeting is run
Convene: to call a meeting
Co-opt: to elect an individual to a committee for a specific purpose
Democratic: decisions being made, and activities handled, according to the agreement of all the members or an elected representative
Executive: the main office-bearers or those who hold the most senior positions of the meeting or the organisation
Ex officio: when a person is a member of a committee as a result of the position he/she holds in the organisation (e.g. the managing director may be an ex officio member of every subcommittee meeting of the board of directors)
In camera: a meeting held in private—not open to the public
Lapse: when a motion is not seconded
Lie on the table: when the question is held over but may be resumed later in the meeting
Minutes: an official record of the activities of the meeting
Motion: a proposition or suggestion put to a meeting (must be seconded by another member of the meeting before it may be discussed)
Nem con: no-one contradicts
Office-bearers: members elected to run the meeting
Point of order: when a member interrupts to query the conduct of another member according to the rules of the meeting
Poll: a secret vote carried out according to the rules and regulations of the meeting
Proxy: when a member unable to attend the meeting appoints someone else (a proxy) to vote in his/her place
Quorum: the minimum number of people required to attend a meeting according to the rules and regulations of the company
Resolution: formal acknowledgement of a decision made at a meeting
Seconded: a motion must be seconded by a member before it can be discussed

Secretary: a person who keeps a record of all matters discussed and decisions made, on behalf of the chairperson
Table a document: when anyone who wants to see a document has access to it
Teller or scrutineer: a person appointed to count votes at a meeting
Treasurer: person who collects membership fees and is responsible for all money matters for a club or committee
Unanimous: when all members have voted in favour of a resolution
Vote: members are invited to vote in favour of, or against, a motion; a majority either way carries the vote

Activity 13.2

Planning and organising a meeting

Divide the class into small groups. Prepare, plan and organise a meeting as follows (this will take two or three weeks of two-hour classes, with the meeting being held in the last week):

1. Each small group brainstorms ideas for topics to discuss at the meeting. The whole class chooses three or four topics for discussion, which are entered on the agenda.
2. Office-bearers and participants holding the meeting are appointed as follows:
   - **Chairperson.** This will also be the leader of the class. She/he will control the planning and organising of the meeting and coordinate each group, so that all tasks are completed in time to hold the meeting at the designated date and time.
   - **Participants**, who will sit with the chairperson at the meeting table and discuss the topics chosen for the meeting. There should not be more than four or five people as participants.
   - **Minute-taker.** One or two members of the class should volunteer to take the minutes. These would preferably be students who are studying note-taking.
3. One small group will be appointed to prepare the agenda when the topics have been decided. In the first week, details will have to be given to the group preparing the notice of the meeting.
4. One small group will be appointed to prepare and circulate the notice of the meeting (after receiving details of the agenda) to all other members of the class and the tutor/supervisor. This should be done well before the date of the meeting.
5. One small group will be appointed to prepare the room and table on the day the meeting is held, and to clear everything away after the meeting.
6. One small group will be appointed to prepare refreshments (organised with the approval of the tutor/supervisor), and to clear these away after the meeting.

7. Complete the meeting planner sheet (Activity sheet 13.2) in your workbook. The room, and date and time, for the meeting will be arranged with your tutor/supervisor.

Hold your meeting at the planned date and time, observed by your tutor/supervisor, who will assess your meeting procedures.

Summary

- Meetings are a valuable group communication process. If meetings are held using the correct procedures they can be more effective than communicating individually. The synergy and brainstorming process of a meeting can also provide many more effective solutions to problems than one person can provide alone.
- A meeting is not necessary if the problem can be solved by one or two people; if the information or objectives could be achieved in another way; or if the authority to make a decision is not held by any participant of the meeting.
- There are many terms given to numbers of people meeting, including convention, conference, seminar and videoconferencing.
- In business there are basically two types of meeting, informal and formal. Informal meetings have no rules or regulations; they may be called on the spur of the moment and held in any location. Formal meetings have predetermined rules and regulations, are held in a prearranged location at a prearranged time, and official records are made of the proceedings.
- There are many different styles of meetings, examples of which are brainstorming, information, opinion-seeking, motivational or team-building, sales, and mutual interest.
- Planning and organising a formal meeting consists of five stages: (1) planning, which includes choosing the location for the meeting and setting the time; (2) informing the participants and noting this in your diary; (3) preparing, which includes setting the date and organising the notice and agenda; (4) controlling, which is achieved when all members are allocated a role with clearly defined duties; (5) summarising and recording the meeting by taking minutes.
- On the day of the meeting the meeting room should be prepared with name cards, meeting supplies, reference material and refreshments. The receptionist should be advised, so that participants can be directed to the correct location and any telephone calls diverted.

Review questions

(These may be used for group discussion.)

1. Why are meetings necessary?
2. When should a meeting not be held?
3. There are two types of business meetings. What are they?
4. List and describe briefly the various styles of meetings.
5. List the five stages of planning and organising a meeting.
6. Describe briefly how you would set the date for a meeting.
7. List and describe briefly the documentation required for a meeting.
8. List the duties of the chairperson, the meeting members and the minute-taker.
9. Describe what minutes are and their purpose.
10. List the items required to be entered in the minutes.
11. How would you prepare for the meeting on the day?
12. What is required to be done after the meeting?
13. List the rules and protocol for a formal meeting.

Competency self-assessment and/or tutor/supervisor assessment

When you have finished reading this chapter, and have completed the activities chosen by your tutor/supervisor (**practical in class, at work or theoretical**), fill in this self-assessment. Then hand it to your tutor/supervisor for official assessment.

Learning outcomes	**Self-assessment**		**Tutor/supervisor assessment**	
I have successfully completed the following learning outcomes	Yes	No	Yes	No
	(please tick)		(please tick)	
1. Procedures for different types of business meetings				
1.1 different types of business meetings and procedures	☐	☐	☐	☐
1.2 meeting terminology	☐	☐	☐	☐
1.3 documentation used for business meetings	☐	☐	☐	☐
2. Plan and manage the diaries of another/others				
2.1 establish priorities of the person/group	☐	☐	☐	☐
2.2 identify contact/time	☐	☐	☐	☐

2.3 make and record appointment times	☐	☐	☐	☐
2.4 advise executives of appointment times in advance	☐	☐	☐	☐
3. Prepare documentation for meetings				
3.1 prepare a notice of meeting	☐	☐	☐	☐
3.2 prepare agenda and identify any other relevant material	☐	☐	☐	☐
3.3 check documentation for accuracy, ensuring that all errors are corrected	☐	☐	☐	☐
3.4 despatch documentation to participants within designated timelines	☐	☐	☐	☐
3.5 prepare a spare master set of documents	☐	☐	☐	☐
4. Make meeting arrangements				
4.1 identify requirements of meeting and participants	☐	☐	☐	☐
4.2 prepare meeting room with supplies and documents	☐	☐	☐	☐
4.3 advise participants of any changes to original meeting details	☐	☐	☐	☐
5. Record minutes of meeting				
5.1 take concise and appropriate notes to ensure an accurate record of meeting	☐	☐	☐	☐
5.2 prepare minutes so that they reflect a true and accurate account of meeting	☐	☐	☐	☐
5.3 format minutes in accordance with organisational procedures	☐	☐	☐	☐
5.4 check minutes for accuracy and submit for approval by nominated person	☐	☐	☐	☐
5.5 despatch minutes to the nominated persons within designated timelines	☐	☐	☐	☐
Date:	Tutor/supervisor signature:			

CHAPTER

14

TRAVEL AND SCHEDULING APPOINTMENTS

ORG401, NOS127, 220, ORG302A, NOS249.V3, 7365CN

objectives

The purpose of this chapter is to help you to organise business travel arrangements, in order to complete the following learning outcomes:

1. **Make travel and accommodation arrangements and bookings from a given itinerary in a workplace-related context**
   - 1.1 clarify purpose of travel and itinerary
   - 1.2 identify travel and accommodation requirements
   - 1.3 make bookings in line with identified needs
   - 1.4 check documentation and rectify discrepancies
   - 1.5 finalise itinerary and forward details of arrangements to traveller within the designated timelines
2. **Scheduling appointments**

INTRODUCTION

With improved telecommunications, business has become more globalised. Because of this, many staff or consultants have overseas postings, or need to travel to other countries to keep up with developments in international trade. International organisations with operations in Australia often exchange staff between countries. Over the next decade there will probably be more trade between Asia and Australia. All these factors mean that travel, either overseas or interstate, is increasing. At some stage in your career it is likely that you will have to arrange travel for yourself or your boss. If you are a personal assistant to someone who travels overseas every few months, you may be involved in the planning and organising of overseas trips regularly. Whether the business trip is international, interstate or intrastate (within your own state), if it is to be successful there are vital steps to take in the planning of the trip.

Clarifying the purpose of the travel arrangements

The first stage in planning business travel is to clarify why the person or persons want to travel. You will need to know how many people are travelling, who they are and what is their destination. Are they visiting an interstate office of your company? Are they going to a conference or attending a series of meetings?

Perhaps your organisation has a regular travel agent that makes all your bookings. In this case all you may have to do is type up a more comprehensive itinerary than the one supplied by the agency.

DOMESTIC TRAVEL

Planning an interstate business trip for two or three days is likely to involve organising the following:

- car-rental voucher
- airline tickets
- hotel accommodation
- appointments for meetings
- a travel itinerary
- travel insurance.

The person travelling may require a detailed itinerary, including not only flight details, hotel accommodation and car hire but times of business meetings and any other information needed to make a well-organised trip. You will need to be aware of time management in order to arrange appointments or meetings with enough time in between for travelling from the hotel or from one appointment to another.

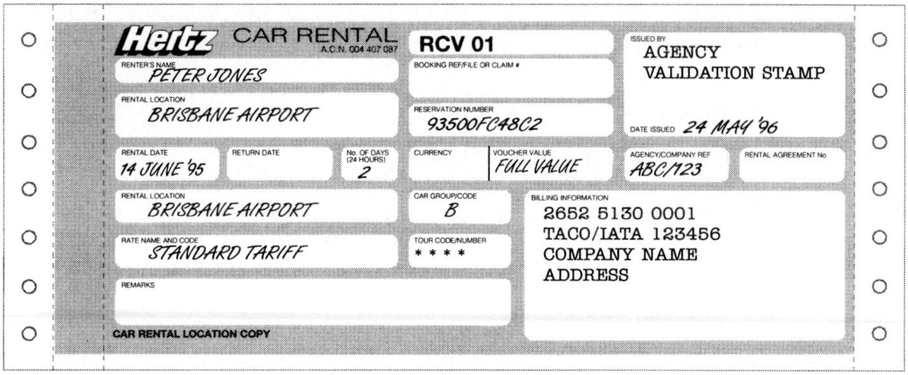

Hertz CAR RENTAL A.C.N. 004 407 087
RCV 01
RENTER'S NAME PETER JONES
BOOKING REF/FILE OR CLAIM #
ISSUED BY AGENCY VALIDATION STAMP
RENTAL LOCATION BRISBANE AIRPORT
RESERVATION NUMBER 93500FC48C2
DATE ISSUED 24 MAY '96
RENTAL DATE 14 JUNE '95
RETURN DATE
No. OF DAYS (24 HOURS) 2
CURRENCY
VOUCHER VALUE FULL VALUE
AGENCY/COMPANY REF ABC/123
RENTAL AGREEMENT No
RENTAL LOCATION BRISBANE AIRPORT
CAR GROUP/CODE B
BILLING INFORMATION
2652 5130 0001
TACO/IATA 123456
COMPANY NAME
ADDRESS
RATE NAME AND CODE STANDARD TARIFF
TOUR CODE/NUMBER * * * *
REMARKS
CAR RENTAL LOCATION COPY

Fig. 14.1 Car-rental voucher

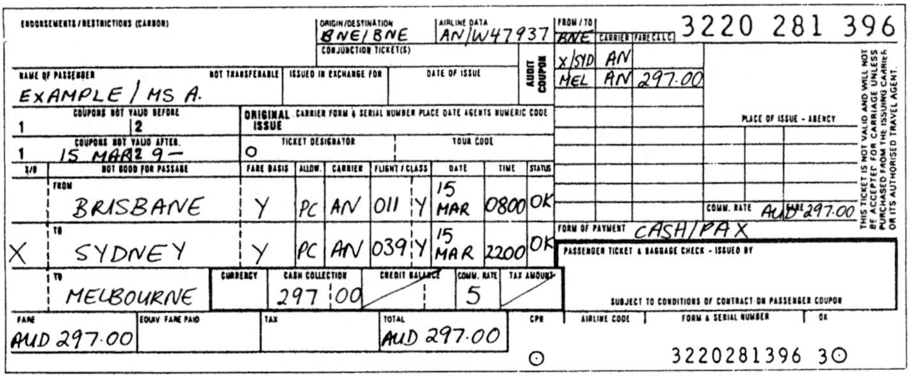

ENDORSEMENTS/RESTRICTIONS (CARBON)
ORIGIN/DESTINATION BNE/BNE
AIRLINE DATA AN/W47937
FROM/TO BNE
3220 281 396
CONJUNCTION TICKET(S)
NAME OF PASSENGER EXAMPLE / MS A.
NOT TRANSFERABLE
ISSUED IN EXCHANGE FOR
DATE OF ISSUE
AUDIT COUPON
X/SYD AN
MEL AN 297.00
COUPONS NOT VALID BEFORE 1 2
COUPONS NOT VALID AFTER 1 15 MAR 2 9—
ORIGINAL ISSUE
TICKET DESIGNATOR
TOUR CODE
PLACE OF ISSUE - AGENCY
X/O NOT GOOD FOR PASSAGE
FARE BASIS ALLOW. CARRIER FLIGHT/CLASS DATE TIME STATUS
FROM BRISBANE Y PC AN 011 Y 15 MAR 0800 OK
TO X SYDNEY Y PC AN 039 Y 15 MAR 2200 OK
TO MELBOURNE
CURRENCY
CASH COLLECTION 297.00
CREDIT BALANCE
COMM. RATE 5
TAX AMOUNT
COMM. RATE AUD 297.00
FORM OF PAYMENT CASH/PAX
PASSENGER TICKET & BAGGAGE CHECK - ISSUED BY
SUBJECT TO CONDITIONS OF CONTRACT ON PASSENGER COUPON
FARE AUD 297.00
EQUIV FARE PAID
TAX
TOTAL AUD 297.00
CPN
AIRLINE CODE
FORM & SERIAL NUMBER
CK
3220281396 3 0
THIS TICKET IS NOT VALID AND WILL NOT BE ACCEPTED FOR CARRIAGE UNLESS PURCHASED FROM THE ISSUING CARRIER OR ITS AUTHORISED TRAVEL AGENT.

Fig. 14.2 Airline ticket

SCHEDULING APPOINTMENTS

Depending on the position you hold in the organisation, you may be responsible for organising appointments for those visiting one or more people within your organisation, or organising appointments for your boss when she/he is travelling. If you are a receptionist in a busy dental or doctor's surgery, or a veterinary clinic, appointments will need to be carefully scheduled at predetermined intervals, according to the particular business.

As mentioned in Chapter 7, there are three types of visitors to a business:

- those expected with appointments;
- those expected without appointments (maintenance and service people);
- those not expected (without appointments).

How to deal with visitors is dealt with in Chapter 7, also how to deal with problem visitors and scheduling appointments with the aid of a paper diary system or a computer diary system.

Electronic scheduling of appointments

There are many computer aids to scheduling appointments, meetings and events. These can be very useful, and have noise and other reminder features.

Appointments, meetings and events

Appointments are activities which you block time for in your calendar and which do not involve inviting other people or resources.

Appointments can have reminders attached to them. You can specify how others view your calendar by designating the time an appointment takes as ***busy, free, tentative*** or ***out of office.*** You can schedule ***recurring*** appointments. Clicking an appointment selects it, while double-clicking the left move handle opens it. You can view your appointments by day, week or month. There may be an autodate functionality, which allows you to type text such as ***next Tuesday*** or ***noon*** instead of typing a date or time.

You can schedule an appointment in your own calendar, and other users (such as your boss) can give you permission to schedule or make changes to appointments in their calendars. Appointments can also be made ***private.***

A meeting is an appointment to which you invite people and resources.

You use the Meeting Planner to create and send meeting requests and to reserve resources. When you create a meeting, you identify the people and resources to invite and pick a meeting time. Responses to your meeting request appear in your ***Inbox.*** You can track responses by clicking ***Show attendee status*** on the ***Meeting planner*** tab in the ***Appointment*** dialogue box. You can also add people to an existing meeting or reschedule a meeting.

An event is an activity that lasts 24 hours or longer.

Examples include a trade show, a vacation, a conference or a seminar. An annual event, such as a birthday or anniversary, occurs yearly on a specific date, while an event occurs once and can last for one day or several days. Events and annual events do not occupy blocks of time in the ***calendar***; instead, they appear in ***banners.*** An all-day appointment shows time as busy, while an event or annual event shows time as free when viewed by others.

Figures 11.4 and 11.5 of Chapter 11 illustrate an electronic planner and organiser which help you to plan appointments and organise meetings and conferences, and Figure 14.3 shows Microsoft Outlook's calendar, which also organises and reschedules appointments with a click on your mouse.

E-mail appointments. You can create an appointment from an e-mail message. Using a software package such as ***Microsoft Outlook,*** you simply drag the e-mail message you want to use to create an appointment onto ***Calendar,*** and enter the location in the ***Location*** box.

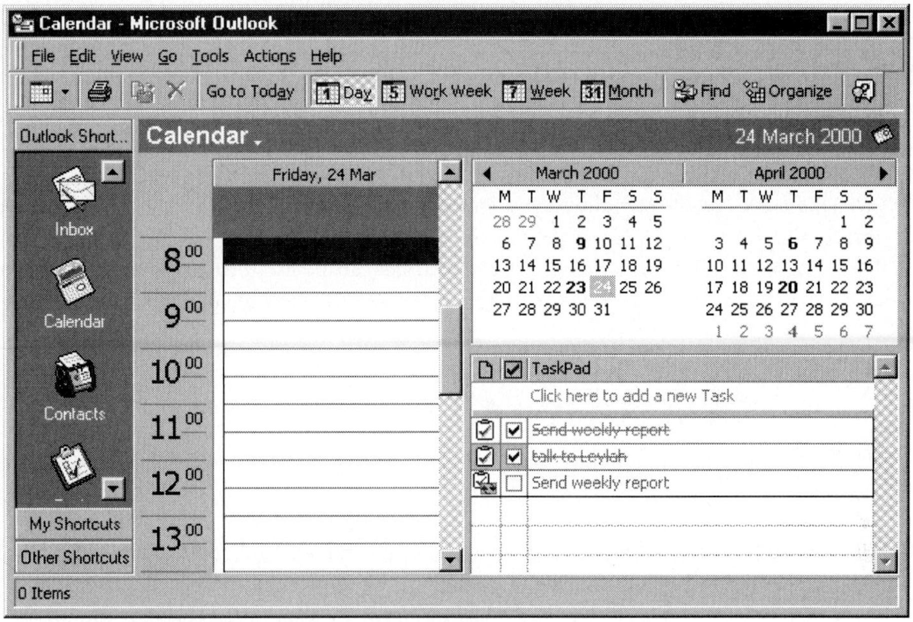

Fig. 14.3 Microsoft Outlook Calendar

You can also set an ***appointment reminder***. Select the ***Reminder*** check box, and then enter the amount of time before the appointment when you want the reminder to occur. You can customise the reminder action and then select the sound you want played.

Record an activity in 'Journal'. You can use 'Journal' to record interactions with important contacts, to record items or documents that are significant to you, and to record activities of all types. For example, you can have all e-mail messages automatically recorded in 'Journal' for the contacts you specify, and tasks, appointments and meetings can be recorded.

Electronic organisers

As pointed out in Chapter 4, office roles are changing and blurring. Since the introduction of IT machines, many managers and executives send their own messages to e-mail. Many of them type their own letters on computers or laptops, and organise their own appointments through e-mail or electronic organisers linked to the main office computer.

Electronic organisers are used by managers and executives to meet all their needs, from simple diary replacements to high-powered pocket personal computers. They are used by a broad cross-section of society—students, personal users and businesspeople—as the price range makes them available to most people. The price goes from under $100 up to $1000.

The low-cost organisers have a single telephone list for storing names, addresses and phone numbers, but the more expensive ones often have multiple 'books' divided into personal, business and specialist categories, with a variety of functions:

- The *scheduler function* is an appointment minder telling you what you are doing on each day, as well as where you should be doing it and with whom.
- Details of *daily, weekly and monthly events* can be stored and retrieved as and when required.
- An *alarm* can sound before each appointment as a reminder.
- A *to-do* function lists tasks to be done today, next week or in the future.
- A *memo pad* or *note-taker* allows you to keep or make notes, which can sometimes be indexed and linked to your scheduler or contacts book.
- Some organisers use an *infrared beam*, and links can be established between any two IrDA devices at a range of 3 metres and can be used to exchange data or drive printers.

Case study 14.1

Travel arrangements

You are the personal assistant to the sales manager of Business Stationery Supplies Pty Ltd. Your boss, Judy Chan, is visiting two potential new and important customers in Cairns next week. The bookings have all been made through the usual travel agent, and she wants you to type up a clear itinerary so that she can see at a glance where she should be at the times specified, and how she will travel to appointments. Her handwritten notes to you are given on Case study sheet 14.1 in your workbook. From these notes, type a clear itinerary in the correct format (see also Figure 14.1).

Activity 14.1

Documents for domestic travel

In Case study 14.1 you organised and typed an itinerary for the sales manager, Judy Chan, for her business trip to Cairns.

On a separate sheet of paper, list the documents you would include in Judy's travel folder for this business trip. If time permits, you could make up a travel folder containing simulated documents.

PLANNING AND ORGANISING AN OVERSEAS BUSINESS TRIP

The first thing you should do when planning and organising a business trip overseas is to make a checklist of what is required for advance planning so that nothing is forgotten. More time and effort will be involved than when planning a domestic trip. Case study sheet 14.2 in your workbook is a checklist of required documents and useful information.

Advance planning

All meetings, appointments or conference attendance should have been confirmed weeks before the departure of the traveller, and details incorporated in the itinerary. Other aspects of advance planning include completing documentation.

Documentation

All overseas travellers require documentation to enter, live in and leave foreign countries. You should allow plenty of time in your advance planning to make sure these documents are all current and valid.

Passport

Has the traveller been overseas before, and does he/she possess a current passport? All overseas travellers require a valid passport from the country of their citizenship. This is a vital document for entry into any foreign country. It is required when checking into hotels and cashing travellers' cheques, or it may be needed for identification. Applications for a passport are made through the Department of Foreign Affairs. You will need to obtain a passport application form from a post office. This will be completed by the person travelling and then taken to the post office by that person, together with two recent photographs. If the application is straightforward the passport will be received by the applicant after two weeks. Take this time into account when you are planning the overseas trip.

Visas

A visa may be required for entry to some countries. A visa is a stamp entered in a passport by the consulate of another country to allow a visitor to stay for a period of time in that country. Each country has its own visa regulations and you will need to check these well in advance as part of your advance planning. There are tourist, work and permanent visas, depending on the purpose of the travel. Sometimes entry is dependent on onward bookings and sufficient finances to pay for the period of the visit. The visa is checked at immigration checkpoints when entering the country, stamped with the date of arrival and again with the date of departure. To work in a foreign country for a long period of time, a work permit is required.

Health regulations

Vaccinations against diseases prevalent in other countries are often necessary, and in some cases are a legal obligation to prevent the spread of infectious diseases from one country to another. These are based on recommendations of the World Health Organization. Your travel agent will tell you what is required for the country your traveller is visiting and how soon vaccination needs to be done before the trip.

Travel and health insurance

Sometimes when you have organised the travel, at the last minute it has to be cancelled due to unforeseen events or illness. Many insurance companies offer cover against such contingencies as:

- loss of deposit for cancellation of advance airline bookings;
- accidents when driving a hire car;
- overseas medical and dental expenses;
- loss of personal property;
- accidental death;
- additional expenses because of delayed or cancelled flights.

Consideration should be given to taking out health insurance if the traveller's stay in the overseas country is for a long period, in case of serious accident or illness requiring hospitalisation (which is not covered in normal travel insurance). Australian Medicare covers some overseas health care, but extended health cover may require a qualifying period and have to be taken out two months before departure. So this must also be taken into account when you are planning an overseas trip.

Making travel reservations

Once you have all the details you need from your advance planning, you can make the bookings through a travel agent. The details your travel agent will require are:

- the name/s of the traveller/s;
- the address, business telephone, and home telephone numbers of the traveller/s;
- the departure and return dates;
- the final destination and any preferred stopover locations;
- the standard of accommodation and facilities required;
- the type and class of air travel.

Air travel

Various types of air travel are available.

First class. This is the most expensive type of air travel, with higher-standard meals, drinks included, and more comfortable and spacious seating. Only top executives use this class.

Business class. This is aimed at business travellers, and is not quite as expensive as first class. The seating is more comfortable and there are more extras than in economy class.

Economy class. This is the cheapest class of air travel. It can be uncomfortable and noisy, and is not suitable for business travellers who want to work during the flight.

Discount air travel. It is a good idea to take advantage of any special offers from different airline companies. Sometimes there is a price war between airlines and very competitive prices are offered. Discounts are usually offered for:

- advance purchase, where a return ticket is booked and paid for 14 days before departure with no minimum period of stay;
- group fares, which can be arranged for business, sporting, charity or social groups;
- 'Kangaroo airpass' fares, which allow air travel around Australia for a lower rate if a minimum number of kilometres is booked (6000 to 10 000).

Special offers. Airlines often make special offers of:

- free car-hire;
- stopovers with free accommodation and car-hire;
- special rates at different times of the year (e.g. cheaper rates for flying from Australia to Europe when it is winter in Europe).

Conditions apply to these special offers and discount fares, and should be carefully noted when making the bookings. However the airline flights have been booked, you should confirm these 24 hours before the traveller's departure, and a contact number should be left with the airline in case of delays or cancellations.

Identifying accommodation requirements

Whether accommodation is booked through a travel agent or with the hotel/motel direct, you will need to find out from the traveller:

- whether there are any special requirements, and specify these to the motel/hotel;
- what standard of accommodation is required;
- any special facilities required (sports and health, working area, communication links, meeting or conference areas);
- any special dietary requirements;
- the type of room and number of guests (single, double, with or without private bathroom facilities);
- the length of stay (including time of arrival and time of departure);
- the method of payment (through a travel agent, by credit card, or through a corporate account);
- whether pick-up from the airport is needed.

Motel and hotel rating definitions are as follows:

*****	International style offering superior standard of appointment
****	Exceptionally well-appointed establishments with high-quality furnishings
***	Well-appointed establishments offering a comfortable standard of accommodation

** Well-maintained establishments offering an average standard of accommodation

* Establishments offering a basic standard of accommodation.

You should confirm hotel bookings before the traveller leaves home.

Using travel agents

The main advantage of using a travel agent for booking your overseas trip is that travel agents can use a computerised reservation system for all bookings. This provides an electronic equivalent of a travel marketplace: travel agents can compare airline and travel services and choose the best one for their customers. They key in a travel request and the system lists the flights available, the best connections, the most suitable flight times, and seat availability. The agent makes the booking, with instant confirmation. The system then automatically issues a ticket, invoices the customer, and updates the airline records.

The two major systems in use in Australia are **Galileo** and **Sabre**. The database stores information on the following things, as well as other travel data:

- flight schedules
- prices
- hotel accommodation
- theatre bookings
- routes
- baggage allowance
- hire cars
- sightseeing tours
- departure times
- flight connections
- travel insurance

A travel agent will also prepare an itinerary and can advise you on health regulations for the country being visited and the most convenient stopovers.

Frequent business travellers will have an account profile with a travel agent. Details are recorded of a traveller's preference for class of air travel, car-rental company, and hotel category. Bookings are made by the travel agent and payment is made by credit card.

The disadvantage of using a travel agent is that it can be expensive. There are things that you may still have to do, such as:

- confirming hotel bookings
- typing up a comprehensive itinerary
- confirming flight bookings
- getting useful information about the country being visited.

Planning and organising an itinerary

When you have organised all the details of the trip, you should prepare a draft itinerary, which can be finalised and handed to the traveller a day or so before the date of departure in case changes or alterations are necessary. An itinerary is a summary of the entire trip, which sets out clearly and concisely:

- departure and arrival times with flight numbers;
- name and address of motel/hotel accommodation;
- times and addresses of appointments and name of the person being met;
- dates and times for relaxation and holiday, if relevant.

Effective time management of a business trip

It is important when planning and organising an itinerary to make allowances for the following:

- jetlag—if the trip is long (e.g. to Europe or the USA), a rest day on arrival should be allowed if possible to recover from the air travel;
- time spent travelling from the airport to the motel/hotel, to appointments and in between appointments;
- the fact that it often takes longer to achieve things in an unfamilar environment.

What, where, why and how

Appointments or meeting details are often set out separately at the end of the itinerary, but it is often better to incorporate them in the itinerary so that everything is set out next to the times in diary format. It is then easier for the traveller to see:

- what has to be done (e.g. attending a meeting or appointment);
- where it is going to be done (name of place, address and location);
- why it has to be done (e.g. to discuss products);
- how to get there (e.g. by taxi, hire car).

An example of part of an overseas trip itinerary is given in Figure 14.4. Appointments and meetings or attendance at conferences should be confirmed in writing well before the traveller's departure time.

Methods available for payment of business trip

There are various methods of paying for the trip. If you work for a large organisation, payment is often made by the corporate travel department to the travel agent for such items as air tickets, hotel accommodation and car hire. But for personal spending on arrival in the overseas country, other means of payment are needed.

Cash. A certain amount of cash (a limit is sometimes specified by the country and the amount you take is usually noted in the passport) may be required for expenses (e.g. taking taxis immediately on arrival in the country).

Credit cards. It is common these days for overseas travellers to use credit cards. The advantages are that they reduce the amount of cash and travellers' cheques that needs to be carried, and the traveller's source of money in Australia can be accessed immediately if there is a sudden need for extra finances.

Travellers' cheques. These are an alternative if access to credit card facilities is not available in the country being visited. Travellers' cheques can be bought from banks or through a travel agent prior to leaving. Travellers' cheques can be cashed by banks, hotels and travel agents. The disadvantage of travellers' cheques is that they can be lost or stolen and take time to replace.

Reimbursement of expenses. There are always some expenses that require reimbursement after the trip. A record should be kept by the traveller of business expenses and relevant personal expenses, which will later be reimbursed. These may include:

ITINERARY FOR Edward Gerrard
Adelaide–Singapore–Malaysia
6 July–8 July, 199X

JULY

<u>Thursday 6</u>

1230 hours	Arrive ADELAIDE INTERNATIONAL AIRPORT terminal and check your baggage into QANTAS AIRWAYS counter for FLIGHT QF95 to SINGAPORE Proceed to the departure lounge to obtain your boarding pass. Check your seat allocation.
1430 hours	Depart ADELAIDE
2005 hours	Arrive SINGAPORE On arrival in SINGAPORE, you will be met at the airport and transferred to your accommodation at: HOLIDAY INN SINGAPORE 25 Scotts Road SINGAPORE 0922 Tel 7377966
Friday 7	Collect your AVIS hire car and drive to MERSING, where accommodation has been reserved for you at: MERSING MERLIN INN 11/2 Mile Jalan Endau JOHORE, MALAYSIA Tel (072) 7913123
Saturday 8	Proceed to KUANTAN. Accommodation has been booked for you at: MERIDIEN LODGE 1219 Kuantan Road KUANTAN, MALAYSIA Tel (9) 49879

Fig. 14.4 Part of an overseas trip itinerary

- local transportation
- meals and snacks
- postage
- telecommunications expenses
- laundry services
- entertainment.

Activity 14.2

Foreign currency and exchange rates

When organising methods for payment of a business trip, an important aspect is the currency exchange rate. How much buying power is your traveller going to have when Australian dollars are exchanged into the currency of the overseas country she/he is visiting?

Below are examples of exchange rates in major overseas countries. These rates change every day and the current exchange rate may be obtained from major newspapers and banks.

Select 10 countries from the list below and calculate the quantity of foreign currency you would receive in exchange for 500 Australian dollars.

Currency exchange rates

One Australian dollar will buy:	
USA, dollar	0.7845
UK, pound	0.5083
Canada, dollar	1.0818
China, yuan	6.4218
Denmark, krone	4.5347
Europe, Curr Unit	0.6262
Fiji, dollar	1.1055
Finland, markka	3.5359
France, franc	4.0093
Germany, mark	1.1753
Greece, drachma	186.470
Hong Kong, dollar	6.0997
India, rupee	27.7710
Indonesia, rupiah	1859.0
Ireland, punt	0.4909
Italy, lira	1197.0
Japan, yen	85.2200
Malaysia, ringgit	1.9683
Norway, krone	5.0754
Singapore, dollar	1.1124
South Africa, rand	3.6316
Spain, peseta	99.1500
Switzerland, franc	0.9517
Thailand, baht	19.9500

Work in the office while the traveller is away

Routine work can be handled by you while your boss is away, and you can catch up with the work you put on one side while you were organising the trip. Very urgent mail will be sent to your boss by fax, and you will be able to keep in touch on urgent matters by telephone or e-mail. Make a note of the matters you have dealt with and the action you have taken.

The travel folder

All documentation and information required by the traveller should be placed in a folder for use on the trip. This includes:

- air tickets
- writing materials
- departure tax stamps
- health documents
- confirmation of hotel reservations
- business cards
- confirmation of car hire
- passport with entry visa stamp or document
- health insurance and vaccination certificates
- travel itinerary
- travel funds (travellers' cheques, credit cards, cash)
- excess baggage vouchers
- international driver's licence
- travel insurance document
- letters, memos, conference notes and other working documents
- list of appointments and meetings, or details of conference (if not on itinerary)
- list of home and overseas telephone/fax numbers
- list of addresses and special instructions.

Useful information

In addition to the documents and itinerary in the folder, attached to the itinerary should be useful information about the country being visited; for example:

- the language used
- business hours
- public holidays (in case banks and other public utilities are closed)
- essential services (opening hours of pharmacies, doctors, post offices)
- essential telephone numbers (e.g. airports, Australian Consulate, ambulance, police, taxi)
- radio and TV service information
- climate
- clothing and dress requirements
- whether smoking is prohibited in public places

- customs regulations
- cultural information. (Recognition should be made of cultural and religious practices when visiting another country so as not to offend the people you are visiting.)

Case study 14.2

Travel planning

You are the personal assistant to the general manager of Business Stationery Supplies Pty Ltd. Your boss, Edward Gerrard, is planning an overseas trip to Singapore and Malaysia, as he is hoping to expand the business into the Asian market. He is attending a conference for four days in Singapore, the theme of which is 'Expanding markets in Asia'. He is then taking a touring holiday in Malaysia for four days. He has asked you to plan and organise the trip for him. You are requested to produce a travel folder containing all documents and information for the trip.

Obtain a travel brochure for Singapore and Malaysia from a travel agency and use this to plan Edward Gerrard's trip.

1. Use the checklist on Case study sheet 14.2 of your workbook to help you plan this trip in advance.
2. Organise a travel folder for the trip containing the following documents (the title page, contents page, itinerary and information sheet must be typed):
   - title page
   - contents page
   - an itinerary for the trip, including
     —flight details (obtain information from a travel agent)
     —departure and arrival times
     —accommodation details (use the brochure)
     —details of times of conference sessions for the four days
     —taxi/car-hire
     —tour/holiday details (from brochure)

   (Make sure you organise the daily itinerary details and activities effectively so that there is enough time for conference sessions, meals and other activities.)
   - an information sheet that will be useful to Edward while on the trip
   - the travel documents (passport/health etc.) that will be required for the trip.
3. Identify what kind of arrangements you would make for the payment of hotel accommodation, for spending money, for booking and paying for air tickets.
4. Identify the documents you would need to obtain from the bank for credit transfers, currency and travellers' cheques.
5. State the methods you would use to organise and make bookings for this overseas trip.

Summary

- The first stage in planning business travel is to clarify why the person or persons want to travel, who is going and where they are going.
- Planning an interstate business trip for two or three days is likely to involve organising the airline tickets; car-rental voucher; hotel accommodation; appointments for meetings; a travel itinerary; and travel insurance.
- When planning and organising an overseas business trip, advance planning is required so that nothing is forgotten. This includes: confirmation of meetings and appointments or conference attendance; obtaining documentation such as passport, visas and health documents (if required) and travel insurance.
- Bookings may be made through a travel agent, and you will need to specify the class of air travel: first class (which is the most expensive), business class (which is aimed at business travellers), or economy class, which is the least expensive but can be noisy and uncomfortable for business travellers wanting to work during the flight.
- Discounts may be offered when buying air tickets for: advance purchase, where a return ticket is booked and paid for 14 days prior to departure; group fares, which can be arranged for people travelling in a group; and special offers, which include free car-hire, stopovers, and special rates at different times of the year.
- Before booking accommodation you will need to find out what standard of hotel/motel is required and whether there are any special requirements for the traveller.
- An itinerary for the traveller should set out clearly and concisely departure and arrival times with flight numbers; the name and address of motel/hotel accommodation; times and addresses of appointments and name of the person being met; and the dates and times for relaxation and holiday, if relevant.
- It is important when planning and organising an itinerary to make allowances for jetlag, time for travelling from the airport, and time in between appointments. Appointments or meeting details are often set out separately at the end of the itinerary, but it is often better to incorporate them in the itinerary so that everything is set out clearly in diary format.
- The advantages of using a travel agent for booking an overseas trip are that everything can be done electronically and the bookings are made with instant confirmation. The system automatically issues a ticket and invoices the customer.
- Methods available for payment of a business trip include cash for small expenses in the country visited, credit cards, travellers' cheques. Many expenses will require reimbursement after the trip.
- All documentation and information required by the traveller should be placed in a travel folder for use on the trip.

Review questions

(These may be used for group discussion.)

1. What is the first stage in the planning of business travel?
2. What advance planning do you need to make for an overseas business trip?
3. List the documentation you would need for an overseas business trip.
4. Name the different classes of air travel.
5. List special accommodation requirements which may be required by a traveller.
6. List the different motel/hotel rating definitions.
7. Name the two major computer reservation systems in Australia.
8. What information do these systems store?
9. What information do you need to include in an itinerary?
10. Explain briefly the different methods of paying for a business trip.
11. What is a travel folder? List the documentation and information placed in the folder for use by the traveller on the trip.
12. Describe briefly information that may be useful to a traveller when staying in an overseas country.

Competency self-assessment and/or tutor/supervisor assessment

When you have finished reading this chapter, and have completed the activities chosen by your tutor/supervisor (**practical in class, at work or theoretical**), fill in this self-assessment. Then hand it to your tutor/supervisor for official assessment.

Learning outcomes	**Self-assessment**		**Tutor/supervisor assessment**	
I have successfully completed the following learning outcomes	**Yes** (please tick)	**No**	**Yes** (please tick)	**No**
1. Make travel and accommodation arrangements and bookings from a given itinerary in a workplace-related context				
1.1 clarify purpose of travel and itinerary	☐	☐	☐	☐
1.2 identify travel and accommodation requirements	☐	☐	☐	☐
1.3 make bookings in line with identified needs	☐	☐	☐	☐
1.4 check documentation and rectify discrepancies	☐	☐	☐	☐
1.5 finalise itinerary and forward details of arrangements to traveller within the designated timelines	☐	☐	☐	☐
2. Scheduling appointments	☐	☐	☐	☐
Date:	Tutor/supervisor signature:			

section five

FINANCE

chapter 15

Financial source documents

chapter 16

Petty cash

chapter 17

Bank deposits

CHAPTER

15

FINANCIAL SOURCE DOCUMENTS

FIN101A, NOS134.V2, 7365U

objectives

The purpose of this chapter is to provide you with the knowledge and skill to identify, prepare and check business financial source documents, in order to complete the following learning outcomes:

1. **Identify, prepare and record business financial documents for the purchase of goods/services according to enterprise procedures**
   1.1 complete purchase/internal stock requisition for authorisation by nominated person according to enterprise guidelines
   1.2 identify and prepare appropriate documents for the purchase of goods/services according to enterprise procedures
   1.3 identify discrepancies between incoming invoices and supporting documents and report to nominated person/section for resolution
   1.4 check incoming invoices for accuracy and identify errors to be reported to nominated person/section for correction/resolution
   1.5 rectify discrepancies and errors on invoices as directed
   1.6 prepare payment documentation for correct and authorised invoices within designated timelines
   1.7 copy and file documents for auditing purposes
   1.8 resolve creditor enquiries or refer to nominated person/section for resolution

2 **Identify, prepare and record business financial documents for the sale of goods/services according to enterprise procedures**
   2.1 identify and prepare accurate documents for the sale of goods/services according to enterprise procedures
   2.2 check outgoing documents against relevant documentation to ensure accuracy of contents
   2.3 distribute invoices to nominated person/section for verification
   2.4 despatch verified invoices and other sales documentation to debtors within designated timelines
   2.5 complete receipts for payment of debtors' invoices and present to nominated person for authorisation
   2.6 copy and file documents for auditing purposes
   2.7 resolve debtors' enquiries or refer to nominated person

ORGANISATIONAL POLICIES AND PROCEDURES

To function effectively, businesses should have **organisational policies and procedures manuals** that establish, for example, the way in which cheques, invoices and other business items are to be prepared. Without well-documented directions, a change of staff could mean a change of procedure to that desired by the new staff member, and this can be confusing for other members of the organisation who have become familiar with the existing methods. There are many different processes for ordering goods. Standardised, well-documented procedures ensure that every member of staff knows the way they should be carried out.

This chapter uses source documents from two companies, A1 Secretarial Services and Business Stationery Supplies Pty Ltd. Such companies would have examples of all forms, together with descriptions for their completion, in their detailed policy manuals. For your purposes, assume that the document descriptions given in the following pages contain the detailed descriptions from the policy manuals of these two companies. Their policy manuals would also state the timelines within which invoices, credit notes and statements must be prepared and sent to customers. For the purpose of this chapter, it will be assumed that invoices and credit notes will be sent to the accounting department of the receiving firm on the same day as the goods are sent or returned. Statements must be posted no later than five working days after the end of each month.

A business would also have a range of policies that affect the way employees operate. For example, the occupational health and safety policy would ensure that you are working in a safe environment.

Therefore, as an employee, you must familiarise yourself with the policies and procedures peculiar to the section or department in which you operate. Some of these will apply to the business as a whole, whereas others will be specific to one section.

AUSTRALIAN ACCOUNTING AND AUDITING STANDARDS

Australian Accounting and Auditing Standards set certain requirements that must be adhered to by an enterprise. Many of the documents presented in this chapter are what are referred to as **source documents**. A source document is one that is used to enter information into the records of an enterprise. These documents are 'accountable' documents, meaning that they must be accurate, and kept for a period of seven years for auditing purposes. For example, invoices are source documents, and therefore they are also accountable documents. Guidelines given in this chapter adhere to the requirements of the Australian Accounting and Auditing Standards.

PREPARATION OF FORMS

When preparing forms used in this chapter, you must be aware that styles will differ from one business to the next. However, regardless of individual style, each form should be identifiable by the content.

What will also differ between one business and another is the number and type of departments. A large organisation might have separate departments for purchasing, receiving and accounting; on the other hand, smaller organisations may combine the receiving and purchasing departments into one. Greater accountability is obtained when separation of duties is possible; however, it is often impossible for small businesses, with low staff numbers and small budgets, to do this.

REPORTING MECHANISMS/SYSTEMS

Reporting mechanisms/systems will differ, depending on the needs of the enterprise. Some will use a paper-based system while others will use electronic systems. No matter what system is used, consideration will need to be given to the accounting system used by the enterprise.

Paper-based systems versus electronic systems

The financial source documents presented in this chapter use a paper-based system. In other words, the documents will be prepared by hand.

Many large organisations use electronic means, particularly computer technology, to generate source documents. These computer-based systems are often tied in via computer programs to other accounting systems. For example, a computer-based system for generating **receipts** of money might automatically update the cash receipts records of an organisation, removing the need to physically enter extra information.

Enterprise accounting systems

Organisations will develop their own reporting mechanisms, which will involve the use of **journals** to record like transactions. The source documents prepared in this module form the basis for further study in journal preparation.

Source documents are an important link in the chain of events that leads to reporting because it is here that the information collected is recorded and checked, and this forms the basis of reports. An example might be seen in an organisation that wants to use a standardised **invoice**, but has three different departments (for example, it might have a margarine department that sells all its margarine, a chicken department that sells all its chickens, and a fertiliser department that sells all its fertiliser). This firm might want to prepare reports that detail the individual operations of each of these departments; in this case, any invoice used must clearly indicate which department each recorded item comes from. This could be done with coding (each department having a specific code, recorded on the invoice) or—possibly more easily—with the use of different columns for the amounts relating to individual departments. The construction of the source document must meet the needs of the reporting functions and be easily understood by the customer.

Security

Security must be taken seriously in all enterprises. When preparing financial documents, one aspect of security that must be practised relates to separation of duty. Separation of duty means that an employee performing one task within an integrated series of tasks must not also perform other duties related to the same sequence of events. An example would be of an employee who receives cash, and writes the receipt, but should not write the information into a cash receipts journal.

Separation of duties makes it less likely that theft or fraud will occur, because if, for example, a receipt is incorrectly written, another person would find the error while recording the information.

Accountability of financial source documents

Preprinting source documents assists with accountability by providing a means whereby each document can be recorded in numerical sequence. Cancelled documents should also be recorded, and listed as 'cancelled'. The original and duplicate pages of documents should be cancelled and filed in accordance with organisational procedures.

Activity 15.1

Theory review

Write your answers to the following on a separate sheet of paper:

(a) Why does a business need to have policies and procedures manual?
(b) What is the relationship between the information obtained from source documents and the reports a business might want to generate?
(c) If a business uses a paper-based system for the preparation of source documents, how would such documents be prepared?
(d) What is the most commonly used electronic system for source document preparation?
(e) Discuss the need for security and accountability of financial source documents.

IDENTIFYING AND PREPARING APPROPRIATE SOURCE DOCUMENTS FOR PURCHASES, SALES, DELIVERIES AND REMITTANCES

Purchasing goods from the point of view of the purchaser

Look at the flow chart under the heading *Tracing documents in the purchase and sale of goods* (page 413). It documents the process of purchase, supply and payment of goods through departments of both the purchasing and the supplying firm. Refer to this chart as you progress through the chapter.

The documents that must be prepared by the **purchaser** when ordering goods are a purchase requisition and an order form. The seller will prepare a delivery docket and an invoice. If required, **credit notes** will also be prepared by the seller.

The original invoice received from the seller, together with original credit notes if required, will be used as the source documents for recording purchases, purchase returns and allowances. The documents prepared by the purchaser will be used to ensure that deliveries are correct, according to the order, and that documentation received from the seller is accurate.

Prior to commencing the practical activities you should be aware that 'liquid paper' and erasers are not permitted in accounting practices. This is because any corrections altered in this way could be seen to be covering up fraudulent acts.

Purchase requisitions

Every department in a business needs to keep a check on how quickly inventories (stock or goods) are being sold. If a business were to run out of saleable stock, or to have stock that had deteriorated in quality or value owing to age or obsolescence, then profits would ultimately fall because either consumers would not want to buy its products or they would not be prepared to pay reasonable prices for them. All departments should have a **reorder point** for each product, which is simply the point at which it will be reordered. This avoids having to store too many inventories at once, but ensures that supplies of the product do not run out. The reorder point becomes apparent over time and is determined by such things as:

- how quickly the product turns over (is sold);
- how long it takes for the product to arrive after it has been ordered (lead time);
- the method of transport used (air transport is fast, rail is slower).

When the reorder point is reached the department requesting the purchase will generate a *purchase requisition*, which is an *internal document* that advises the department responsible for ordering goods (the purchasing department) which goods are required.

A purchase requisition is illustrated below. Note that it contains:

1. the number of the requisition;
2. the order date;
3. the date goods are required by;

4. the department or section requiring goods;
5. the name of the person requesting goods;
6. particulars of the goods required.

No. 1076 (1)	**Purchase Requisition**	Date: *2 May 19XX* (2)
	A1 SECRETARIAL SERVICES	

Date required: *16 May 19XX* (3)

Supply to: *Miscellaneous Department* (4)

Requested by: *H. Black, Manager* (5)

	Quantity / unit	Description
(6)	*1 doz*	*Biros, blue*
(6)	*3 doz*	*Lead pencils, HB*
(6)	*3 boxes*	*Highlighter pens, green*
		Per: (x) ....

Note: (x) indicates that a signature is required here, and is used throughout this chapter.

Purchase orders

After approval and the selection of a supplier the purchasing department will raise a purchase order (order form), which is an *external document.* (An external document is one that is sent to an individual or organisation 'external to' the organisation that prepared the document.) This is sent to the supplier of the product, advising that the goods are required.

Suppliers are selected on the basis of price, quality, availability of inventories, terms, discounts, the date required and their believed or known reliability and/or after-sales services.

As illustrated in the following example, the purchase order should contain:

1. name and address of the buyer (letterhead);
2. a purchase order number;
3. name and address of the seller or supplier (the boxed section);
4. full details of the goods being ordered, including size, type, colour and code number (if known);
5. number of days before payment is required (nett means payment in full—no discount);
6. delivery instructions;
7. date goods required by;
8. transport to be used;
9. signature of an *authorised person*);
10. date of preparation;
11. statement indicating quoted price includes GST.

(1) **A 1 SECRETARIAL SERVICES**
52 Letterfield Lane
CANNON HILL QLD 4170

Purchase order
No. 33 (2)

To:
(3) *Business Stationery Supplies Pty Ltd*
2109 Old English Road
COOLANGATTA QLD 4225

Please supply the following goods and charge to our account.

	Quantity	Product number	Description	Unit price $ (incl. GST)	Amount $
(4)	*1 doz*	*A3*	*Biros, blue*	*8.00 doz*	*8.00*
	3 doz	*C3*	*Lead pencils, HB*	*3.00 doz*	*9.00*
	3 boxes	*A15*	*Highlighter pens, green*	*5.00 doz*	*15.00*

(5) Terms: *Nett 30 days*
(6) Deliver to: *33 Raidy Road, Uralunga* (9) Per: (x) ....
(7) Required by: *16 May 19XX* (10) Date: *14 May 19XX*
(8) Transport: *Tranters Transport* (11) GST included in quoted prices.

Comments

- The product number will be included if it is known (Appendix 15.1 price list).
- The price (Appendix 15.1 price list) is usually included on the purchase order. If it has been raised since a quote was made, the supplier is not obliged to sell at the price quoted but should contact the customer and advise of the change, and the reason for it. A new purchase order reflecting the change might be requested. To cover price rises, suppliers should specify a time limit on quotes, say 30 days.
- Any special price or condition should be noted on the purchase order. The terms might state that a discount is to be given if payment of the full amount is received within a certain period (such as 30 days). Another example of a term is 2.5 per cent if paid within 60 days. In other words, a discount of 2.5 per cent of the total will be allowed if payment is received within 60 days.
- The 'Deliver to' line might state a department, or even a different location.
- Note that transport costs will be added by the supplier if they apply, but this should be advised when orders are placed.

Skill development 15.1

Purchase requisitions and purchase orders

Use the purchase requisitions that follow to prepare purchase orders in your workbook. Note that each purchase requisition number requires a separate purchase order. There are six different purchase orders that must be completed.

(a) The purchasing department of A1 Secretarial Services has chosen Business Stationery Supplies Pty Ltd, 2109 Old English Road, Coolangatta, QLD 4225, as supplier for the items requisitioned on the following purchase requisitions.

Complete the purchase orders in your workbook using this additional information:

- dates are as per purchase requisitions;
- terms are nett 30 days;
- Gosford's Transport is to be used for delivery;
- the purchase orders begin at no. 34.

No. 1077	**Purchase Requisition**	Date: *17 May 19XX*
	A1 SECRETARIAL SERVICES	
Date required: *24 May 19XX*		
Supply to: *Retail Office Department*		
Requested by: *J. Jones*		
Quantity / unit	Description	
10 only	*Calculators, Tandy*	
1 doz	*Rulers, plastic (30 cm)*	
	Per: (x) *J. Jones*	

No. 1078	Purchase Requisition	Date: 18 May 19XX
	A1 SECRETARIAL SERVICES	

Date required: 25 May 19XX

Supply to: Retail Office Department

Requested by: J. Jones

Quantity / unit	Description
2 boxes	Highlighter pens, green
3 boxes	Fountain pens, black
	Per: (x) J. Jones

No. 1079	Purchase Requisition	Date: 19 May 19XX
	A1 SECRETARIAL SERVICES	

Date required: 26 May 19XX

Supply to: Retail Office Department

Requested by: J. Jones

Quantity / unit	Description
5 boxes	Felt tip pens, black
10 reams	Paper, A4, pink
	Per: (x) J. Jones

No. 1080	Purchase Requisition	Date: 23 May 19XX
	A1 SECRETARIAL SERVICES	

Date required: 30 May 19XX

Supply to: Retail Office Department

Requested by: J. Jones

Quantity / unit	Description
4 reams	Beige marble paper (200 sheets)
1 roll	Computer paper, standard, continuous (1000 sheets)
3 only	Pencil sharpeners, desktop
	Per: (x) J. Jones

No. 1081	Purchase Requisition	Date: 24 May 19XX
	A1 SECRETARIAL SERVICES	

Date required: 31 May 19XX

Supply to: Retail Office Department

Requested by: J. Jones

Quantity / unit	Description
5 boxes	Highlighter pens, yellow
10 boxes	Computer disks $3\frac{1}{2}''$
10 only	Rulers, wood (30cm)
	Per: (x) J. Jones

No. 1082	Purchase Requisition A1 SECRETARIAL SERVICES	Date: *30 May 19XX*
Date required: *6 June 19XX* Supply to: *Retail Office Department* Requested by: *J. Jones*		
Quantity / unit	Description	
10 pkts *5 pkts* *1 only*	*Envelopes, 228 x 101 mm (100)* *Envelopes, beige, logo, 228 x 101 mm (100)* *2-drawer filing cabinet* Per: (x) *J. Jones*	

Check the purchase orders before continuing; you will be required to use them in the next Skill development.

(b) The table below provides extension exercises for which solutions are given for invoices only. These need be completed only if you require further reinforcement. However, you may complete all documents and ask your supervisor to check your responses.

(Retain all documents, as many are used in subsequent Activities or Skill developments.)

Goods required

Purchase order no.	*Date required by*	*Quantity/ unit*	*Description*
40	6 June	3 boxes	Fountain pens, black
		1 only	Computer paper, standard, continuous (1000 sheets)
		10 boxes	Guides for files (50 per box)
41	6 June	20 only	Pencil sharpeners, plastic
		6 boxes	Papermate pens, black
		3 only	Staplers, front-loading
42	7 June	10 boxes	Manila folders, plain
		10 boxes	Manila folders, yellow
		20 boxes	File inserts
		5 boxes	Suspension files
43	7 June	2 only	Filing cabinet, 4-drawer
		10 reams	A4 paper, pink
		5 only	Disk containers (3½ inch)

Selling goods from the point of view of the supplier

Documents that must be prepared by the seller include delivery dockets, invoices and, if applicable, credit notes. The purchaser, as has been explained, will prepare purchase requisitions and order forms.

Duplicate invoices and *duplicate credit notes* will be the source documents for recording sales and sales returns and allowances. Original invoices and credit notes will have been sent to the purchaser.

Delivery dockets

A *delivery docket* (delivery note or delivery slip) is an *external document*; it might not be numbered, and is made out by the *seller*. It usually accompanies the invoice, which is made out at the same time. Delivery dockets are made out for invoices only when goods are being delivered.

The delivery docket is used by the transport company, and contains only limited information. A brief description of the goods being delivered may be included, but sometimes—to avoid the temptation to steal—goods may only be listed as, for example, '1 carton'. However, firms generally itemise goods being delivered.

A delivery docket completed for the sample purchase order (on page 390) is given below. Note that it contains:

1. the delivery date;
2. the name of the buyer;
3. the name of the seller;
4. a brief description of the goods;
5. the docket number (usually preprinted).

Delivery Docket (1)

Date. *16 May 19XX*

To: *A1 Secretarial Services*
(2) *52 Letterfield Lane*
CANNON HILL QLD 4170

Received from: *Business Stationery Supplies Pty Ltd*
(3) *2109 Old English Road*
COOLANGATTA QLD 4225

In good order and condition:

(4)
1 doz biros, blue
3 doz lead pencils, HB
3 boxes highlighter pens, yellow

No. 123 (5) Received by: (x) ....

Comments

- The 'Received by' line is not filled in. The person who accepted these goods in the receiving department would sign to state that they were in good order and condition (which implies that they must be opened and checked) and that they 'match' the purchase order. (You will note that the colour of the highlighter pens is incorrect. Ignore this point for the moment.)
- Where differences occurred, these would be noted on the delivery docket prior to signing it.

Skill development 15.2

Delivery dockets

Business Stationery Supplies requires you to complete delivery dockets in your workbook, using the purchase orders prepared in Skill development 15.1. The dates on the delivery dockets will be the required delivery dates, and the delivery docket numbers will begin at 124. (Check your delivery dockets before continuing.)

Invoices

All incoming goods go first to the receiving department. The purchase order would accompany them, and a 'goods received' slip would be prepared.

An **invoice** (similar to the example on page 398) would be prepared by the supplier and the original (top copy) sent to the purchaser; it might arrive with the goods or it might be sent directly to the accounting department, where it would be used as the source document for the purchases journal. Either way, a copy of the *purchase order*, the *delivery docket* and the *goods received slip* would also be sent to the accounting department, which would compare them with the invoice.

The example on page 398 shows that an invoice should contain:

1. the name and address of the seller (letterhead);
2. the name and address of the buyer (the boxed section);
3. the invoice number (preprinted);
4. the delivery date;
5. the number of the purchase order to which it applies;
6. transport details;
7. full details of the goods being sold;
8. any extra delivery instructions;
9. terms and conditions (such as discount) that may apply to the transaction;
10. the rate and amount of trade discount given (if applicable) (usually given to tradespeople who use the product in the providing of services to make a profit);
11. delivery charges, if applicable;
12. the final total or amount owing;
13. the GST calculation (see page 15).

Note that all calculations are rounded to the nearest cent. Note also that, depending on company policy, all calculations may be made to the nearest 5 cents.

Invoice

BUSINESS STATIONERY SUPPLIES PTY LTD
(1) **2109 Old English Road**
COOLANGATTA QLD 4225
ABN 19 112 309 432

(2) *A1 Secretarial Services*
52 Letterfield Lane
CANNON HILL QLD 4170

Inv. no. (3) **201**

TAX INVOICE

	Date	**Purchase order no.**	**Method of delivery**
(4)	*16 May 19XX*	*33* (5)	(6) *Tranters Transport*

	Item	**Qty ordered**	**Description**	**Unit price**		**per**	**Extension**	
(7)	*1*	*1 doz*	*Biros, blue*	*8*	00	*doz*	*8*	00
	2	*3 doz*	*Lead pencils, HB*	3	00	*doz*	9	00
	3	*3 boxes*	*Highlighter pens, yellow*	*5*	00	*box*	*15*	00

	Subtotal	32	00	
(8) Deliver to: *33 Raidy Road Uralunga Qld*	10% Trade discount	3	20	(10)
	Subtotal after trade discount	*28*	*80*	
(9)	Plus delivery charges			(11)
Terms: *Nett within 30 days*	Invoice total $	*28*	*80*	(12)

(13) The total price includes GST of $*2.62*

Comments

- The date is when the invoice is prepared.
- 'Purchase order no.' is the number on the purchase order (look back to the example under the heading *Purchase orders*).
- 'Method of delivery' will be taken from information on the purchase order, but a business may choose its own method if this is not stated.
- Any special details will be noted in the space at the bottom left-hand corner (e.g. 'Deliver to 33 Raidy Road, Uralunga, QLD). This section is not used to direct delivery to a particular department, because goods would be delivered initially to the receiving department, which would forward them to internal departments. Rather, this section is used where, for example, the firm has a branch at another location.

  If *urgent delivery* is required, this might be indicated in large print across the bottom of the invoice.
- Trade discount is not recorded in the sales or purchases journals, but is used as a way to encourage tradespeople to purchase goods from a particular company. It is calculated as a percentage of the listed price and is deducted from the amount owing.

- 'Nett within 30 days' means that the account must be paid in full within 30 days. Sometimes a discount is given if it is paid within a set period.

Skill development 15.3

Invoices

Business Stationery Supplies requires you to complete invoices (see your workbook) using the purchase orders prepared in Skill development 15.1. (Use the price list in Appendix 15.1, page 443, in preparing the invoices.)

The following information is given:

(a) The date of each invoice will be the required delivery date.
(b) Trade discount is to be 10 per cent on all invoices. (Note, however, that different discount percentages might apply.)
(c) Invoices are to be numbered sequentially, beginning at 202.

(Check the invoices before continuing; you will be required to use them in other Skill developments.)

Goods and Services Tax

The Goods and Services Tax is a tax levied by the federal government under the *Goods and Services Tax Act.* Most goods incur this tax, which is paid by the consumer. Some products are GST-free, for example basic food, cars for the disabled, child care, some travel and transport.

Many businesses will collect and forward the tax on behalf of the government. Each business with a turnover of $50 000 or more must register and will receive an Australian business number (ABN). The ABN will need to be quoted when this consumer tax is forwarded to the government, and the number will be quoted when a business with an ABN purchases products for resale. It will be most important to these ABN businesses to keep accurate records of all GST inputs into their business.

The GST is usually included in the listed price of a product, and will state 'includes GST'. A 10 per cent GST is to be charged on all items, unless GST-free. This will result in 1/11th of the invoiced cost being the GST component. For example, if an article which costs $100.00 prior to GST has a 10 per cent GST added, its new cost will be $110.00 ($100 plus $10). If $10 out of a total cost of $110 is the GST component, then it is equal to 10/110. Mathematically, this converts to 1/11th of the final cost.

When discounts are given, the GST is calculated on the actual price paid (i.e. the discounted price). However, for your purposes, the GST is included in the listed price of the product, and no extra calculation is required by you.

Activity 15.2

Goods and Services Tax

- Calculate the GST component on each of the invoices you prepared in Skill development 15.3. Remember, it will be one-eleventh of the final invoice total (discounted or otherwise). To find this, you will need to divide the invoice total by 11 and your result will be the GST cost.
- Write the value on your invoice, in the space provided.

Ask your supervisor/teacher to verify your calculations before continuing.

Credit notes

A credit note will be prepared by the seller if goods received are damaged or incorrect in some way; for example, the colour, type or size may be wrong, an incorrect amount may be delivered or an inadvertent overcharge made on an invoice. Note, however, that an *undercharge* (a pricing error which results in an invoice total that is less than it should be) would require an additional invoice to cover it.

A credit note is very similar to an invoice in appearance and in content. It will be clearly titled 'Credit note', and either the paper or the ink used in printing should be in a different colour (preferably red) to distinguish it easily from an invoice.

As illustrated on page 401, information on a credit note should include:

1. the credit note number, usually preprinted;
2. the number and date of the invoice to which it relates;
3. the date of the credit note;
4. the reason for credit;
5. the GST value.

Note the following additional points:

- The calculations for a credit note are performed in exactly the same way as for an invoice. The trade discount applicable to the product being returned must also be the same. Note that if a GST is charged on the invoice, it must also be applicable to the credit note.
- Credit notes do not have delivery instructions. If goods are incorrect and/or faulty they will usually be returned with the transport company that delivered them.
- The credit note shown relates to the sample invoice on page 398. The highlighter pens ordered (on page 391) were 'green', and 'yellow' pens were delivered in error (see delivery docket page 396).

Credit Note

BUSINESS STATIONERY SUPPLIES PTY LTD
2109 Old English Road
COOLANGATTA QLD 4225
ABN 19 112 309 432

A1 Secretarial Services 52 Letterfield Lane CANNON HILL QLD 4170	**Credit note no.** ① 99

TAX CREDIT NOTE

Date of invoice	Invoice no.	Date
② 16 May 19XX	201 ②	16 June 19XX ③

Item	Qty returned	Description	Unit price		per	Extension	
1	3 boxes	Highlighter pens, yellow	5	00	box	15	00
Reason for credit: Incorrect colour ④			Subtotal			15	00
			10% Trade discount			1	50
			Subtotal after trade discount			13	50
			Credit note total		$	13	50

The total price includes GST of $1.23

Skill development 15.4

Credit notes

Credit notes are required (see your workbook) with regard to the following orders. (Note that these relate to items in previous Skill developments.)

Date	*Order no.*	*Amount*	*Type*	*Reason for credit*
4	36	1 box	Felt-tip pens, black	Packages damaged in transit
6	35	1 box	Highlighter pens, green	Incorrect number ordered in error
8	37	1 only	Pencil sharpener, desktop	Slightly scratched ($2 allowance given)

Credit note numbers are to begin at 100; the date is now June 19XX.

Statements

When selling goods to customers on credit, a business will each month send a **statement** that details the *balance outstanding* and the status of any amounts outstanding from the previous month. Only basic documents will be considered here.

Information contained in statements will be used by customers to determine how much money should be sent in payment of their debts.

The sample statement opposite will be used as the basis for writing a receipt. Note that a statement will contain:

1. the full name and address of the sender, including ABN;
2. the full name and address of the customer;
3. the month to which the statement refers;
4. an opening balance, if applicable (balance outstanding from previous statements, referred to as A/c rendered);
5. transactions during current month;
6. a section at the bottom showing the final balance and, if necessary, a listing of outstanding balances (e.g. 30 days, or 1 month, 60 days, or 2 months, overdue);
7. the statement 'terms'. For example, '2.5% discount if paid in 30 days' means the customer will deduct 2.5 per cent from the final balance if paid in the 30 days, as stated.

Statement

(1) **A1 SECRETARIAL SERVICES**
52 Letterfield Lane
CANNON HILL QLD 4170
Phone: (07) 3893 1277 Fax: (07) 3893 1778
ABN 17 119 207 198

(2) *Example Pty Limited*
187 Old World Road
FALCON WA 6210

(3) Statement for:
May 19XX

Do not detach—please return this stataement with remittance.

Particulars			Debit	Credit	Balance
May	*1*	*A/c rendered*			*700.00 Dr* (4)
(5)	*7*	*Invoice no. 208*	*50.00*		*750.00 Dr*
	12	*Invoice no. 230*	*60.00*		*810.00 Dr*
	15	*Receipt no. 40*		*600.00*	*210.00 Dr*
	20	*Interest charged*	*20.00*		*230.00 Dr*
	24	*Invoice no. 301*	*70.00*		*300.00 Dr*
	26	*Credit note no. 107*		*50.00*	*250.00 Dr*
	28	*Receipt no. 52*		*100.00*	*150.00 Dr*

Outstanding				
3 months and over	2 months	1 month	Current	Balance
			$150.00	*$150.00* (6)

TERMS: *2.5% DISCOUNT IF PAID IN 30 DAYS* (7) AMOUNT DUE

This statement is made up to the end of the month.
Any payments received or credits issued after this date will be credited on next month's statement.
A receipt will not be forwarded unless specifically requested.

Notice that the 'current' balance includes all values in the statement except amounts of cash that have been paid off 'A/c rendered' amounts. For example, the amount of cash received was $700 (receipt nos 40 and 52). This amount cancels out the 'A/c rendered' balance; therefore, all other values must be current: in other words, they should relate to the month of the statement.

Skill development 15.5

Statements

Complete statements in your workbook, using the following details relating to A1 Secretarial Services for the month of May 19XX. Terms are to be 2.5 per cent discount if paid in 30 days.

B. Boston
236 Main Road
Capalaba, QLD 4157

Date	*Particulars*	*Amount ($)*
1	Account rendered	500
7	Invoice no. 32	100
10	Invoice no. 50	200
15	Interest charged	5
20	Cash received, rec. no. 30	500
25	Invoice no. 62	100
27	Returns, credit note no. 7	20

S. Smith
22 Brown Street
Capalaba, QLD 4157

Date	*Particulars*	*Amount ($)*
1	Account rendered	200
5	Invoice no. 44	300
7	Invoice no. 45	150
10	Returns, credit note no. 9	10
12	Cash received, rec. no. 30	200
20	Returns, credit note no. 12	20
25	Invoice no. 55	100

Cheque requisition vouchers, cheques and cheque remittance advice from the point of view of the purchaser

Cheque requisition vouchers

When a cheque is required, all relevant documentation to prove the validity of the request is given to an authorised supervisor. The supervisor checks the presented documents, and when satisfied issues a cheque requisition voucher (see opposite page). The cheque requisition voucher can be presented to the cashier as authorisation to obtain a cheque for the amount and details stated.

A cheque requisition should contain:

1. the name of the person requesting the cheque;

2. the cheque amount;
3. the name of the payee;
4. the signature of an authorised person.

CHEQUE REQUISITION VOUCHER
(1) Authorisation is hereby given for *B. Smythe* to obtain a cheque to the value of *$480.00* (2) being for *Business Stationery Supplies Pty Ltd* (3) (payment of *May* statement).
Signed: *A Supervisor* (4)

Cheques

Businesses must keep accurate records of all monies paid out. This ensures that items paid for (in the form of expenses) throughout a year can be offset against revenue to calculate profit (or loss). It also ensures that all records relating to cash payments can be maintained.

For security purposes, a cheque book should be kept in a locked cabinet. It is preferable to lock it in a fireproof safe when not in use; it should not be left lying unattended on a desk.

There are three parties to a cheque that must be known:

- the **drawer**, who owns the account (the one who signs the cheque);
- the **payee**, who will receive the cheque and bank it;
- the **drawee**, the bank at which the cheque account is held.

Another term you will need to know is *bearer*: this is the person who will present the cheque to the drawee (bank) for payment.

A **cheque butt** is used as the **source document** for the cash payment records of a business. It is therefore very important that it—as well as the cheque itself—is completed accurately and fully.

Remember that 'cash' refers to either cash or cheques.

Cheques can be of three types: cash, crossed, and crossed 'not negotiable'.

Cash cheques

A cash cheque—that is, one on which the payee is recorded as 'cash'—is the least secure form, because it can be readily cashed by anyone over the counter of a bank. This type of cheque is used for wages, drawings and petty cash. Businesses sometimes need to obtain cash: for example, they may need to pay employees using cash (not cheques). Of course, with the advancement in electronic banking for payroll, the need for cash has diminished greatly.

Crossed cheques

If a cheque is crossed with two parallel lines drawn from top to bottom it can then be passed to another 'payee', who becomes the 'bearer'. This is still an unsafe form, because if it is lost another person can bank it in his or her own account and withdraw the amount in cash. However, it is safer than a cash cheque, because the name of the bearer can be identified via his or her bank account.

Crossed 'not negotiable' cheques

This is the preferred method of writing cheques. Many cheque forms include the printed words 'Not negotiable', and this is a useful safeguard. A cheque that is crossed and not negotiable should be banked in the account of the bearer, as identified on the cheque. 'Not negotiable' draws attention to the fact that the person whose name is written on the cheque is the intended recipient of the cheque. The words 'Account payee only' can also be added to the crossed section, which implies that the cheque cannot be passed to another bearer. Such a cheque eliminates a lot of the security problems that are evident with the use of cash cheques or crossed cheques. It is considered to be reasonably safe; however, no cheques should be left lying around for someone to 'pick up'.

A **cheque** should contain:

1. the cheque number;
2. the date;
3. the name of the drawee (bank);
4. the name of the payee;
5. the name of the drawer;
6. the signature of the drawer;
7. the amount in words and figures;
8. a detailed description of the cheque's purpose.

Hint: Always complete the cheque butt before writing out the cheque.

Study the following examples showing (a) a cash cheque and (b) a crossed 'not negotiable' cheque.

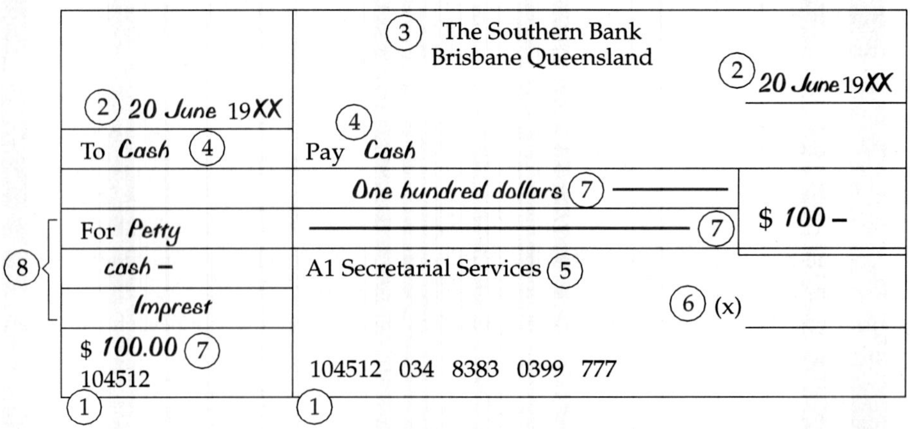

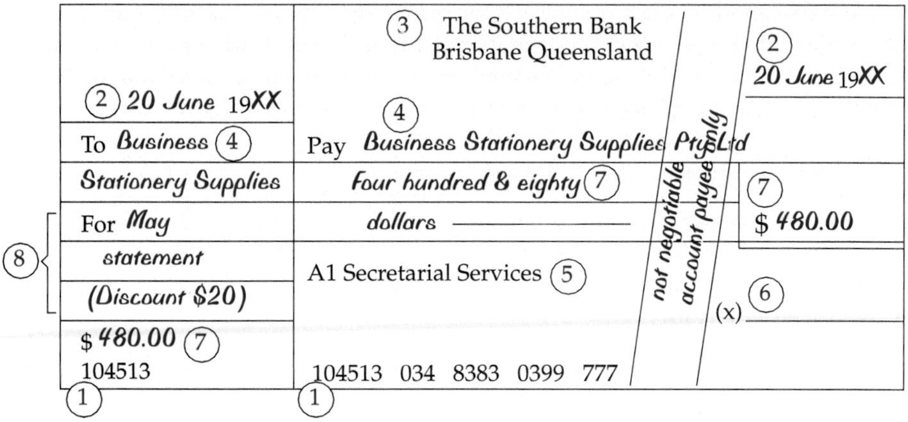

Skill development 15.6

Cheques

A1 Secretarial Services has asked you to prepare cheques (see your workbook) according to the information in the table below. Note that 'Amount' is the actual amount on the cheque; discount will be recorded separately. The date is May 19XX.

Date	*Cheque no.*	*Payee*	*Particulars*	*Amount ($) *Incl. GST*	*Disc. ($)*
2	109	K. Kash	April a/c	270	30
7	110	Cash	Petty cash imprest	100	
8	111	Cash	Drawings	200	
15	112	J. Johns	April a/c	360	40
20	113	Cash	Wages, w/e 20/5	350	
20	114	Electric Company	Electricity	*120	
22	115	KC's Garage	Petrol	*150	
27	116	Cash	Wages, w/e 27/5	350	
27	117	Cash	Drawings	100	
30	118	F. Fred	Vehicle repairs	*55	

Cheque remittance advice

A cheque remittance advice slip is a small document that must by law accompany a cheque when it is forwarded through the mail.

In the example on page 408, which is a cheque remittance advice for the crossed 'not negotiable' cheque shown above, you can see that A1 Secretarial Services has paid an amount of $480. The account was for $500, less a $20 discount that was given for prompt payment. The discount is shown in the adjustment section.

Nowadays most invoices are prepared on a computer. In this case each one has a tear-off section that acts as the cheque remittance advice. When writing a cheque, you may know the value of the cheque amount and the discount. In this instance you would need to add the value of the cheque ($480) to the discount ($20) so that you could enter the statement balance on the cheque remittance advice.

A cheque remittance advice slip should contain:

1. the name and address of the sender;
2. the name and address of the receiver;
3. the details of payment;
4. the sender's signature.

Note that the actual amount of the cheque is the value that will be recorded at the bottom.

Cheque Remittance Advice

(1) **A1 SECRETARIAL SERVICES**
52 Letterfield Lane
CANNON HILL QLD 4170

(2) *Business Stationery Supplies Pty Ltd*
2109 Old English Road
COOLANGATTA QLD 4225

Enclosed please find cheque for payment of account as under:

Statement for month of:[1]
May

Less adjustments

(3)

	$	c
	500	00
Discount	20	00
Cheque total:	480	00

(4) Signature (x) _______________

[1] If the cheque is for something other than the payment of an account (for example, it may be payment of a service fee, where no statement has been received), the reason for the payment will still be written underneath 'Statement for month of'.

Skill development 15.7

Cheque remittance advice

Prepare remittance advice slips (see your workbook) for the cheques completed in Skill development 15.6. Remember that remittance advice slips are prepared only for cheques that will be posted (not for cash cheques).

You will require the following addresses:

- K. Kash, 118 Brode Road, Coorparoo, QLD 4578
- J. Johns, 120 Fence Street, Woolloongabba, QLD 4102
- Electric Company, 239 Redland Bay Road, Capalaba, QLD 4157
- KC's Garage, 335 Forbes Place, Capalaba, QLD 4157
- F. Fred, 377 Ipswich Road, Ipswich, QLD 4305

Receipt form from the point of view of the supplier

Receipts

A receipt is a document that is written out by the *seller*. It acknowledges that an amount of money has been received from (or paid by) the purchaser.

For internal control purposes, to ensure that the correct amounts of money are banked and recorded, *all money received should be evidenced either by a receipt or, if appropriate, by a cash register summary, and the receipt should be prepared at the time goods are paid for* (preparation should not be left until a more convenient time). Receipt forms might be specially printed for a particular business, or a receipt book bought from a stationery supply firm might be used. To be considered an 'official' receipt, however, the document must be correctly prepared to include all pertinent details, either typed or handwritten, and must bear the signature of an authorised person.

The receipts will be used as the source documents for a business's record of cash received. Note that the use of the word 'cash' in business refers to the receipt either of cash (notes or coins) or of cheques.

The example below shows that a receipt should contain:

1. the receipt number;
2. the date;
3. the name of the seller;
4. the name of the buyer;
5. information about the transaction;
6. the discount, if applicable;
7. the amount received;
8. an authorised signature.

A1 SECRETARIAL SERVICES (1)
52 Letterfield Lane (3)
CANNON HILL QLD 4170

Official Receipt
No. 108

Date	Received from	Particulars	Discount	By Cash/Cheque
(2) *15 June 19XX*	*Example Pty Ltd* (4)	*Payment of account* (5)	(6) *$3 75*	(7) *$146 25*

per: (x).. (8) (Signature)

Comments

- 'Received from' will usually be filled in with the name of the person or firm making the payment. However, if the receipt is for a cash sale, the word 'cash' will usually be recorded here, because for accounting purposes it is not necessary to have a record of each person or firm involved in cash transactions.
- The particulars will identify why the money is being received. If it is for a cash sale, the word 'sale' will be inserted. If it is for payment of account (as per the example), this must also be clear for accounting purposes.

Skill development 15.8

Receipts

The following cheques have been received by A1 Secretarial Services during the month of May 19XX. Write a receipt for each payment (see your workbook). Receipt numbers are to begin at no. 5.

The Southern Bank
Brisbane Queensland

not negotiable *account payee only*

4 May 19*XX*

Pay *A1 Secretarial Services*

The sum of *Twenty-five dollars*

$ 25.00

B. Smith

(x) *B. Smith*

1178 999 689

B. Smith paid commission of $25 (includes GST).

The Western Bank
Harrisville New South Wales

not negotiable *account payee only*

7 May 19*XX*

Pay *A1 Secretarial Services*

The sum of *Ninety dollars*

$ 90.00

Ace Pty Ltd

(x) *A. Ace*

978663 00983 999

Ace Pty Ltd paid their April account. Discount of $10 was also given.

The Southern Bank
Brisbane Queensland

not negotiable *account payee only*

12 May 19*XX*

Pay *A1 Secretarial Services*

The sum of *One hundred and fifty dollars*

$ *150.00*

J. Jones

(x) *J. Jones*

078378 997 083

J. Jones paid $150 for the purchase of goods (includes GST).

The Southern Bank
Brisbane Queensland

not negotiable *account payee only*

15 May 19*XX*

Pay *A1 Secretarial Services*

The sum of *Twenty dollars*

$ 20.00

S. Smyth

(x) *S. Smyth*

98767308 973 073

S. Smyth paid a service fee of $20 (includes GST).

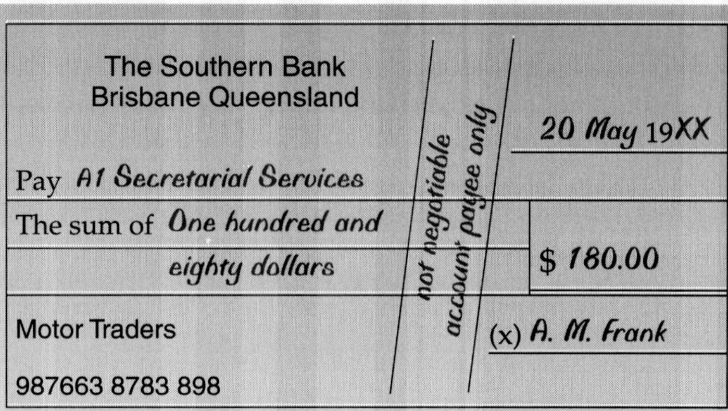

The Southern Bank
Brisbane Queensland

20 May 19XX

Pay A1 Secretarial Services

The sum of One hundred and eighty dollars

$ 180.00

not negotiable account payee only

Motor Traders

987663 8783 898

(x) A. M. Frank

Motor Traders paid their April account. Discount of $20 was also given.

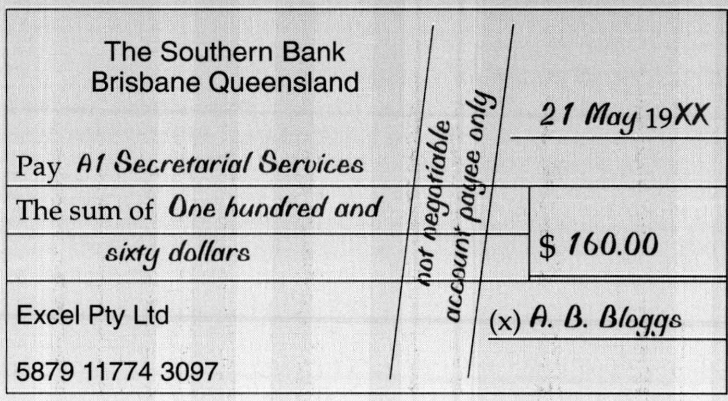

The Southern Bank
Brisbane Queensland

21 May 19XX

Pay A1 Secretarial Services

The sum of One hundred and sixty dollars

$ 160.00

not negotiable account payee only

Excel Pty Ltd

5879 11774 3097

(x) A. B. Bloggs

Excel Pty Ltd paid their March account.

DOCUMENT FLOW CHART

Tracing documents in the purchase and sale of goods

Documentation process for the purchase and sale of goods

The chart opposite summarises the flow of documents through the system in the purchase and sale of goods.

The documentation process for the purchase and sale of goods

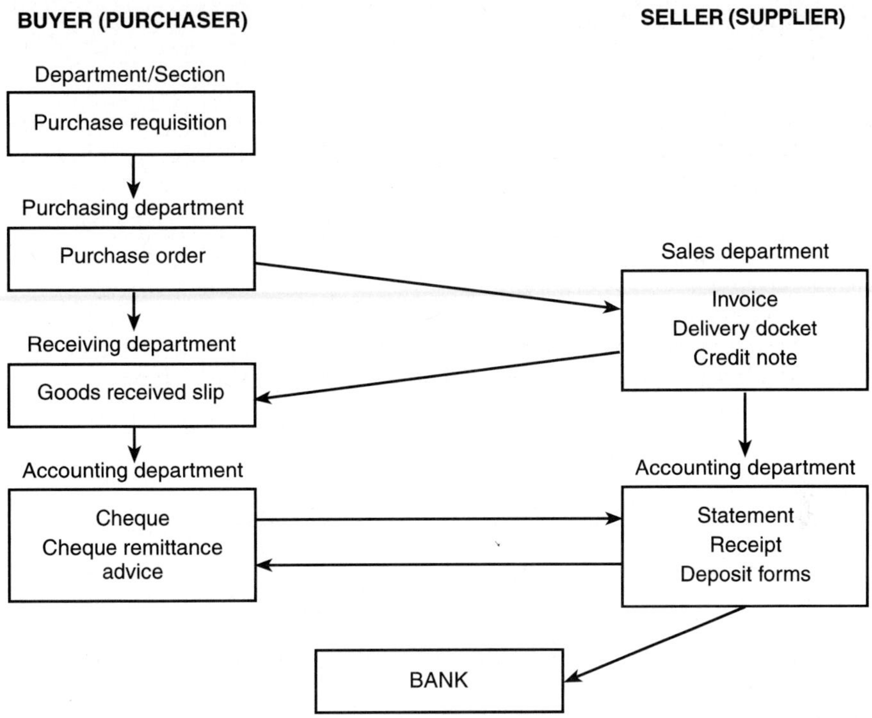

- The *section* or *department* requiring goods prepares a purchase requisition, which is forwarded to the purchasing department.
- The *purchasing department* selects a suitable supplier and prepares a purchase order, which is forwarded to the supplier.
- The *supplier's sales department* prepares an invoice and a delivery docket (and a credit note, if applicable), which are sent to the purchaser's receiving department.
- This *receiving department* checks the goods and forwards the invoice and the delivery docket to the accounting department.
- The *supplier* sends a statement to the purchaser's accounting department, for payment of amounts due.
- The purchaser's *accounting department* checks that the invoice, the delivery docket and a duplicate copy of the order form all match before entering in the purchasing records and paying on the statement.
- The *supplier* forwards a receipt for money received in payment of the account.
- Once cash is received, a bank deposit form is completed and the cash received is banked.

Remember that a duplicate copy of every form would be retained and filed in the department that generated it, as proof of completion. Extra copies might also be required for other departments; however, the main sequence of events is detailed in the above chart.

Note: debt collection procedures may occur if the payment is not made.

COPYING AND FILING DOCUMENTS FOR AUDITING PURPOSES

Documents today are typically prepared using computers. Consequently, many source documents do not automatically have duplicate copies. It is necessary to have copies (either duplicates or photocopies) of all financial source documents. Documents are required by law to be accessible for a period of seven years for auditing purposes.

Auditing is the process of checking the accuracy and reliability of the accounting process. An audit may be performed by internal auditors, which means being checked by people within the organisation, who then report their findings to management. Alternatively, an external audit may be performed, which usually requires accountants from an external and separate organisation to check the accounting records of the company and report to its shareholders.

CHECKING AND DISTRIBUTING SOURCE DOCUMENTS

Validity of incoming and outgoing transactions

You now have the skills to prepare documents that would be either received by or prepared by your business. You must be able to check their contents for:

- accuracy of contents, quantities and cost;
- authorisation by supervisor/manager with approval to do so.

Incoming documents requiring checking include:

- invoices;
- credit notes;
- delivery dockets;
- statements.

Checking of documents is an important aspect of bookkeeping and is an ongoing process.

Source documents (both incoming and outgoing) need to be checked to ensure that they are valid (or correct). For example, any invoice received by a firm should be checked to ensure that: (1) a suitably authorised order has been placed for the product or services being invoiced; (2) the product or services have, in fact, been received; and (3) all contents, quantities and costs are accurate. In the sample source documents prepared by you, validity of the invoices could be justified by matching purchase orders and delivery dockets that relate to them.

Similarly, invoices and statements being sent to customers must be checked to ensure that all aspects of their content are accurate. This will include ensuring that all dates, names and addresses, numerical calculations and additional information are accurate.

CORRECTING IRREGULARITIES

An important aspect of this chapter has been the correction of source documents related to one section prior to moving on to the next. However, a business must also have a system of checking and correcting irregularities that may occur. For example, if an incoming invoice that has been overcharged is allowed to progress through the system, it is possible that an overpayment will occur and go unnoticed. A business cannot afford to have this happen; therefore, as documentation is prepared and received, procedures must be in place that correct irregularities before they progress through the system. Section supervisors or managers need to be advised if any irregularity is noticed.

The example below shows how you would compare a delivery docket with its matching invoice. Assume that the delivery docket has already been signed to acknowledge that the items that appear on it are the items received. Assume also that you have the authority to make changes to the invoice to enter the correct details in the business records. Note, however, that the firm that sent the invoice would have to be notified and a corrected invoice sent out.

Look at the delivery docket and the invoice that follow. Use the policy extract below, together with the sample corrected invoices, as the basis of your checking procedure.

Staff who are authorised to check the accuracy of invoices and credit notes must perform the following tasks:

- **Check that the amount/unit, description (including colour, etc.) details are correct.**
- **Check with the price list to see that the unit cost is correct; if any items are incorrect, insert the correct price.**
- **Check that all calculations are correct, including trade discount and all totals; if any are incorrect, alter the invoice accordingly.**
- **If any errors or omissions have appeared, advise your section supervisor or manager immediately.**

Delivery Docket

Date. *16 May 19XX*

To: *A1 Secretarial Services*
52 Letterfield Lane
CANNON HILL QLD 4170

Received from: *Business Stationery Supplies Pty Ltd*
2109 Old English Road
COOLANGATTA QLD 4225

In good order and condition:

1 doz biros, blue
3 doz lead pencils, HB
3 boxes highlighter pens, yellow

No. 123 Received by: (x) ....

Invoice

BUSINESS STATIONERY SUPPLIES PTY LTD
2109 Old English Road
COOLANGATTA QLD 4225
ABN 19 112 309 432

A1 Secretarial Services *52 Letterfield Lane* *CANNON HILL QLD 4170*	**Inv. no.** **201**

TAX INVOICE

Date	Purchase order no.	Method of delivery
14 May 19XX	*33*	*Tranters Transport*

Item	Qty ordered	Description	Unit price		per	Extension	
1	*2 doz*	*Biros, red*	*4*	*00*	*doz*	*8*	*00*
2	*3 doz*	*Lead pencils, HB*	*2*	*50*	*doz*	*7*	*50*
3	*3 boxes*	*Highlighter pens, yellow*	*6*	*00*	*box*	*18*	*00*

	Subtotal	*33*	*50*
Deliver to: *33 Raidy Road* *Uralunga Qld*	10% Trade discount	*3*	*35*
	Subtotal after trade discount	*30*	*15*
	Plus delivery charges		
Terms:	Invoice total $	*30*	*15*

The total price includes GST of $ *2.74*

Note: You should notify your supervisor immediately where errors exist in invoices.

Look at the corrected invoice opposite. It indicates how corrections to incorrect invoices must be completed. Note that all corrections have been initialled.

Invoice

BUSINESS STATIONERY SUPPLIES PTY LTD
2109 Old English Road
COOLANGATTA QLD 4225
ABN 19 112 309 432

	Inv. no.
A1 Secretarial Services 52 Letterfield Lane CANNON HILL QLD 4170	201

TAX INVOICE

Date	Purchase order no.	Method of delivery
16 ~~14~~ May 19XX IL	33	Tranters Transport

Item	Qty ordered	Description	Unit price		per	Extension	
1	1 ~~2~~ doz IL	Biros, blue ~~red~~ IL	8 ~~4~~	00 ~~00~~	IL doz	8	00
2	3 doz	Lead pencils, HB	3 ~~2~~	00 ~~50~~	IL doz	IL 9 ~~7~~	00 ~~50~~
3	3 boxes	Highlighter pens, yellow	5 ~~6~~	00 ~~00~~	IL box	IL 15 ~~18~~	00 ~~00~~

	Subtotal	IL 32 ~~33~~	00 ~~50~~
Deliver to:	10% Trade discount	IL 3 ~~3~~	20 ~~35~~
33 Raidy Road Uralunga Qld	Subtotal after trade discount	IL 28 ~~30~~	80 ~~15~~
	Plus delivery charges		
Terms:	Invoice total $	IL 28 ~~30~~	80 ~~15~~

The total price includes GST of $~~2.74~~ 2.61 IL

Skill development 15.9

Correcting irregularities

(a) Assume that you are working for A1 Secretarial Services. Compare the following delivery docket and invoice. The contents of the delivery docket have been checked and are correct. Alter the invoice on page 419 in ink so that it is correct. Trade discount is 10 per cent.

Use Appendix 15.1 to check the prices.

Delivery Docket

Date. *14 May 19XX*

To: *A1 Secretarial Services*
52 Letterfield Lane
CANNON HILL QLD 4170

Received from: *Business Stationery Supplies Pty Ltd*
2109 Old English Road
COOLANGATTA QLD 4225

In good order and condition:
3 only pencil sharpeners, plastic
2 doz lead pencils, 2B
3 doz Papermate pens, blue
(Order no. 54)

No. 123 Received by: (x) ....

Invoice

BUSINESS STATIONERY SUPPLIES PTY LTD
2109 Old English Road
COOLANGATTA QLD 4225
ABN 19 112 309 432

A1 Secretarial Services *52 Letterfield Lane* *CANNON HILL QLD 4170*	**Inv. no.** 201

TAX INVOICE

Date	Purchase order no.	Method of delivery
14 May 19XX	*54*	*Tranters Transport*

Item	Qty ordered	Description	Unit price		per	Extension	
1	*4 only*	*Pencil sharpeners, plastic*	2	00	*each*	*8*	*00*
2	*2 doz*	*Lead pencils, 2B*	2	50	*doz*	*5*	*50*
3	*3 doz*	*Papermate pens black*	8	00	*doz*	*24*	*00*

	Subtotal	*37*	*50*
Deliver to: *33 Raidy Road* *Uralunga Qld*	10% Trade discount	*3*	*35*
	Subtotal after trade disc.	*41*	*25*
	Plus delivery charges		
Terms:			
	Invoice total $	*41*	*25*

The total price includes GST of $*3.75*

On a separate sheet of paper, answer the following questions.

(b) Who should you advise if you find any errors in the documents in part (a)?

(c) When should you advise them of these errors?

(d) Discuss the importance of despatching invoices within designated timelines.

Despatching debtors' invoices within designated timelines

Documents must be sorted and batched into like transactions before distribution, for processing within designated timelines.

Timelines are important in business because they ensure, for example, that documents are recorded in a timely fashion and accounts are paid by due dates.

The policy manual for Business Stationery Supplies Pty Ltd states:

> **Invoices and credit notes will be sent to the accounting department of the receiving firm on the same day as goods are sent or returned.**

Preparing payment documentation for correct and authorised invoices within designated timelines

Once an invoice is received and verified, a cheque requisition voucher is obtained from the authorised officer, and the cheque and cheque remittance advice can be prepared. This should be completed within designated timelines, in order to take advantage of any discounts offered for prompt payment, and also in order to maintain a good credit rating. For these reasons, payment of invoices by the due date is very important to an organisation.

Resolving or referring enquiries to nominated person/section

As an employee, there may be instances when financial source document enquiries are directed to you. It is the responsibility of all employees to familiarise themselves with the required procedure for the different types of enquiries. Each of these enquiries will require the attention of a particular person or department, and must be dealt with correctly by referring the enquirer to the nominated person, according to company policy.

Alternatively, any enquiry to which you are authorised to respond should be dealt with in a polite and efficient manner. If you are unsure, ask the enquirer to wait, and refer them to your supervisor. Alternatively, take their details and ask the nominated person to contact them forthwith.

Activity 15.3

Ask your tutor/supervisor whether you are to photocopy the following documents. You could number the documents: for example, number 1 would show each related document—purchase order, delivery docket, invoice and credit note.

Photocopy the source documents you prepared in Skill developments 15.1 (purchase orders), 15.2 (delivery dockets), 15.3 (invoices) and 15.4 (credit notes). They must now be collated ready for entering in the business's purchasing records.

Note that the invoices and credit notes relate to purchases because they were sent to A1 Secretarial Services, whose name appears in the boxed section towards the top of the invoice.

The seller of the goods is Business Stationery Supplies, because this name appears in the letterhead on each invoice and credit note. (Business Stationery Supplies sent these documents, and requires payment.)

- Sort each delivery docket, invoice and credit note according to the purchase order to which it relates; make sure the purchase order number given on the invoice matches the number on the order form. These documents will be used in A1 Secretarial Services' purchases journal.
- Staple together the invoice, delivery docket and purchase order with the invoice on top and the purchase order at the bottom.
- Sort the documents according to the invoice dates, placing the earliest on top and the most recent at the bottom.
- Staple the credit notes together in date order, with the most recent date on the bottom, ready for entering in the purchase returns and allowances journal.

Present all the documents to your supervisor for checking.

Skill development 15.10

Review

The following series of activities are designed to review documents presented in this chapter. Use the blank documents in your workbook.

Note: for all calculations, round your answer to the nearest cent.

Question 1

(a) Mr John Black, manager of the customer service department of A1 Secretarial Services, 52 Letterfield Lane, Cannon Hill, QLD 4170, requires the following items:

- 4 reams A4 beige marble paper (200 sheets)
- 10 pkts envelopes 228 × 101 mm (100)
- 1 only 2-drawer filing cabinet
- 5 bxs highlighter pens, green

Prepare requisition number 107, dated 20 March 19XX. The items are required exactly seven days from the date ordered.

(b) The purchasing department of A1 Secretarial Services has chosen Business Stationery Supplies Pty Ltd, 2109 Old English Road, Coolangatta, QLD 4225, as the supplier for the purchase requisition completed in part (a). Complete a purchase order, noting the following additional information:

- the purchase order will be number 40;
- terms include 2.5% discount if the account is paid within 30 days;
- Tranters Transport is to be used for delivery.

(c) On behalf of Business Stationery Supplies, complete the invoice relating to the above purchase order and prepare a delivery docket. The invoice number is 208 and the delivery docket is number 44. To price the invoice, use the price list in Appendix 15.1.

The following further information is given:

- The required delivery date is to be used for the invoice and the delivery docket.
- Trade discount is to be 10%.

(d) On 12 March 19XX, A1 Secretarial Services received a cheque valued at $400 for payment of the February account of Amy Productions Pty Ltd. Discount of $20 was also given for prompt payment of this account. Write out receipt number 50 for this amount.

(e) The invoice that was prepared in part (c) is now ready for payment. A1 Secretarial Services has been given a 2.5% discount for prompt payment of this amount. Prepare cheque number 1078, dated 15 April 19XX, and the remittance advice that will accompany it.

On a separate sheet of paper, write your answers to the following questions.

(f) Debtors' invoices must be dispatched within designated timelines. Give two reasons why this is necessary in a business.

(g) You have received a query in relation to the invoice prepared in part (c). Explain how you should handle this enquiry.

Question 2

The invoice and delivery docket that follow have been received by A1 Secretarial Services. To check that all details are correct, you must:

- compare the delivery docket with the invoice;
- correct any irregularities on the invoice so that it matches the delivery docket;
- correct the pricing structure of the invoice, if necessary (use the price list in Appendix 15.1).

Invoice

BUSINESS STATIONERY SUPPLIES PTY LTD
2109 Old English Road
COOLANGATTA QLD 4225
ABN 19 112 309 432

A1 Secretarial Services
52 Letterfield Lane
CANNON HILL QLD 4170

Inv. no. 203

TAX INVOICE

Date	Purchase order no.	Method of delivery
25 May 19XX	*45*	*Gosford's Transport*

Item	Qty ordered	Description	Unit price		per	Extension	
1	*3 bxs*	*Highlighter pens, green*	*5*	*00*	*box*	*15*	*00*
2	*3 bxs*	*Fountain pens, black*	*16*	*00*	*box*	*48*	*00*

	Subtotal	63	00
Deliver to:	10% Trade discount	6	30
	Subtotal after trade discount	*56*	*70*
	Plus delivery charges		
Terms:	Invoice total $	*56*	*70*

The total price includes GST of $*5.15*

Delivery Docket

Date: *25 May 19XX*

To: *A1 Secretarial Services*
52 Letterfield Lane
CANNON HILL QLD 4170

Received from: *Business Stationery Supplies Pty Ltd*
2109 Old English Road
COOLANGATTA QLD 4225

In good order and condition:

2 bxs highlighter pens, green
3 bxs fountain pns, black
1 only 2-drawer filing cabinet

(Order no. 45)

No. 125 Received by: (x) ....

Question 3

In response to the three purchase orders given below, A1 Secretarial Services has received the delivery dockets and invoices that follow them. The documents must now be sorted. To do this, photocopy and cut the relevant pages.

(a) Staple together the matching delivery dockets, purchase orders and invoices ready for processing in the business's records.

(b) Sort the invoices in date order ready for processing, placing the latest date at the bottom.

(c) The invoices given in this question must be for either purchases or sales. State which type of transaction each one relates to (from the point of view of A1 Secretarial Services), and explain how you can tell this from the information it contains. Give your answer on a separate sheet of paper.

A1 SECRETARIAL SERVICES
52 Letterfield Lane
CANNON HILL QLD 4170

Purchase order
No. 34

to:
Business Stationery Supplies Pty Ltd
2109 Old English Road
COOLANGATTA QLD 4225

Please supply the following goods and charge to our account.

Quantity	Product number	Description	Unit price	Amount
10 only		*Calculators, Tandy*	$60.00	$600.00
1 doz		*Rulers, plastic, 30 cm*	$0.50	$6.00

Terms: *Nett 30 days*
Deliver to: *Retail Office Department*
Required by: *24 May 19XX*
Transport: *Gosford's Transport*

Per: (x) ....
Date: *22 May 19XX*
GST included in quoted prices

A1 SECRETARIAL SERVICES
52 Letterfield Lane
CANNON HILL QLD 4170

Purchase order
No. 36

to:
Business Stationery Supplies Pty Ltd
2109 Old English Road
COOLANGATTA QLD 4225

Please supply the following goods and charge to our account.

Quantity	Product number	Description	Unit price	Amount
5 bxs		*Felt-tip pens, black*	$9.00	$45.00
10 reams		*Paper, A4 pink*	$15.00	$150.00

Terms: *Nett 30 days*
Deliver to: *Retail Office Department*
Required by: *26 May 19XX*
Transport: *Gosford's Transport*

Per: (x) ....
Date: *24 May 19XX*
GST included in quoted prices

A1 SECRETARIAL SERVICES
52 Letterfield Lane
CANNON HILL QLD 4170

Purchase order
No. 38

to: *Business Stationery Supplies Pty Ltd*
2109 Old English Road
COOLANGATTA QLD 4225

Please supply the following goods and charge to our account.

Quantity	Product number	Description	Unit price	Amount
5 bxs		*Highlighter pen, yellow*	*$5.00*	*$25.00*
10 bxs		*Computer disks ($3\frac{1}{2}$")*	*$20.00*	*$200.00*
10 only		*Rulers, wood, 30 cm*	*$ 0.30*	*$3.00*

Terms: *Nett 30 days*
Deliver to: *Retail Office Department*
Required by: *31 May 19XX*
Transport: *Gosford's Transport*

Per: (x) ....
Date: *29 May 19XX*
GST included in quoted prices

Delivery Docket

Date: *26 May 19XX*

To: *A1 Secretarial Services*
52 Letterfield Lane
CANNON HILL QLD 4170

Received from: *Business Stationery Supplies Pty Ltd*
2109 Old English Road
COOLANGATTA QLD 4225

In good order and condition:

5 bxs felt-tip pens, black
10 reams paper, A4, pink

No. 126

Received by: (x) ....

Delivery Docket

Date: *24 May 19XX*

To: *A1 Secretarial Services*
52 Letterfield Lane
CANNON HILL QLD 4170

Received from: *Business Stationery Supplies Pty Ltd*
2109 Old English Road
COOLANGATTA QLD 4225

In good order and condition:

10 only calculators, Tandy
1 doz rulers, plastic, 30 cm

No. 124 Received by: (x) ....

Delivery Docket

Date: *31 May 19XX*

To: *A1 Secretarial Services*
52 Letterfield Lane
CANNON HILL QLD 4170

Received from: *Business Stationery Supplies Pty Ltd*
2109 Old English Road
COOLANGATTA QLD 4225

In good order and condition:

5 bxs highlighter pens, yellow
10 bxs computer disks (3 $\frac{1}{2}$")
10 only rulers

No. 128 Received by: (x) ....

Invoice

BUSINESS STATIONERY SUPPLIES PTY LTD
2109 Old English Road
COOLANGATTA QLD 4225
ABN 19 112 309 432

A1 Secretarial Services
52 Letterfield Lane
CANNON HILL QLD 4170

Inv. no.
202

TAX INVOICE

Date	**Purchase order no.**	**Method of delivery**
24 May 19XX	*34*	*Gosford's Transport*

Item	**Qty ordered**	**Description**	**Unit price**		**per**	**Extension**	
1	*10 only*	*Calculators, Tandy*	60	00	*item*	600	00
2	*1 doz*	*Rulers, plastic, 30 cm*	0	50	*doz*	6	00

	Subtotal		606	00
Deliver to:	10% Trade discount		60	60
	Subtotal after trade disc.		*545*	*40*
	Plus delivery charges			
Terms:	Invoice total	$	*545*	*40*

The total price includes GST of $*49.58*

Invoice

BUSINESS STATIONERY SUPPLIES PTY LTD
2109 Old English Road
COOLANGATTA QLD 4225
ABN 19 112 309 432

A1 Secretarial Services
52 Letterfield Lane
CANNON HILL QLD 4170

Inv. no.
204

TAX INVOICE

Date	Purchase order no.	Method of delivery
26 May 19XX	36	Gosford's Transport

Item	Qty ordered	Description	Unit price		per	Extension	
1	5 bxs	Felt-tip pens, black	9	00	box	45	00
2	10 reams	Paper, A4, pink	15	00	ream	150	00

	Subtotal	195	00
Deliver to:	10% Trade discount	19	50
	Subtotal after trade disc.	175	50
	Plus delivery charges		
Terms:	Invoice total $	175	50

The total price includes GST of $15.95

Invoice

BUSINESS STATIONERY SUPPLIES PTY LTD
2109 Old English Road
COOLANGATTA QLD 4225
ABN 19 112 309 432

A1 Secretarial Services
52 Letterfield Lane
CANNON HILL QLD 4170

Inv. no.
206

TAX INVOICE

Date	**Purchase order no.**	**Method of delivery**
31 May 19XX	*38*	*Gosford's Transport*

Item	**Qty ordered**	**Description**	**Unit price**		**per**	**Extension**	
1	*5 bxs*	*Highlighter pens, yellow*	*5*	*00*	*box*	*25*	*00*
2	*10 bxs*	*Computer disks* ($3\frac{1}{2}''$)	*20*	*00*	*box*	*200*	*00*
3	*10 only*	*Rulers, wood (30 cm)*	*0*	*30*	*each*	*3*	*00*

	Subtotal		*228*	*00*
Deliver to:	10% Trade discount		*22*	*80*
	Subtotal after trade disc.		*205*	*20*
	Plus delivery charges			
Terms:	Invoice total	$	*205*	*20*

The total price includes GST of $*18.68*

Summary

- Purchase requisitions are prepared by the department that requires goods, which advises the department responsible for ordering goods that stock is required.
- A purchase order is prepared by the purchasing department when ordering inventories.
- The supplier prepares a delivery docket. This is sent with the goods and is signed by the purchaser as proof of delivery.
- Invoices and credit notes, if they are necessary, will be prepared by the supplier. Invoices usually accompany the goods, but may be sent direct to the purchasing firm's accounting department ready for payment of the amount owed. Credit notes are usually sent direct to the accounting department of the purchasing firm.
- Calculation of trade discount and GST, if applicable, must be performed when an invoice is prepared. Trade discount is deducted from the value of the invoice, whereas GST will increase the amount owed.
- A statement is prepared by the supplier to show how much a purchaser owes at the end of the month. The statement will be forwarded to the purchaser.
- Cheques and cheque remittance advice slips are prepared by the purchaser, and are sent when the statement is due for payment. Cheques will also be written for goods and services that are not documented by a statement. There are a number of different types of cheques, including cash, crossed and crossed 'not negotiable'.
- When the cash (cheque) is received, a receipt is issued by the supplier to indicate that the amount owed, as recorded on the statement, has been paid.
- In banking receipts of cash, a bank deposit form is used as proof of the transaction.
- The source documents used in this chapter include:
  - duplicate invoices for the sales journal;
  - duplicate credit notes for the sales returns and allowances journal;
  - original invoices for the purchases journal;
  - original credit notes for the purchase returns and allowances journal;
  - cheque butts or duplicate cheques for the cash payments journal;
  - receipts for the cash receipts journal.

  These source documents are forwarded to the firm's accounting department to be entered in the business's records.

Review questions

(This integrated activity may be used for self-assessment or for formal assessment.)

Question 1

(a) Mr John Black, manager of the customer service department of A1 Secretarial Services, 52 Letterfield Lane, Cannon Hill, QLD 4170, requires the following items:

No. 201 **Purchase Requisition** **Date:** *20 March 19XX*

Date required: *27 March 19XX*

Supply to: *Customer Service Department*

Requested by: *J. Black, Manager*

Quantity/unit	Description
5 boxes	*Highlighter pens, yellow*
4 reams	*Beige marble paper (200 sheets)*
1 roll	*Computer paper, continuous (1000 sheets)*

Per: (x) *J. Black*

The purchasing department of A1 Secretarial Services has chosen Business Stationery Supplies, 2109 Old English Road, Coolangatta, QLD 4225, as the supplier for the purchase requisition. Complete a purchase order, noting the following additional information:

- the purchase order will be number 40;
- terms include 2.5% discount if the account is paid within 30 days;
- Tranters Transport is to be used for delivery.

(b) On behalf of Business Stationery Supplies, complete the invoice relating to the above purchase order and prepare a delivery docket. The invoice number is 208 and the delivery docket number is 44. To price the invoice, use the price list in Appendix 15.1.

The following further information is given.

- The required delivery date is to be used for the invoice and the delivery docket.
- Trade discount is to be 10%.

(c) On March 15 19XX A1 Secretarial Services received $650 for payment of the February account of Amy Productions Pty Ltd. Discount of $35 was given for prompt payment of this account. Write out receipt number 60 for this amount.

(d) The invoice that was prepared in part (b) is now ready for payment. A1 Secretarial Services has been given a 2.5% discount for prompt payment of this amount. Prepare cheque number 1978, dated 25 April 19XX, and the remittance advice that will accompany it.

(e) Debtors' invoices must be dispatched within designated timelines. On a separate sheet of paper, give two reasons why this is necessary in a business.

(f) You have received a query in relation to the invoice prepared in part (b). Explain on a separate sheet of paper how you would determine whether you or someone else should handle this enquiry.

Question 2

The invoice and delivery docket that follow have been received by A1 Secretarial Services. To check that all details are correct, you must:

- compare the delivery docket with the invoice;
- correct any irregularities on the invoice so that it matches the delivery docket;
- correct the pricing structure of the invoice, if necessary (refer to the price list in Appendix 15.1).

Invoice
BUSINESS STATIONERY SUPPLIES PTY LTD
2109 Old English Road
COOLANGATTA QLD 4225
ABN 19 112 309 432

A1 Secretarial Services
52 Letterfield Lane
CANNON HILL QLD 4170

Inv. no.
206

TAX INVOICE

Date	**Purchase order no.**	**Method of delivery**
31 May 19XX	*40*	*Gosford's Transport*

Item	**Qty ordered**	**Description**	**Unit price**		**per**	**Extension**	
1	*5 bxs*	*Highlighter pens, yellow*	*5*	*00*	*box*	*25*	*00*
2	*11 bxs*	*Computer disks ($3\frac{1}{2}''$)*	*25*	*00*	*box*	*275*	*00*
3	*9 only*	*Rulers, wood (30 cm)*	*0*	*30*	*each*		*30*

	Subtotal	*300*	*00*
Deliver to:	10% Trade discount	*30*	*30*
	Subtotal after trade disc.	*270*	*00*
Terms:	Plus delivery charges		
	Invoice total $	*270*	*00*

The total price includes GST of $*24.55*

Delivery Docket

Date: *31 May 19XX*

To: *A1 Secretarial Services*
52 Letterfield Lane
CANNON HILL QLD 4170

Received from: *Business Stationery Supplies Pty Ltd*
2109 Old English Road
COOLANGATTA QLD 4225

In good order and condition:

5 bxs highlighter pens, green
10 bxs computer disks ($3\frac{1}{2}$")
10 only rulers, wood (30 cm)

No. 128 Received by: ....

Question 3

In response to the three purchase orders that follow, A1 Secretarial Services has received the delivery dockets and invoices that follow them. The documents must now be sorted. To do this, photocopy and cut the relevant pages.

(a) Staple together the matching delivery dockets, purchase orders and invoices ready for processing into the business's records.

(b) Sort the invoices in date order ready for processing, placing the latest date at the bottom.

(c) The invoices given in this question must be for either purchases or sales. State which type of transaction each one relates to (from the point of view of A1 Secretarial Services), and explain on a separate sheet of paper how you can tell this from the information it contains.

(d) The invoices you have sorted must be copied and filed. Explain on a separate sheet of paper why this is necessary.

A1 SECRETARIAL SERVICES
52 Letterfield Lane
CANNON HILL QLD 4170

Purchase order
No. 37

To:
Business Stationery Supplies Pty Ltd
2109 Old English Road
COOLANGATTA QLD 4225

Please supply the following goods and charge to our account.

Quantity	Product number	Description	Unit price	Amount
4 reams		*Beige marble paper (200 sheets)*	*$16.50*	*$66.00*
1 roll		*Computer paper (standard continuous, 1000 sheets)*	*$25.00*	*$25.00*
3 only		*Pencil sharpeners, desktop*	*$20.00*	*$60.00*

Terms: *Nett 30 days*
Deliver to: *Retail Office Department*
Required by: *30 May 19XX*
Transport: *Gosford's Transport*

Per: (x) ....
Date: *25 May 19XX*
GST included in quoted prices

A1 SECRETARIAL SERVICES
52 Letterfield Lane
CANNON HILL QLD 4170

Purchase order
No. 35

to:
Business Stationery Supplies Pty Ltd
2109 Old English Road
COOLANGATTA QLD 4225

Please supply the following goods and charge to our account.

Quantity	Product number	Description	Unit price	Amount
2 bxs		*Highlighter pens, green*	*$5.00*	*$10.00*
3 bxs		*Fountain pens, black*	*$15.00*	*$45.00*

Terms: *Nett 30 days*
Deliver to: *Retail Office Department*
Required by: *26 May 19XX*
Transport: *Gosford's Transport*

Per: (x) ....
Date: *25 May 19XX*
GST included in quoted prices

A1 SECRETARIAL SERVICES
52 Letterfield Lane
CANNON HILL QLD 4170

Purchase order
No. 39

to: *Business Stationery Supplies Pty Ltd*
2109 Old English Road
COOLANGATTA QLD 4225

Please supply the following goods and charge to our account.

Quantity	Product number	Description	Unit price	Amount
10 pkts		*Envelopes, 228 x 101 mm (100)*	*$5.00*	*$50.00*
5 pkts		*Envelopes, beige logo, 228 x 101 mm (100)*	*$4.00*	*$20.00*
1 only		*2-drawer filing cabinet*	*$165.00*	*$165.00*

Terms: *Nett 30 days*
Deliver to: *Retail Office Department*
Required by: *6 June 19XX*
Transport: *Gosford's Transport*

Per: (x) ....
Date: *5 June 19XX*
GST included in quoted prices

Delivery Docket

Date: *25 May 19XX*

To: *A1 Secretarial Services*
52 Letterfield Lane
CANNON HILL QLD 4170

Received from: *Business Stationery Supplies Pty Ltd*
2109 Old English Road
COOLANGATTA QLD 4225

In good order and condition:

2 bxs highlighter pens, green
3 bxs fountain pens, black

No. 125

Received by: (x) ....

Delivery Docket

Date: 6 *June 19XX*

To: *A1 Secretarial Services*
52 Letterfield Lane
CANNON HILL QLD 4170

Received from: *Business Stationery Supplies Pty Ltd*
2109 Old English Road
COOLANGATTA QLD 4225

In good order and condition:

10 pkts envelopes (228 x 101) x 100
5 pkts envelopes, beige logo (228 x 101) x 100
1 only 2-drawer filing cabinet

No. 129 Received by: (x) ....

Delivery Docket

Date: *30 May 19XX*

To: *A1 Secretarial Services*
52 Letterfield Lane
CANNON HILL QLD 4170

Received from: *Business Stationery Supplies Pty Ltd*
2109 Old English Road
COOLANGATTA QLD 4225

In good order and condition:

4 reams beige marble paper (200 sheets)
1 roll computer paper (standard, continuous x 1000 sheets)
3 only pencil sharpeners, desktop

No. 127 Received by: (x) ....

Invoice

BUSINESS STATIONERY SUPPLIES PTY LTD
2109 Old English Road
COOLANGATTA QLD 4225
ABN 19 112 309 432

A1 Secretarial Services
52 Letterfield Lane
CANNON HILL QLD 4170

Inv. no. 203

TAX INVOICE

Date	Purchase order no.	Method of delivery
25 May 19XX	35	Gosford's Transport

Item	Qty ordered	Description	Unit price		per	Extension	
1	2 bxs	Highlighter pens, green	5	00	box	10	00
2	3 bxs	Fountain pens, black	15	00	box	45	00

	Subtotal	55	00
Deliver to: Terms:	10% Trade discount	5	50
	Subtotal after trade disc.	49	50
	Plus delivery charges		
	Invoice total $	49	50

The total price includes GST of $4.50

Invoice

BUSINESS STATIONERY SUPPLIES PTY LTD
2109 Old English Road
COOLANGATTA QLD 4225
ABN 19 112 309 432

A1 Secretarial Services *52 Letterfield Lane* *CANNON HILL QLD 4170*	**Inv. no.** **205**

TAX INVOICE

Date	Purchase order no.	Method of delivery
30 May 19XX	*37*	*Gosford's Transport*

Item	Qty ordered	Description	Unit price		per	Extension	
1	*4 reams*	*Beige marble paper*	*16*	*50*	*ream*	*66*	*00*
2	*1 roll*	*Computer paper (continuous, 1000 sheets)*	*25*	*00*	*roll*	*25*	*00*
3	*3 only*	*Pencil sharpeners, desktop*	*20*	*00*	*each*	*60*	*00*

	Subtotal	*151*	*00*
	10% Trade discount	*15*	*10*
Deliver to:	Subtotal after trade disc.	*135*	*90*
Terms:	Plus delivery charges		
	Invoice total $	*135*	*90*

The total price includes GST of $*12.35*

Invoice

BUSINESS STATIONERY SUPPLIES PTY LTD
2109 Old English Road
COOLANGATTA QLD 4225
ABN 19 112 309 432

A1 Secretarial Services
52 Letterfield Lane
CANNON HILL QLD 4170

Inv. no.
207

TAX INVOICE

Date	**Purchase order no.**	**Method of delivery**
6 June 19XX	*39*	*Gosford's Transport*

Item	**Qty ordered**	**Description**	**Unit price**		**per**	**Extension**	
1	*10 pkts*	*Envelopes (228 x 101)*	*5*	*00*	*pkt*	*50*	*00*
2	*5 pkts*	*Envelopes (228 x 101), beige logo*	*4*	*00*	*pkt*	*20*	*00*
3	*1 only*	*2-drawer filing cabinet*	*165*	*00*	*each*	*165*	*00*

	Subtotal	*235*	*00*
	10% Trade discount	*23*	*50*
Deliver to:	Subtotal after trade disc.	*211*	*50*
Terms:	Plus delivery charges		
	Invoice total $	*211*	*50*

The total price includes GST of $*19.23*

Competency self-assessment and/or tutor/supervisor assessment

When you have finished studying the chapter, and have completed all of the Activities and Skill development exercises, complete the self-assessment part and then hand it to your tutor/supervisor to complete.

Learning outcomes	Self-assessment		Tutor/supervisor assessment	
I have successfully completed the following learning outcomes	**Yes** (please tick)	**No**	**Yes** (please tick)	**No**
1. Identify, prepare and record business financial documents for the purchase of goods/services according to enterprise procedures				
1.1 complete purchase/internal stock requisition for authorisation by nominated person according to enterprise guidelines	☐	☐	☐	☐
1.2 identify and prepare appropriate documents for the purchase of goods/services according to enterprise procedures	☐	☐	☐	☐
1.3 identify discrepancies between incoming invoices and supporting documents and report to nominated person/section for resolution	☐	☐	☐	☐
1.4 check incoming invoices for accuracy and identify errors to be reported to nominated person/section for correction/resolution	☐	☐	☐	☐
1.5 rectify discrepancies and errors on invoices as directed	☐	☐	☐	☐
1.6 prepare payment documentation for correct and authorised invoices within designated timelines	☐	☐	☐	☐
1.7 copy and file documents for auditing purposes	☐	☐	☐	☐
1.8 resolve creditor enquiries or refer to nominated person/section for resolution	☐	☐	☐	☐
2. Identify, prepare and record business financial documents for the sale of goods/services according to enterprise procedures				
2.1 identify and prepare accurate documents for the sale of goods/services according to enterprise procedures	☐	☐	☐	☐

2.2	check outgoing documents against relevant documentation to ensure accuracy of contents	☐	☐	☐	☐
2.3	distribute invoices to nominated person/section for verification	☐	☐	☐	☐
2.4	despatch verified invoices and other sales documentation to debtors within designated timelines	☐	☐	☐	☐
2.5	complete receipts for payment of debtors' invoices and present to nominated person for authorisation	☐	☐	☐	☐
2.6	copy and file documents for auditing purposes	☐	☐	☐	☐
2.7	resolve debtors' enquiries or refer to nominated person	☐	☐	☐	☐
Date:		Tutor/supervisor signature:			

APPENDIX 15.1

Price list

Product number	Item	Retail Price (Incl. GST)
	Section 1	
	Biros (all 12 per box)	per box:
A1	Biros, mixed colours	4.00
A2	BIC, black	8.00
A3	BIC, blue	8.00
A4	BIC, red	8.00
A5	Papermate, blue	12.00
A6	Papermate, black	12.00
A7	Papermate, red	12.00
A8	Fountain pen, black	15.00
A9	Fountain pen, blue	15.00
A10	Fountain pen, red	15.00
A11	Felt tip, blue	9.00
A12	Felt tip, black	9.00
A13	Felt tip, red	9.00
	Highlighter pens	per box:
A14	Multicoloured (6 per box)	6.00
A15	Green (10 per box)	5.00
A16	Pink (10 per box)	5.00
A17	Yellow (10 per box)	5.00

Product number	Item	Retail price (Incl. GST)
	Section 1 (cont.)	
A18	Pencil eraser, 'Multi' metal .... each	6.80
A19	Pencil sharpeners, desktop .... each	20.00
A20	Pencil sharpeners, plastic .... each	2.00
	Section 2	
	Paper	per ream:
B1	A4 ream, white (photocopying)	10.00
B2	A4 ream, blue	15.00
B3	A4 ream, yellow	15.00
B4	A4 ream, pink	15.00
B5	A4 blue-grey marble (200 sheets)	16.50
B6	A4 beige marble (200 sheets)	16.50
	Computer paper	
B7	Standard continuous (1000 sheets)	25.00
	Computer	
B8	Disks, 5¼ .... per box 10	13.00
B9	Disks, 3½ .... per box 10	20.00
B10	Disk container (holds 5¼) .... each	7.00
B11	Disk container (holds 3½) .... each	8.50
	Envelopes	
B12	228 × 101 mm (100)	5.00
B13	146 × 90 mm (100)	3.00
	Envelopes, beige, logo	
B14	228 × 101 mm (100 per pkt) .... per pkt	4.00
B15	146 × 90 mm (100 per pkt) .... per pkt	3.50
B16	Plastic inserts A4 .... per box 50	9.00
	Manila folders	
B17	Plain .... per box 100	8.00
B18	Red .... per box 100	9.50
B19	Blue .... per box 100	9.50
B20	Beige .... per box 100	9.50
B21	Yellow .... per box 100	9.50
	Section 3	
	Miscellaneous	
C1	Stapler, front loading .... each	55.00
C2	Calculators, Tandy .... each	60.00
C3	Lead pencils, HB (12 per box) .... per box	3.00
C4	Lead pencils, 2B (12 per box) .... per box	2.50
C5	Rulers, plastic (30 cm) .... each	0.50
C6	Rulers, wood (30 cm) .... each	0.30
	Filing	
C7	File tray, stackable, blue .... each	4.00
C8	Suspension files (50 per box) .... per box	12.00
C9	Guides for files (50 per box) .... per box	2.50
C10	File inserts (50 per box) .... per box	0.50
C11	Filing cabinets, 2-drawer .... each	165.00
C12	Filing cabinets, 4-drawer .... each	230.00

CHAPTER

16

PETTY CASH

FIN101A, NOS151, 8872B

objectives

The purpose of this chapter is to provide you with the knowledge and skills to operate a petty cash system, using the imprest method according to organisational procedures, in order to complete the following learning outcomes:

1. **Claim reimbursement from the petty cash fund according to enterprise procedures**
   - 1.1 discuss the need for accountability, security and control when dealing with money
   - 1.2 complete petty cash voucher for expenditure in accordance with enterprise procedures and present to authorised person for reimbursement
2. **Record and balance petty cash transactions for financial records according to organisational procedures**
   - 2.1 check petty cash claims/vouchers for accuracy and authenticity before processing
   - 2.2 refer irregularities to nominated person/section
   - 2.3 distribute petty cash according to enterprise guidelines
   - 2.4 record, total and balance transactions in petty cash book
   - 2.5 present records of transactions to relevant authority for checking within designated timelines
   - 2.6 Prepare documents for reimbursement of the petty cash fund according to enterprise guidelines

---

This chapter is written in the form of a petty cash system, designed from the beginning, for Business Stationery Supplies Pty Ltd. The chapter also includes petty cash examples prepared for a number of stores at different locations, in order to consolidate understanding. It is structured to allow you to proceed at your own pace. There is no prerequisite for this chapter.
(At the end of the chapter, when you have achieved all the above outcomes, you and/or your tutor/supervisor will be able to evaluate your performance for assessment.)

THE IMPREST SYSTEM OF PETTY CASH

Petty cash is a system that allows for the payment of items that are individually too small to require payment by cheque. It would be impractical for a $1.50 bus ticket to be paid using a cheque. Generally a firm would have an amount of between $50 and approximately $150 placed in a separate cash drawer or tin, referred to as the **petty cash fund.** The amount that is required in the petty cash fund is dependent on the needs of the particular enterprise.

The imprest system

The easiest system for recording petty cash is referred to as the **imprest system.** This system requires that a cheque be drawn to establish the petty cash fund. At intervals, on the production of supporting evidence or proof of expenditure, the amount paid during a period is reimbursed by the drawing of another cheque, referred to as the reimbursement cheque. For example, if the imprest amount is $100.00, and an amount of $55.00 is paid out during a period, the reimbursement cheque would need to be for $45.00 ($100 less $55).

In an imprest system, the amount of the cash on hand in the petty cash tin or drawer, added to the proof of expenditure documents, would always add up to the imprest amount: that is, in the example above it must add up to $100.00.

Petty cash advance

When using the imprest system for petty cash, the cash that is obtained to begin the system is said to be an 'advance' to the system. An account is set up to record the amount of the advance, and this account is called the **petty cash advance**. Some enterprises also refer to it as the 'petty cash imprest' account.

ENTERPRISE POLICIES AND PROCEDURES

Larger enterprises will generally have a policies and procedures manual that directs employees in all aspects of petty cash. Employees performing duties relating to petty cash should have a working knowledge of all policies and procedures relevant to their position, and should strictly adhere to these requirements.

Policies and procedures relating to petty cash should include:

- specified duties of the petty cashier;
- procedures that ensure accountability, security and control over petty cash;
- the amount of the petty cash fund (called the **imprest amount**);
- account names that will be used in the petty cash book;
- example petty cash documents;

- other relevant procedures relating to petty cash, required by the specific enterprise.

The policies and procedures manual should also detail occupational health and safety aspects relating to the safety of employees performing petty cash duties. To deter theft, aspects to be considered include that:

- cash should not be openly displayed;
- the cash fund should not be too large;
- cash should not be collected routinely from the bank;
- different routes should be used to and from the bank when collecting petty cash reimbursements;
- the amount and location of the petty cash fund should not be disclosed.

Clearly defined policies and procedures should ensure that processes relating to petty cash are easy to follow, and ensure that theft and fraud cannot occur.

Activity 16.1

Policies and procedures manual

Write your answers to the following on a separate sheet of paper:

1. List four policies and/or procedures relating to petty cash that should be included in an enterprise's policies and procedures manual.
2. Discuss the prevention of at least four hazards that could affect the safety of employees when dealing with petty cash.

THE NEED FOR ACCOUNTABILITY, SECURITY AND CONTROL WHEN DEALING WITH MONEY

The petty cashier

An employee who is solely responsible for the petty cash operations generally carries out most of the duties relating to petty cash. This person is referred to as the **petty cashier**. The petty cashier has a position of responsibility within an enterprise. Because the position requires dealing with cash it is important that they be trustworthy and have the ability to complete their tasks within the guidelines specified in the enterprise's policies and procedures manual.

Accountability for petty cash

Accountability refers to being accountable for one's actions within an enterprise. The person who is responsible for recording of petty cash is the petty cashier. The petty cashier is accountable for ensuring that guidelines outlined in the enterprise's policies and procedures manual are maintained for the petty cash process. Accountability also refers to being accountable to someone else within an enterprise. A petty cashier would generally be accountable to his/her supervisor, who should check the validity of the petty cash processes.

Security measures over petty cash

Security relates to ensuring that access to petty cash documents and the petty cash fund is restricted. Security over petty cash should include the following.

- The petty cash fund should be kept in a safe place, under lock and key, preferably in a fireproof safe.
- Access to the petty cash fund should be limited to one person, the petty cashier, with provision being made for unexpected absences (e.g. sick leave).
- Keys to the petty cash tin/drawer should be kept in a secure place, with restricted access.
- Petty cash documents should be kept under lock and key, and should be accessed only by the petty cashier.

Control measures over petty cash

Cash is easily stolen, and therefore reliable internal controls are required to deter or detect employees with dishonest intentions. A good system of internal control over petty cash should include the following.

- The petty cashier should be solely responsible for the recording of the petty cash book.
- The petty cashier should ensure that each petty cash expenditure is accompanied by proof of expenditure in the form of, for example, bus tickets, checkout dockets, cab receipts.
- The petty cashier should not be responsible for the recording of petty cash in records of the enterprise, other than the petty cash book. This is called separation of duties, and makes it more difficult for an employee to steal cash.
- Rotation of duties should be practised regularly. This means another person on a rotational basis, and without prior notice, should undertake the petty cashier position. Furthermore, the petty cashier should be required to take annual holidays. Another person performing the petty cash duties would discover any unusual entries. Employees are less likely to commit theft if they know their actions may be detected.
- Petty cash documents should be serially numbered, and used sequentially.
- Internal auditing of petty cash records should be randomly performed.
- Balancing of the petty cash fund should be done on a regular basis. This should be performed according to the guidelines in the enterprise's policies and procedures manual, and checked by the supervisor.

Activity 16.2

The need for accountability, security and control when dealing with cash

List your answers to the following on a separate sheet of paper:

1. Accountability for a petty cashier means two things. Explain in detail these two aspects of accountability.
2. Security is important when dealing with cash. Discuss four aspects of security that are required when dealing with petty cash.
3. Why is it important to have good internal control when dealing with petty cash?
4. Show how you would write a section of the enterprise's policies and procedures manual for Business Stationery Supplies Pty Ltd that gives instructions relating to internal control over petty cash. List as many options as you can.

COMPLETE A PETTY CASH VOUCHER FOR EXPENDITURE IN ACCORDANCE WITH ENTERPRISE PROCEDURES

Petty cash expenditure

Small expenditures can be reimbursed to employees out of the petty cash fund. The types of expenditure commonly claimed through petty cash include:

- bus and taxi fares;
- parking tickets;
- bridge tolls;
- stationery items;
- coffee, tea and biscuits;
- cleaning items;
- magazines for the reception;
- other work-related expenses.

There would generally be guidelines relating to the maximum amount allowed for individual claims using petty cash: for example, expenditure items over $20 may be paid by cheque.

Validation of expenditure

Petty cash expenditure items must be validated by proof of expenditure. Examples of documentation would include:

- receipts;
- bus, ferry, train tickets;
- cash register dockets;
- EFTPOS receipts.

Preparing petty cash vouchers

A petty cash voucher is prepared for each item of expenditure claimed through the petty cash process. Petty cash vouchers must be prepared in accordance with the enterprise's policies and procedures manual.

When a petty cash voucher is to be prepared, proof of expenditure must first be given and then attached to the back of the petty cash voucher. The voucher may be prepared by the person requesting the reimbursement, with assistance from the petty cashier.

Transaction. On June 20 19XX, K. French requested a petty cash reimbursement for the sum of $20.00. Proof of expenditure in the form of a cash register summary, showing a breakdown for stamps, express post, parcel delivery and envelopes, was presented. Authorisation for this expenditure had previously been given.

When completed, the voucher looks like this:

Petty Cash Voucher

No. 10 (1)			(2) *20 June* 19*XX*
Debit *Postage* (3)			
Particulars	*Stamps, Express Post,* *Parcel to Maryborough* *Envelopes*	(4)	*$18.18*
	GST		*$1.82*
Total: *$20.00* (5)			
	Signed: *K. French* (6)		
	Authorisation: *P. Cashier* (7)		

Comments

1. Number (no.) is usually preprinted on the voucher. The vouchers are in book form, and they are used sequentially (e.g. number 1 is followed by number 2).
2. Date is the date of the reimbursement.
3. Debit relates to the account name to which the expenditure will be directed. It would be too complex to have an account for each item (e.g. one for stamps and another for parcels), so expenditure items are grouped together in a logical manner. If an item grouping is unclear, the petty cashier should obtain assistance from the supervisor relating to the account (debit). The policies and procedures manual may state which account names will be used for that enterprise, and all other expenditure, which does not 'fit' into a particular group, will be placed in an account called 'Sundries'. Every enterprise would require a 'Sundries' account name for infrequent transactions.
4. Particulars gives a fuller explanation of the expenditure. Available space requires that it be kept short.
5. The total at the end of the petty cash voucher is the amount being reimbursed.
6. The voucher must be signed by the person seeking a petty cash reimbursement.
7. Some enterprises also require that the petty cashier sign to authorise the petty cash reimbursement.

Skill development 16.1(i)

Complete petty cash vouchers for expenditure in accordance with enterprise procedures

Assume that you are the petty cashier for Business Stationery Supplies Pty Ltd. The petty cash fund operates on the imprest system. You have been asked to complete the following petty cash vouchers in your workbook.

1. Accounts to be used include GST, postage, stationery and sundries.
2. Voucher numbers are to begin at 1, and are sequential (use one voucher for each line of information).
3. Proof of expenditure has been given.
4. Assume that vouchers have been signed (indicate using a 'x'), and authorise the expenditure.
5. GST is not included in prices.
   - July 12 paid $10.00 for tea and $2.73 for biscuits, GST .27¢
   - July 16 paid $8.64 for stamps, $9.09 for express post, and $2.73 for proof of delivery, GST $2.04
   - July 16 paid $9.09 for stamps, $3.64 for express post, GST $1.27
   - July 18 paid $6.36 for cab fare, GST .64¢
   - July 18 paid $2.27 for biscuits, $12.00 for coffee, GST .23¢
   - July 19 paid $4.55 for magazines and $6.36 for flowers for the reception area, GST$1.09

Present petty cash voucher to authorised person for reimbursement

Once completed, with supporting proof attached, the petty cash voucher is presented to the petty cashier for reimbursement of the amount expended.

Activity 16.3

Validation and authorisation following proof of purchase

Write your answers to the following on a separate sheet of paper.

1. Look at the voucher information in Skill development 16.1(i) above, and discuss the types of proof of purchase that you might expect to receive for each of these items.
2. Who usually has approval to authorise reimbursement of the amount on the petty cash voucher?

Petty cash tin or drawer

Generally, once a petty cash voucher is authorised and paid, it is placed in the petty cash tin or drawer, sequentially with other petty cash vouchers for this period and together with the remaining cash. Note that the total of the cash and vouchers in a petty cash fund must always add up to the original imprest amount.

For example, if the imprest amount is \$50.00, after reimbursement of the voucher shown previously (voucher number 10, for \$20, on page 452), the petty cash fund would have vouchers and cash to the value of \$50.00 as follows:

Vouchers	\$20.00
Cash	\$30.00
Total	\$50.00 (the imprest amount)

Skill development 16.1(ii)

Petty cash tin or drawer

For each of the following, use the table in your workbook to show the total value of vouchers and cash that should be in the petty cash tin for Business Stationery Supplies Pty Ltd on each day. The imprest amount is \$100.00.

- July 12 19XX, voucher number 1, for \$13.00
- July 16 19XX, voucher number 2, \$22.50, and number 3, \$14.00
- July 18 19XX, voucher number 4, \$2.00, and number 5, \$14.50
- July 19 19XX, voucher number 6, \$12.00

Look back to Skill development 16.1(i) in your workbook. Note that you have used the information from your prepared vouchers to determine the value of daily cash remaining.

RECORDING PETTY CASH TRANSACTIONS

Check vouchers for accuracy and authenticity before processing

The amounts for reimbursement on petty cash vouchers must be checked by the petty cashier to ensure they agree with the proof of expenditures attached. They must also be checked for accuracy and authenticity. When checking, ensure that:

- the proof of expenditure amount agrees with the amount being claimed;
- the proof of expenditure has not been altered in any way;
- the voucher date is the reimbursement date (not necessarily the date the expense was incurred);
- the debit (account name) is accurate, and reflects the actual transaction;
- particulars briefly describe the transaction;
- the voucher is signed by the person seeking reimbursement.

The voucher and summary of cash sales printout below was presented to the petty cashier by S. Smythe for reimbursement, and has been returned for correction. Study the voucher and proof of expenditure details and identify the errors. Note that today's date is 25 December 19XX. The enterprise has accounts for travel, postage, stationery and sundries.

Petty Cash Voucher

No. 14 19

Debit

Particulars

Total:

Signed:

Authorisation:

The cash register printout, dated 18 December 19XX and attached to the petty cash voucher, shows:

GST	$1.46	
10 stamps @	$4.55	
Express Post @	$4.09	
Envelopes @	$6.00	$16.10

Comments

1. The voucher date is incorrect; it should state the reimbursement date.
2. The account debit is incorrect; it should be postage.
3. The amount on the voucher is incorrect; it does not agree with the cash register printout.
4. The voucher is unsigned.

The corrected voucher should look like this:

Petty Cash Voucher

No. *14* S.S. *25* ~~18~~ *December* 19 *XX*

Debit *Postage* ~~Travel~~ S.S.

Particulars *Stamps, Express Post, Envelopes* *$14.64*

GST *$1.46*

Total: *$16.10* ~~$16.00~~ S.S. Signed: *S. Smythe*

Authorisation: *P. Cashier*

Note that the petty cash voucher is now authorised by the petty cashier.

Skill development 16.2

Check claims/vouchers for accuracy

The vouchers and proof of expenditure in your workbook were presented to the petty cashier for reimbursement, and have been returned for correction. Study the vouchers and proof of expenditure details and identify and correct the errors.

Authenticity points to note include:

- Authenticity must be checked prior to giving a reimbursement for petty cash.
- An enterprise's policies and procedures manual should state how authenticity is to be verified. For example, a signed authorisation slip may be required, prior to incurring the expenditure.
- Proof of expenditure documentation may include other items that do not form part of the petty cash reimbursement. If so, the relevant items must be circled or highlighted to clearly indicate agreement between the petty cash voucher and proof of expenditure.
- If authenticity cannot be verified, the matter must be referred to the supervisor.

Once accuracy and authenticity have been determined, the petty cashier can authorise the petty cash voucher and give the cash reimbursement.

Activity 16.4

Check petty cash claims/vouchers for authenticity

You are the petty cashier for Business Stationery Supplies Pty Ltd and have been presented with a voucher and supporting proof of purchase for $15.00 for a candle burner and oil for use in the reception area. As you have never had authorisation to approve expenditure of this type, and there is no accompanying authorisation slip, what should you do? List your ideas on a separate sheet of paper.

REFER IRREGULARITIES TO NOMINATED PERSON/SUPERVISOR

The person nominated for referral of irregularities should be documented in an enterprise's policies and procedures manual. In the absence of this, any irregularities should be referred to the petty cashier's supervisor. Irregularities may include:

- supporting evidence not given;
- inconsistent dates on supporting documents (some variation of dates will occur, but a discrepancy of some months may be questioned);
- amounts on supporting documents that appear to have been altered;
- proof of expenditure authorisation not verifiable.

Activity 16.5

Refer irregularities to nominated person

You have been presented with two vouchers for reimbursement:

- One voucher did not have supporting proof of expenditure, but did have an authorisation slip. The person seeking reimbursement has promised to fill out a statutory declaration stating that the expense was incurred.
- The second voucher had supporting proof of purchase dated 9 months prior to the reimbursement request, and did not have an authorisation slip.

What should you do? List your ideas on a separate sheet of paper.

DISTRIBUTE PETTY CASH ACCORDING TO ENTERPRISE GUIDELINES

Distribution of petty cash funds includes the type of system that will be used for petty cash, either paper-based or electronic, and all of the processes, from commencing the fund with the initial imprest cheque, right through to the reimbursement of amounts paid during a particular period.

Prepare petty cash using paper-based or electronic methods

The choice of either paper-based or electronic methods of completing petty cash will be determined by each enterprise, and should be clearly documented in its policies and procedures manual.

Many enterprises choose to use a manual or paper-based system for recording the vouchers and petty cash book, but may use a computerised accounting package, such as 'Mind Your Own Business' (MYOB), for recording associated cheques.

Programs such as Excel lend themselves to the preparation of petty cash books. These can easily be programmed to perform all of the mathematical functions relating to balancing of the petty cash book; however, security measures must be taken when using an electronic system to ensure that changes are not made once the petty cash book has been validated by a supervisor. A printout of each completed page should be kept, and validated, to make theft or fraud more difficult.

Activity 16.6

Paper-based versus electronic systems

Present an argument for using a paper-based system for the petty cash book to your supervisor. Supply at least three valid reasons why you consider a paper-based system would be superior to an electronic system for your needs.

(Alternatively, you may argue the case for an electronic system for petty cash.)

RECORD, TOTAL AND BALANCE TRANSACTIONS IN A PETTY CASH BOOK

Commencing petty cash using the imprest system

The imprest system of petty cash has previously been explained (on page 448). Prior to commencing a petty cash system, it must be determined how much cash is required for the petty cash fund. This will differ according to the needs of the enterprise. Some will require quite large amounts of petty cash, say $200, if they regularly buy items that are individually too small for payment by cheque. An example of an organisation that uses larger amounts of petty cash might be one that repairs motors and requires regular smaller items such as spare parts to be paid for using cash. Other enterprises may use petty cash infrequently, and may require only $50 for small items of stationery, milk and coffee. This amount, once determined, is called the **imprest amount**.

A cheque is prepared and cashed for the amount of the imprest:

Cheque butt	Cheque	
(1) *15 June* 19*XX*	The Northern Bank Brisbane Queensland	(1) *15 June* 19*XX*
To *Cash* (2)	Pay *Cash* (2)	
	The sum of *Fifty Dollars* (4)	$ *50.00* (4)
For *Petty*		
cash –	Business Stationery Supplies Pty Ltd	
Imprest (3)		(x) *P. Cashier* (5)
$ *50.00* (4) 1176 (6)	1176 0999 086754 (6)	

The following must appear on the cheque:

1. the date the cheque was written;
2. the name of the payee. This cheque will be handed over to the bank teller for 'cash', to begin the petty cash fund; therefore, 'cash' is written on both the cheque butt and against the pay instruction on the cheque;
3. petty cash imprest (or advance), indicating the account name in which a record of the transaction will be held;
4. the amount of the cheque, written in words and figures on both the cheque butt and on the cheque;
5. an authorised person's signature;
6. cheque numbers preprinted on the cheque and the cheque butt.

Skill development 16.1(iii)

Prepare the petty cash imprest cheque

Business Stationery Supplies Pty Ltd commenced a petty cash fund using the imprest system on 1 July 19XX. It was decided that an imprest amount of $100 was required to service the needs of the enterprise.

- Prepare the petty cash imprest cheque number 1576.

Choice of notes and coins for the petty cash tin or drawer

Small notes and coins are required for a petty cash fund to enable the payment of small reimbursement amounts. A range of notes and coins is requested, using a change required slip. The cash is placed in a petty cash tin or drawer, under lock and key, and preferably in a fireproof safe.

CHANGE REQUIRED SLIP		
Denomination		Amount $
Notes:	$20.00	
	$10.00	
	$5.00	
TOTAL NOTES		
Coin:	$2.00	
	$1.00	
	$0.50	
	$0.20	
	$0.10	
	$0.05	
TOTAL COINS		
TOTAL CASH		

Recording the imprest cheque

Transaction. An imprest amount of $50, drawn on 15 June 19XX using cheque number 1176 (shown on page 459), will be recorded in the petty cash book as follows:

Petty Cash Book of Business Stationery Supplies Pty Ltd (1)

					Analysis of expenditure			
Amount received	Date	Particulars	voucher no.	Amount paid	Amount of GST	Postage	Stationery	Sundries
$	19XX	(4)		$	$	$	$	$
(2) 50.00	15 June (3)	Petty cash imprest, cheque no. 1176						

Comments

1. The name of the enterprise is recorded at the top of each page.
2. The amount received is the value of the imprest (cash received to start the fund).
3. The date on the cheque must appear in the petty cash book.
4. Write 'petty cash imprest', and list the cheque number. The particulars will always be similar when commencing a petty cash fund for the first time. Note that petty cash imprest can also be referred to as the petty cash advance, because it is the amount 'advanced' to the petty cash fund, to be used for the reimbursement of expenditure.

Skill development 16.1(iv)

Record the imprest cheque

Use the imprest cheque prepared in Skill development 16.1(iii) to begin the petty cash book in your workbook.

Recording petty cash vouchers

The documents that follow have been checked for accuracy and authenticity. Each voucher has proof of expenditure attached. Look carefully at each one, and then refer to the petty cash book that follows them.

Petty Cash Voucher

No. 10 — 20 June 19XX

Debit *Postage*

Particulars *Stamps, Express Post, Parcel to Maryborough, Envelopes*

Total: *$20.00 (includes $1.82 GST)*

Signed: *K. French*
Authorisation: *P. Cashier*

Petty Cash Voucher

No. 11 — 22 June 19XX

Debit *Postage $4.00 & Sundries $8.00*

Particulars *Stamps $3.67, Coffee $8.00 GST .33¢*

Total: *$12.00*

Signed: *B. Ransome*
Authorisation: *P. Cashier*

Petty Cash Voucher

No. 12 — 24 June 19XX

Debit *Stationery*

Particulars *Liquid paper, writing materials, pens, pencils*

Total: *$10.00 (includes .91¢ GST)*

Signed: *K. Carr*
Authorisation: *P. Cashier*

Petty Cash Voucher

No. 13 — 24 June 19XX

Debit *Sundries*

Particulars *Milk and biscuits*

Total: *$5.50 (includes .27¢ GST)*

Signed: *S. Adams*
Authorisation: *P. Cashier*

Petty Cash Book of Business Stationery Supplies Pty Ltd

					Analysis of expenditure (5)			
Amount received	Date	Particulars	Voucher no.	Amount paid	Amount of GST	Postage	Stationery	Sundries
$	19XX			$	$	$	$	$
50.00	June 15	Petty cash advance cheque no. 1176						
	20	Stamps, envelopes, express post, parcel	10	20.00	1.82	18.18		
	22	Stamps & coffee	11	12.00	0.33	3.67		8.00
	24 (1)	Liquid paper, pens, pencils, writing materials	12	10.00	0.91		9.09	
	(7)	Milk & biscuits	13	5.50	0.27			5.23
		(2)	(3)	(4)	(8)	(6)		

Comments

1. Record the voucher date.
2. Record the particulars on the voucher, abbreviated if necessary.
3. Record the voucher number, entered consecutively, from the preprinted voucher.
4. Record the amount paid for reimbursement from each voucher. Note the entry on 22 June, voucher number 11, 19XX. Two items have been bought relating to different accounts, but the total is recorded under 'amount paid'. This price includes GST.
5. The 'analysis of expenditure' columns include all account names required by the enterprise, with a column for sundries. The 'sundries' column will be used for those expenses that occur too infrequently to have their own column heading.
6. Each item of expenditure must be recorded under an analysis of expenditure column.
7. The date is not repeated when it is the same as the date above.
8. The amount of GST.

Skill development 16.1(iv), continued

Record petty cash vouchers

Record details from the petty cash vouchers prepared in Skill development 16.1(i) to update the petty cash book of Business Stationery Supplies Pty Ltd.

Balancing the petty cash book, and drawing a reimbursement cheque

Petty Cash Book of Business Stationery Supplies Pty Ltd

					Analysis of expenditure			
Amount received	Date	Particulars	Voucher no.	Amount paid	Amount of GST	Postage	Stationery	Sundries
$	19XX			$	$	$	$	$
50.00	June 15	Petty cash advance cheque no. 1176						
	20	Stamps, envelopes, express post, parcel	10	20.00	1.82	18.18		
	22	Stamps & coffee	11	12.00	0.33	3.67		8.00
	24	Liquid paper, pens, pencils, writing materials	12	10.00	0.91		9.09	
		Milk & biscuits	13	5.50	0.27			5.23
			(1)	47.50	$3.33	$21.85	$9.09	$13.23 (2)
	(5)	Balance c/d	(4)	2.50				(3)
(6) $50.00			(6)	50.00				
(7) 2.50	25	Balance b/d						
(8) 47.50		Reimbursement cheque no. 1201						
		Bus fare	14	3.60	0.33			3.27
	27	Postage	15	15.00	1.36	13.64		(9)
		Coffee	16	6.00				6.00

Comments

1. Rule a line under the last entries.
2. Total the amount paid and analysis of expenditure columns, and write in the balance date.
3. Rule a double line under the analysis of expenditure columns only.
4. Add up the amount received, deduct the amount paid ($50.00 less $47.50). This is the balance ($2.50).
5. Place this balance on the smaller side, ready to be carried down (c/d), write 'Balance c/d'.
6. Total both sides (look at each side that has been totalled to $50.00. They must always agree). Note the ruling and use of $ signs. Total values are in alignment with each other.
7. Bring the balance down (b/d). As you can see, the balance is brought down into the cash received side, because it represents the amount of cash remaining in the petty cash float. Write 'balance b/d'.
8. Record the amount of the reimbursement cheque, write 'reimbursement cheque' and record the cheque number. The reimbursement amount will be the 'amount paid' out of the petty cash fund. The petty cash book is now ready for you to begin entering vouchers for the next period.
9. Continue entering the next period's vouchers.

Pay particular attention to the use of ruling. Single lines are used when additions are required, and double lines are used once column balancing is completed. Dollar signs are also used to indicate totals that are completed.

The totals of the analysis of expenditure columns, added together, must agree with the total of the amount paid column ($24.00 plus $10.00 plus $13.50 agrees with $47.50).

Skill development 16.1(iv), continued

Balance the petty cash book

Use the numbered instructions above to balance the petty cash book. Complete and carefully check each numbered instruction before continuing to the next. Do not record the reimbursement cheque just yet.

Presenting records of transactions to relevant authority for checking

The policies and procedures manual should detail the checking procedure required for the petty cash book, prior to drawing the reimbursement cheque.

Usually, the petty cashier will give the supervisor all vouchers with supporting evidence, together with the petty cash book, for checking. On completion, the supervisor should verify the reimbursement amount and complete a cheque requisition voucher. This can be used as authorisation for the petty cashier to obtain a petty cash reimbursement cheque from the cashier.

Cheque Requisition Voucher

Authorisation is hereby given for
P. Cashier to obtain a cheque to the
value of *$47.50*, being for *reimbursement of petty cash expenditure.*

Signed: *A. Supervisor*

Complete checking within designated timelines

The process of checking should occur soon after the petty cash book is balanced, and before the reimbursement cheque is drawn. Timelines for this procedure should be clearly documented in the policies and procedures manual.

Skill development 16.1(v)

Cheque requisition voucher

Obtain authorisation of your petty cash procedures by requesting the tutor/supervisor sign the cheque requisition voucher in your workbook, acknowledging the accuracy and authenticity of your procedures. Complete this process within the timeframe designated by your tutor/supervisor.

PREPARE DOCUMENTS FOR REIMBURSEMENT

In the example on page 462 of this chapter, $47.50 has been paid out of the petty cash fund. The petty cashier will give the cashier the signed cheque requisition voucher for a cheque for the amount to reimburse the fund, in order to bring the fund back to the original imprest amount of $50.00. In some enterprises, the petty cashier will be required to prepare the petty cash reimbursement cheque and present it to an authorised person for signature. For a detailed explanation of preparing cheques, refer to page 404 of Chapter 15.

The petty cash cheque, and the cheque butt, should look like this:

	The Northern Bank Brisbane Queensland	
25 June 19XX		25 June 19XX
To Cash	Pay Cash	
	The sum of Forty-Seven Dollars	$ 47.50
For Petty cash	and fifty cents	
Reimbursement (1) GST $3.33 Postage $21.82 Stationery $9.09 Sundries $13.23 $47.50	Business Stationery Supplies Pty Ltd	P. Cashier
1201	1201 0999 086764	

Comments

1. Petty cash reimbursement, followed by details from the dissection columns of the petty cash book, is written on the cheque butt (note that $ amounts are required for each of the analysis of expenditure dissection columns).

For further instruction, refer to page 459 of this chapter.

Look at the petty cash book on page 462 of this chapter, and identify the total from the amount paid column. It is the same as the reimbursement cheque.

Skill development 16.1(vi)

Prepare reimbursement cheque, and update the petty cash book

Use information contained in the petty cash book in your workbook to prepare reimbursement cheque number 1698, dated 19 July 19XX. On completion, record the reimbursement cheque in the petty cash book (Skill development 16.1(iv)) in your workbook.

Starting a new page in the petty cash book

On commencement of a new page in the petty cash book, it is necessary to balance the completed page to obtain balances to be carried forward. These balances must then be brought forward to the next pages. Care must be taken to bring forward correct balances, and to ensure they are brought forward into the correct analysis of expenditure columns.

Using the example previously given, carried forward balancing will look like this:

Petty Cash Book of Business Stationery Supplies Pty Ltd

					Analysis of expenditure			
Amount received	Date	Particulars	Voucher no.	Amount paid	Amount of GST	Postage	Stationery	Sundries
$	June 19XX			$	$	$	$	$
						(previous entries not shown)		
		Milk & biscuits	13	5.50	0.27			5.23
				47.50	$3.33	$21.85	$9.09	$13.23
		Balance c/d		2.50				
$50.00				$50.00				
2.50	25	Balance b/d						
47.50		Reimbursement cheque no. 1201						
		Bus fare	14	3.60	0.33			3.27
	27	Postage	15	15.00	1.36	13.64		
		Coffee	16	6.00				6.00
$50.00		Balance c/f		$24.60	$1.69	$13.64		$9.27

And when they are brought forward to the new page, the new page is as follows:

Petty Cash Book of Business Stationery Supplies Pty Ltd

					Analysis of expenditure			
Amount received	Date	Particulars	Voucher no.	Amount paid	Amount of GST	Postage	Stationery	Sundries
$	19XX			$	$	$	$	$
50.00	27 June	Balance b/f		24.60	1.69	13.64		9.27

Hint: As you write 'balance c/f', think 'Balance to be carried forward', and as you write 'balance b/f' think 'Balance is now brought forward'. This way, you will use the correct abbreviations.

Skill development 16.1(iv), continued

Continue entering the following vouchers, starting with voucher number 7. You are not required to prepare petty cash vouchers.

- August 4 paid $12.50 for bus and cab fares
- August 6 paid $7.50 for tea and biscuits
- August 12 paid $15.00 for pens, pencils and writing paper
- August 12 paid $10.00 for stamps, express post
- August 20 paid $16.90 for cleaning products
- August 25 paid $12.00 for cab fare

Skill development 16.1(vii)

Prepare reimbursement cheque number 1751, dated 26 August 19XX.

Skill development 16.1(iv), continued

Enter vouchers for second month

Balance and reimburse the petty cash book in your workbook.

Increasing the imprest amount

At some time it may be necessary to increase the petty cash imprest. This requires approval from a person indicated in the enterprise's policies and procedures manual, usually at management level. The petty cashier will obtain approval for the increase, and on approval will obtain a cheque requisition voucher requesting a cheque for the higher amount from the cashier. This cheque will be taken to the bank's teller and cashed. The increase in imprest will look like this:

Petty Cash Book of Business Stationery Supplies Pty Ltd

					Analysis of expenditure			
Amount received	Date	Particulars	Voucher no.	Amount paid	Amount of GST	Postage	Stationery	Sundries
$	19XX			$	$	$	$	$
50.00	June 27	Balance b/f		$24.50	1.69	13.64		9.27
(1) 40.00	29 (2)	Petty cash advance (increase—cheque no. 1307) (3)						
	30	Bus & cab fares	17	15.00	1.36			13.64 (4)

Comments

1. Record the amount of the increase.
2. Record the date from the cheque butt (the date of the increase).
3. Write 'petty cash advance (increase)' and record the cheque number.
4. Continue entering the next period's vouchers.

Skill development 16.1(iv), continued

Record imprest cheque

Management has today (4 September 19XX) approved an increase in the petty cash fund, to bring the fund up to $140.00. Record the imprest increase cheque number 1832 by updating your petty cash book in your workbook. You are not required to complete the cheque.

Skill development 16.3

You have today (1 September 19XX) opened a new branch of Business Stationery Supplies Pty Ltd at Wynnum, and have agreed to set up and complete their new petty cash book for the first three months, to assist them.

An imprest amount of $50.00 is sufficient to begin the fund. Prepare and record imprest cheque number 1888. Voucher numbers commence at 60.

Record the following transactions that occurred during the month of September **(all amounts include GST (1/11th) as shown)**:

- Sept 18 paid $8.00 for bus fares (GST .73¢)
- Sept 18 paid $2.50 for parcel express (GST .23¢)
- Sept 20 paid $7.00 for stamps (GST .64¢)
- Sept 28 paid $10.00 for magazines for the reception room (GST .91¢)
- Sept 30 paid $4.50 for refreshments (GST .41¢)
- Sept 30 paid $12.00 for pens (GST $1.09)

- Balance the petty cash book for 1 October and prepare and record reimbursement cheque number 1950. You will need to use columns for postage, stationery, GST and sundries.
- Check the completed vouchers in your workbook for accuracy. If any errors are evident, make corrections (on the vouchers), and initial them. Place a note at the bottom of the page, advising the tutor/supervisor of the error.
- Complete the last two vouchers.
- Present your completed petty cash book and vouchers to your supervisor for validation.

On 4 October approval was given to bring the imprest up to $100.00. Enter the petty cash imprest increase (cheque number 2044) in the petty cash book (you are not required to complete the cheque).

Continue entering the next month's transactions. You are not required to prepare the vouchers or reimbursement cheque. If you need to commence a new page, refer to page 466 of this chapter. The following transactions have occurred during the month of October:

- Oct 16 paid $14.00 for tea and coffee (GST exempt)
- Oct 18 paid $7.00 for express post (GST .64¢)
- Oct 20 paid $20.00 for cab fares (GST $1.82)
- Oct 20 paid $15.50 for flowers for the reception (GST $1.41)
- Oct 21 paid $12.50 for notepads, and telephone message pads (GST $1.14)
- Oct 22 paid $5.50 for bus fares (GST .50¢)

Balance the petty cash book for 31 October and record the reimbursement cheque number 2155.

Continue entering the next month's transactions. You are not required to prepare the vouchers or reimbursement cheque. The following transactions have occurred during the month of November:

- Nov 4 paid $5 for cake and coffee (GST .45¢)
- Nov 4 paid $10.50 for a tap washer (GST .95¢)
- Nov 10 paid $15.00 for cab fares (GST $1.36)
- Nov 20 paid $20.50 for postage (GST $1.86)

Balance the petty cash book for 30 November and record the reimbursement cheque number 2277.

Skill development 16.4

You have today (1 April 19XX) opened a new branch of Business Stationery Supplies Pty Ltd at Richmond, and have agreed to set up and complete their new petty cash book for the first three months to assist them.

An imprest amount of $70.00 is sufficient to begin the fund. Prepare and record imprest cheque number 1777. Voucher numbers start at 20.

Record the following transactions that occurred during the month of April **(all amounts include GST (1/11th) as shown)**:

- April 18 paid $16.90 for nuts and bolts (GST $1.54)
- April 18 paid $4.80 for cab fares (GST .44¢)
- April 20 paid $18.50 for stamps (GST $1.68)
- April 30 paid $6.00 for cab fares (GST .55¢)
- April 30 paid $12.00 for pens and writing paper (GST $1.09)
- April 30 paid $10.00 for magazines for the reception room (GST .91¢)
- You will need to use columns for postage, stationery, GST and sundries.
- Check the completed vouchers in your workbook for accuracy. If any errors are evident, make corrections (on the vouchers), and initial them. Place a note at the bottom of the page, advising the tutor/supervisor of the error.

- Complete the last two vouchers.
- Balance the petty cash book for 30 April and prepare and record reimbursement cheque number 1876.
- Present your completed petty cash book and vouchers to your supervisor for validation.

On 4 May approval was given to bring the imprest up to $100.00. Enter the petty cash imprest increase cheque number 2001 in the petty cash book (you are not required to complete this cheque).

Continue entering the next month's transactions. You are not required to prepare the vouchers or reimbursement cheque. The following transactions have occurred during the month of May:

- May 16 paid $15.00 for tea and coffee (GST exempt)
- May 20 paid $20.00 for cab fares (GST $1.82)
- May 20 paid $15.50 for flowers for the reception (GST $1.44)
- May 21 paid $12.50 for notepads and writing paper (GST $1.14)
- May 28 paid $17.00 for express post (GST $1.55)
- May 28 paid $9.00 for prepaid postage bags (GST .82¢)

Balance the petty cash book on 31 May and record reimbursement cheque number 2100.

Continue entering next month's transactions. You are not required to prepare the vouchers or reimbursement cheque.

The following transactions have occurred during the month of June:

- June 4 paid $10.50 for a taxi fare (GST .95¢)
- June 10 paid $15.00 for cab fares (GST $1.36)
- June 20 paid $20.50 for postage (GST $1.86)
- June 23 paid $4.50 for a birthday card (GST .41¢)
- June 25 paid $9.00 for rubbers, pencils and writing paper (GST .82¢)

Balance the petty cash book for 30 June and record reimbursement cheque number 2179.

Activity 16.7

Review theory

1. Write a short paragraph explaining the following statement: 'A petty cashier is accountable for the petty cash'.
2. Give two security procedures designed to protect staff and cash when dealing with petty cash.
3. Internal controls over petty cash are important. Why are they important?
4. Give three internal controls that should be practised when handling petty cash.
5. Why is it necessary to have specific policies and procedures relating to petty cash?
6. How should the petty cashier use the policies and procedures manual?
7. List at least three irregularities that you should check for when reimbursing from petty cash vouchers.
8. To whom should you refer irregularities?

Summary

In this chapter you have learnt how to prepare petty cash vouchers in order to claim reimbursement according to enterprise procedures. You have also learnt how to record and balance petty cash transactions, following the enterprise's policies and procedures guidelines.

You should now be able to:

- understand the importance of the job of the petty cashier;
- see that the petty cashier must be trustworthy and be accountable for his/her actions;
- follow security procedures to ensure the safety of yourself and other employees when dealing with cash;
- complete vouchers accurately, and detect errors in documents presented by others for reimbursement;
- record, balance, reimburse and increase the petty cash fund, in line with the needs of the enterprise;
- prepare related documents such as cheques in accordance with enterprise guidelines.

Review questions

(This integrated activity may be used for self-assessment or for formal assessment.)

1. You have been asked today (10 January 19XX) to set up and administer the petty cash fund for the local police youth club. They have decided that $50.00 is sufficient for their needs, and have given you approval to complete a petty cash cheque (use cheque number 1899) in readiness for signing.

   You need account headings for postage, cleaning, GST and sundries. Voucher numbers start at 40. **The amount paid includes GST as shown.**

   The following transactions occurred during the month of January:

   - Jan 14 paid $6.00 for Bon Ami cleaning liquid (GST .55¢)
   - Jan 14 paid $2.50 for scouring cloth (GST .23¢)
   - Jan 19 paid $5.90 for stamps (GST .54¢)
   - Jan 19 paid $12.00 for tennis balls (GST $1.09)
   - Jan 30 paid $8.50 for cordial and biscuits (GST .77¢)
   - Jan 30 paid $12.00 for window cleaner (GST $1.09)

   Balance the petty cash book for 30 January and prepare and record reimbursement cheque number 2014.

   Check the completed vouchers in the workbook (Skill development 16.1(ii)) for accuracy. If any errors are evident, make corrections (on the vouchers) and initial them. Place a note at the bottom of the page, advising the supervisor of the error.

   Complete the last two vouchers.

   Present your completed petty cash book and vouchers to your supervisor for validation.

   On 4 February approval was given to increase the imprest by $30.00. Complete the entry in the petty cash book for petty cash imprest increase cheque number 2195 (you are not required to complete this cheque).

   Continue entering the next month's transactions. You are not required to prepare the vouchers or reimbursement cheque.

   The following transactions have occurred during the month of February:

   - Feb 16 paid $14.00 for cordial and chocolate biscuits (GST $1.27)
   - Feb 18 paid $6.75 for Palmolive dishwashing liquid (GST .61¢)
   - Feb 20 paid $20.00 for record of achievement certificates (GST $1.82)
   - Feb 20 paid $5.50 for coffee (GST exempt)
   - Feb 21 paid $13.75 for Kleen & Clear floor cleaner (GST $1.25)
   - Feb 22 paid $5.50 for a birthday cake (GST .50¢)

   Balance the petty cash book for 28 February and record reimbursement cheque number 3002.

2. Accountability, security and control are important when dealing with petty cash. Explain, giving two examples for each term.

3. Explain irregularities that may occur in relation to petty cash vouchers, and identify the person you would refer these irregularities to.

Competency self-assessment and/or tutor/supervisor assessment

When you have finished studying the chapter, and have completed all of the Activities and Skill development exercises, complete the self-assessment and then hand it to your tutor/supervisor to complete.

Learning outcomes	Self-assessment		Tutor/supervisor assessment	
I have successfully completed the following learning outcomes	**Yes** (please tick)	**No**	**Yes** (please tick)	**No**
1. Claim reimbursement from the petty cash fund according to enterprise procedures				
1.1 discuss the need for accountability, security and control when dealing with money	☐	☐	☐	☐
1.2 complete petty cash voucher for expenditure in accordance with enterprise procedures and present to authorised person for reimbursement	☐	☐	☐	☐
2. Record and balance petty cash transactions for financial records according to organisational procedures				
2.1 check petty cash claims/vouchers for accuracy and authenticity before processing	☐	☐	☐	☐
2.2 refer irregularities to nominated person/section	☐	☐	☐	☐
2.3 distribute petty cash according to enterprise guidelines	☐	☐	☐	☐
2.4 record, total and balance transactions in petty cash book	☐	☐	☐	☐
2.5 present records of transactions to relevant authority for checking within designated timelines	☐	☐	☐	☐
2.6 prepare documents for reimbursement of the petty cash fund according to enterprise guidelines	☐	☐	☐	☐
Date:	Tutor/supervisor signature:			

CHAPTER

17

BANK DEPOSITS

FIN101A, NOS152, 8872A

objectives

The purpose of this chapter is to provide you with the knowledge and skills to prepare and process documents for banking, in order to complete the following learning outcomes:

1. Prepare and process documents for banking
   - 1.1 check credit card vouchers and cheques for validity before processing
   - 1.2 count and list cash, cheques and credit card vouchers on relevant pay-in banking forms for depositing funds and total according to banking institution's guidelines
   - 1.3 reconcile pay-in documentation with all money calculations
   - 1.4 bundle cash, cheques and credit card vouchers according to banking institution's guidelines

2. Lodge deposits with banking institution

This chapter is written in the form of the banking policies and procedures manual extract used by Business Stationery Supplies Pty Ltd. It is structured to allow you to proceed at your own pace. There are no prerequisites.

(At the end of the chapter, when you have achieved all the above outcomes, you and/or your tutor/supervisor will be able to evaluate your performance for assessment.)

Banking is a function that is performed by all businesses. Many transactions lead to the eventual receipt of cash. Because cash is easily stolen, adequate procedures must be implemented to ensure that people with dishonest intentions cannot easily commit theft. Procedures must also ensure that if theft and fraud do occur, they are detected quickly. The receipt of cash, cheques and, for example, credit card vouchers must be acknowledged by the preparation of a receipt. All enterprises should practise daily banking, to ensure that cash does not remain overnight on the premises, as this could encourage break-ins.

ORGANISATIONAL POLICIES AND PROCEDURES

Policies and procedures manual

Policies and procedures manuals are books that include detailed examples and explanations showing staff how to perform their related duties.

The policies and procedures manual for bank deposits should include a guide to:

- preparation of banking documents, which comply with the bank's requirements;
- checking procedures to ensure that the correct amount is banked;
- receiving and acknowledging the receipt of money (cash, cheques etc.);
- internal controls over cash.

General security procedures

The following procedures will make it more difficult for employees to steal cash.

- The duties of employees should be arranged such that those who receive cash should not also record the cash. This is called separation of duties.
- When receiving cash in the mail, there should be two employees involved, who collectively record the receipt.
- Cash received should be banked regularly, usually daily, and *intact* (intact meaning that the whole amount received is banked and amounts are not deducted, for example, for small cash expenses).
- A supervisor should regularly check the daily banking against money calculations for the same day.

Occupational health and safety procedures

The manual should also include guidelines for occupational health and safety for employees who handle cash. Because cash is easily stolen, guidelines should be implemented to ensure the safety of those employees. For example:

- Employees should not divulge amounts being banked.
- Different employees should perform banking.

- A different route should be taken to the bank each day.
- Vehicles being used for banking should alternate.
- Cash for banking should not be openly carried in a bank's carrying bag, as these are too obvious (put it into another carrying bag).
- If banking large sums of money, security guards should be employed to accompany staff while banking.
- Security firms should be employed to transport cash to the bank if amounts are excessively large or the area is considered high-risk.

It is important when considering safety and security procedures for employees who handle cash that predictable patterns do not occur. Being unpredictable makes it far more difficult for thieves to plan a robbery.

Accountability

Because cash is easily stolen, it is very important to ensure that employees are accountable for their actions in relation to the following of procedures designed to ensure the security of cash. Employees should be accountable for the correct preparation of banking documents, and this should be detailed in the enterprise's policy manual. They should also be accountable to a more senior employee, usually a supervisor, who ensures that each process is accurately performed. Supervisors in turn are accountable to management. A tiered system of accountability in relation to cash ensures that the policies and procedures of an enterprise are accurately followed; in so doing, theft and fraud can be largely eliminated.

Activity 17.1

Organisational policies and procedures

On a separate sheet of paper, answer the following questions:

1. What is a policies and procedures manual?
2. List four examples of items relating to bank deposits that may be included as a policy or procedure.
3. List two general security procedures that help to guard against theft.
4. What does it mean to be accountable?
5. Explain how accountability can be carried out in relation to bank deposits.

RECEIVING CASH, CHEQUES AND CREDIT CARD INFORMATION

A narrow interpretation of **cash** includes notes and coins; however, in bookkeeping the word cash is taken to include all forms, including notes, coins, cheques and credit card vouchers. Cash includes all forms of negotiable instruments. On receipt of cash, a receipt must be written immediately, documenting the total amount. These receipts can be used as a basis for checking the accuracy of the amount being banked each day. Receipts are source documents that will be used by the business to enter details in their cash receipts records. It is very important that receipts are accurate.

CHECKING THE VALIDITY OF CHEQUES BEFORE PROCESSING

Cheques

When a cheque is received, it must be checked immediately to ensure its validity. If incorrect cheques are banked, the cheque will be dishonoured (not paid by the bank and returned to the drawer). This is expensive, as a dishonoured cheque fee is charged. It is therefore important to ensure that cheques are correct before banking. Assume that today's date is 16 May 19XX. Look at the sample cheque below:

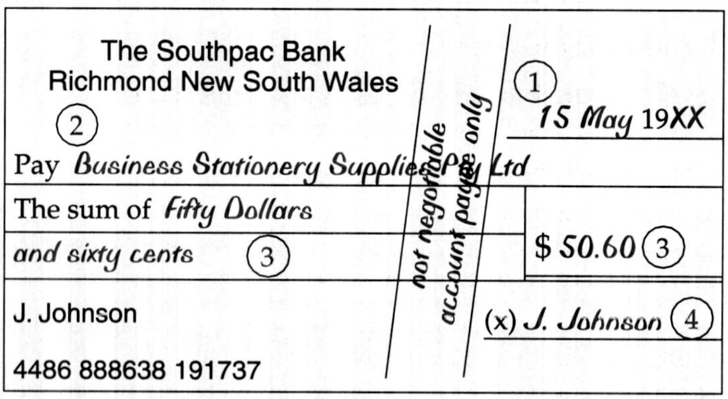

The following checks must be performed:

1. Look at the date: it must not be postdated (a date which has not yet arrived). It is illegal to postdate a cheque. It is particularly important to check for date errors at the beginning of a new year, as it is a common error to enter the previous year's date. If the date is in excess of six months old the cheque may not be honoured (paid) by the bank, as it could be considered 'stale'. If a cheque were nearing six months old, you

would need to contact the bank whose name appears on the cheque to verify that the cheque is still valid.

2. Check that the name indicated as payee is correct. If another name appears as the payee then an endorsement must be written on the back of the cheque, from the payee, stating that the cheque is to be made payable to your enterprise; this must be signed by the payee.
3. The amount in words and figures should be the same. If there is a discrepancy the bank will honour, at its discretion, the written value of the cheque. It would be reasonably safe to bank cheques where small discrepancies occur (e.g. less than one dollar), but the cost of a dishonoured cheque may be such that it would be safer to return the cheque for correction.
4. The cheque must be signed.

Look at the sample cheque that follows and identify the errors (note that today's date is 16 May 19XX):

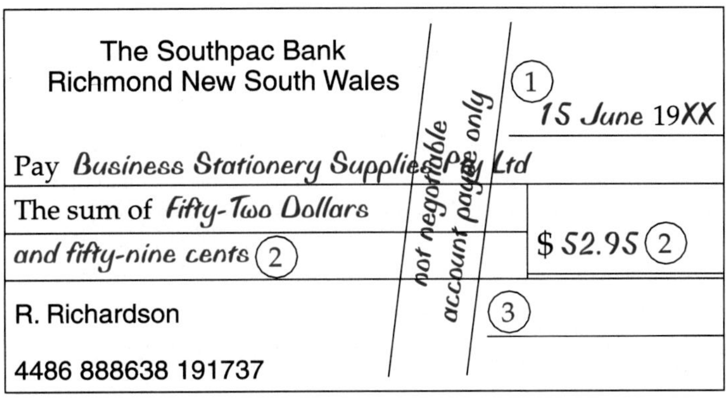

The following errors are observed:

1. The cheque is incorrectly postdated.
2. The numerical value is different from the written value.
3. The cheque is unsigned.

This cheque must be returned for correction before processing.

For the purposes of this chapter, it is the policy of Business Stationery Supplies Pty Ltd to return all cheques, requesting their correction, where errors exist.

Skill development 17.1(i)

Check cheques for validity before processing

Look at the cheques presented in your workbook for banking. There are six cheques. If you find any errors, circle the error, and write a note on a separate piece of paper to advise your tutor/supervisor of the errors. Incorrect cheques are to be returned for correction, in line with the policy of Business Stationery Supplies Pty Ltd. Assume that today's date is 17 May 19XX.

ELECTRONIC BANKING (EFTPOS)

Some organisations offer EFTPOS (Electronic Funds Transfer at Point Of Sale) facilities as a means for customers to pay for goods and services. With EFTPOS the transaction is processed immediately via on-line facilities that link the organisation directly to the EFTPOS department of its bank. When using EFTPOS, the customer swipes a card into the EFT terminal, the merchant enters transaction details, and the customer enters an account type and secret (pin) number (or may choose 'credit', which does not require a pin). The transaction is sent and, if sufficient funds are available, the funds are approved and the customer is given a receipt. This form of banking happens automatically, without the need to physically bank cash.

Activity 17.2

Electronic banking

1. On a separate sheet of paper, write a definition for EFTPOS.
2. Research a local store that uses EFTPOS, and then list as many advantages and disadvantages that you can envisage when using EFTPOS.

CHECK CREDIT CARD VOUCHERS FOR VALIDITY BEFORE PROCESSING

Sales vouchers

Some organisations do not have on-line facilities available and therefore process transactions manually, using a credit card imprinter. In this case (e.g. for Bankcard, MasterCard, Visa) a sales voucher is completed.

Authorisation must be obtained from the credit card section of the merchant's bank for amounts in excess of the allowable store value. This excess amount differs from one organisation to the next, depending on the type of goods and/or services being offered. If authorisation is required, the merchant (storekeeper) will phone his/her bank to obtain an authorisation number.

The merchant requesting authorisation must give relevant information, including:

- the merchant number (this is the number given to the merchant granting him/her authorisation to offer this credit facility to customers);
- the credit card number of the client;
- the expiry date of the credit card; and
- the amount and type of transaction (e.g. mail order, manual sale).

An action code will then be attached to the transaction, and an approval number will be quoted, which must be written in the box at the top of the sales voucher, marked 'authorisation no'.

Although you are not required to complete a sales voucher, you must be able to check that it is correctly completed prior to entering it in a daily transaction summary (merchant summary). A sample credit card voucher appears below.

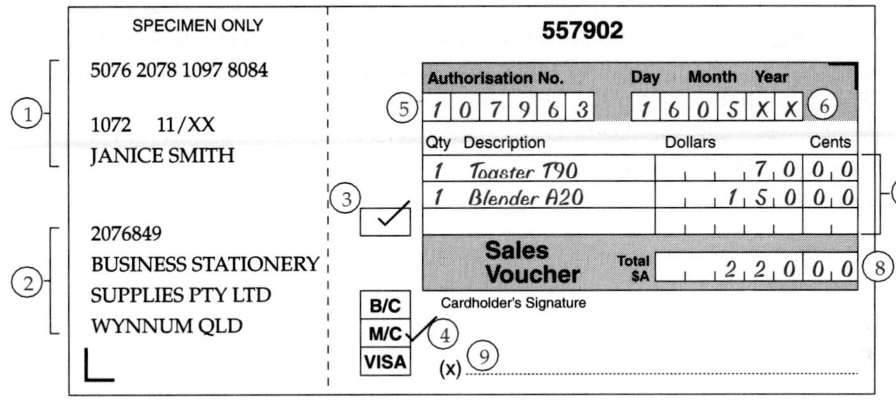

SPECIMEN ONLY

557902

(1) 5076 2078 1097 8084
1072 11/XX
JANICE SMITH

(2) 2076849
BUSINESS STATIONERY
SUPPLIES PTY LTD
WYNNUM QLD

(3) ✓

Authorisation No.	Day	Month	Year
(5) 107963	16	05	XX (6)

Qty	Description	Dollars	Cents
1	Toaster T90	70	00
1	Blender A20	150	00

(7)

Sales Voucher — Total $A 220 00 (8)

B/C
M/C ✓ (4)
VISA

Cardholder's Signature

(x) (9) ....

This voucher must be checked to ensure that it contains:

1. imprinted information from the customer's credit card, giving card number, expiry date and name;
2. imprinted information relating to the merchant, giving merchant number, name and suburb;
3. a 'tick' in the top box to indicate that the card expiry date and validity have been checked—a list of lost and stolen cards should be checked to ensure its validity, and the expiry date on the card should be checked to ensure that it is not out of date;
4. a second 'tick' to indicate which type of credit card is being presented (Bankcard, MasterCard, Visa);
5. an authorisation number, given if the amount exceeds the store's limit for non-authorised transactions;
6. the transaction date;
7. transaction details, including the value;
8. the total value of the sales voucher;
9. the cardholder's/customer's signature.

Skill development 17.1(ii)

Check credit card vouchers for validity before processing

Look at the credit card vouchers presented in your workbook for banking. There are six sales vouchers. If you find any errors, circle them, and write a note on a separate piece of paper to advise your supervisor of the errors. Accuracy in checking sales vouchers is imperative at the point of sale, in line with the policy of Business Stationery Supplies Pty Ltd. (Note that if errors are found once a customer has left the store, the matter must be referred to the tutor/supervisor.) ***Only correct sales vouchers will be banked.***

Note that the store's limit for non-authorised transactions is **$50.00.**

COUNT AND LIST CASH, CHEQUES AND CREDIT CARD VOUCHERS ON BANKING FORMS, ACCORDING TO BANKING INSTITUTION'S GUIDELINES

Merchant summary (transaction summary)

At the end of each day, the merchant must bank all sales vouchers. These are prepared for inclusion in the daily banking by summarising sales vouchers on a merchant summary. A copy of the merchant summary may be used as the source document for entry into the business's cash records.

Look at the merchant summary shown below.

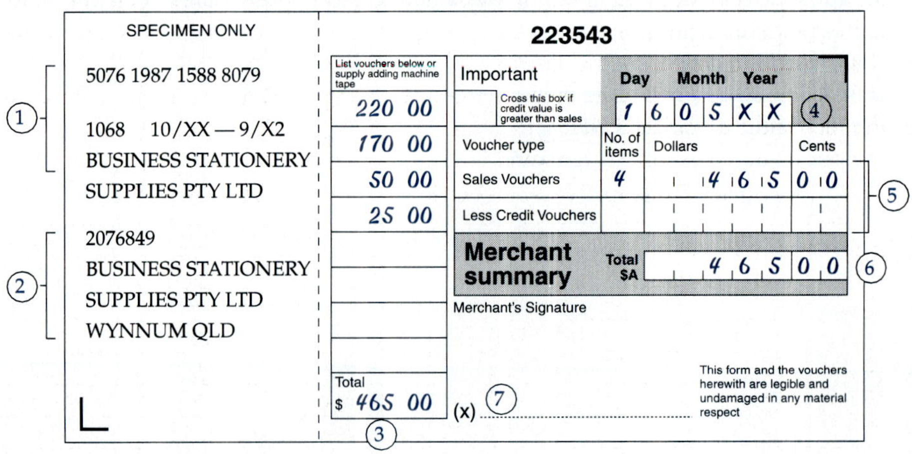
SPECIMEN ONLY

223543

(1) 5076 1987 1588 8079
1068 10/XX—9/X2
BUSINESS STATIONERY
SUPPLIES PTY LTD

(2) 2076849
BUSINESS STATIONERY
SUPPLIES PTY LTD
WYNNUM QLD

List vouchers below or supply adding machine tape

220 00
170 00
50 00
25 00
Total $ 465 00 (3)

Important — Cross this box if credit value is greater than sales

Day Month Year: 16 05 XX (4)

Voucher type	No. of items	Dollars	Cents
Sales Vouchers	4	465	00
Less Credit Vouchers			

(5)

Merchant summary — Total $A 465 00 (6)

Merchant's Signature

(x) (7) ....

This form and the vouchers herewith are legible and undamaged in any material respect

Comments

The following must appear on the merchant summary:

1. the imprint of the merchant's own credit card, in order that the funds can be placed in the required account;
2. the imprint of the merchant number;
3. the value of each sales voucher, added together;
4. the banking date;
5. the total of all sales vouchers, and any credit vouchers, if applicable;
6. the total of the sales vouchers less any credit vouchers, if applicable;
7. the merchant's signature.

On completion of the merchant summary, the amount is recorded on the bank deposit form (see page 485, point 3).

Skill development 17.1(iii)

Merchant summary

Using the credit card vouchers presented in Skill development 17.1(ii), enter the correct sales vouchers only in the merchant summary provided in your workbook.

Counting notes and coins

Before counting notes they should be placed in bundles of 10, folded right-side out, and secured with a rubber band. Coins should be sorted into denomination, in plastic packets available from the bank. Each packet should contain a specific stated amount. The sorting of notes and coins in this manner makes it easier for the enterprise and the bank to avoid counting errors, and therefore simplifies the banking function.

When working out the value of notes and coins being banked, it is useful to draw up a summary sheet to look something like this:

Summary Table for Notes and Coins Received on 16 May 19XX

Denomination	No.	Total	Denomination	No.	Total
$100	23	2300	$2.00	4	8.00
$50	5	250	$1.00	0	0
$20	8	160	0.50 cents	10	5.00
$10	1	10	0.20 cents	16	3.20
$5	2	10	0.10 cents	8	.80
			0.05 cents	12	.60
Total notes		$2730	Total coins		$17.60

Once a total for notes and coins is ascertained, the value is inserted on the bank deposit form (see page 485, points 1 and 2).

Skill development 17.1(iv)

Count notes and coins

Complete the summary sheet in your workbook for the following notes and coins received on 17 May 19XX: 4 × \$100, 5 × \$50, 24 × \$20, 18 × \$10, 35 × \$5, 28 × \$2, 36 × \$1, 20 × \$0.20, 5 × \$0.05.

Use a calculator to add the totals.

LODGE DEPOSITS WITH BANKING INSTITUTION

Bank deposit forms

To record the total of notes, coins, credit card vouchers and cheques for a particular day, the bank uses a bank deposit form. When a business decides to bank with a particular bank, it agrees to abide by the bank's policies and procedures in relation to the presentation of its daily banking. This means that an enterprise must complete the bank deposit form, and lodge notes and coins etc. in the manner required by the bank. A business may be required to bag or roll coins, to bundle notes together, and to complete a merchant summary to summarise credit card vouchers, in order to simplify the banking process. Furthermore, the bank will have its own bank deposit form, which contains particulars relating to the bank, the branch, and other depositing details as required by the bank.

Most business accounts use a large deposit form, as business deposits usually consist of numerous cheques. Small deposit forms can also be used, particularly if a business has a policy of cash rather than cheque receipts. The information contained within large and small deposit forms is similar, the main difference being in the amount of space provided for the insertion of cheque information. The location of information will differ slightly from bank to bank, and also from form to form; however, the forms are similar enough for you to be able to transfer the knowledge learned in this chapter in order to complete other bank deposit forms.

Look at the sample bank deposit form opposite:

Where this deposit is lodged at a bank or branch other than shown below it will be transferred under the bank's internal procedures.

TO>	The State Queensland Bank Limited (11) *16 May 19XX*		Notes	*2730*	*00* (1)
	Branch: *WYNNUM*		Coins	*17*	*60* (2)
	Amount in words: *Four thousand, seven hundred and forty-one dollars and ten cents* (12)	(x) (14) Teller	Bankcard/Credit vouchers	*465*	*00* (3)
	Paid in by (Signature) (x) *A. Jackson* (13)	No. of chqs / Comm.	Cheques	*1528*	*50* (4)
	(Proceeds of cheques, etc. will not be available until cleared)	Account number	*Less* Charges		
	Credit account name (in block letters) *BUSINESS STATIONERY SUPPLIES PTY LTD* (15)	*5594 9849494 888494* (16)	Total $	*4741*	*10* (5)

This bank is not responsible for delays in transmission to branch.

Where proceeds are remitted to another branch, detach cheque listing and retain at initiating branch

Date deposited | Account name and number | Particulars of cheques etc. to be completed by depositor

Drawer (6)	Bank (7)	Branch (8)	Amount (9)	
T. Toms	*NAB*	*Capalaba*	*544*	*50*
K. Karson	*Bank of Qld*	*Cleveland*	*594*	*00*
C. Jones	*CTB*	*Wynnum*	*220*	*00*
O. Mann	*NAB*	*Ormiston*	*170*	*00*
		Carried forward $	*1528*	*50* (10)

Comments

The following must appear on the bank deposit form:

1. the total value of notes, from summary sheet;
2. the total value of coins, from summary sheet;
3. the total of credit card vouchers, taken from merchant summary;
4. the total of cheques, from the carried forward balance in the cheque listing at the bottom of the form;

5. the total value of all notes, coins, credit card vouchers and cheques, less any charges;
6. the 'drawer' of the cheque (from printed name on the cheque);
7. the bank, shown on the cheque (can be abbreviated);
8. the branch, shown on the cheque;
9. the amount of each cheque, itemised;
10. the total value of all cheques being banked;
11. the deposit date;
12. the amount written in words (must be the same value as point 5);
13. the signature of the person depositing the funds;
14. the teller's signature, number of cheques and comm. information for bank use;
15. the account name where funds are being deposited (sometimes preprinted);
16. the account number, relating to the deposit.

Skill development 17.1(v)

Prepare a bank deposit form

- On the basis of the information contained in Skill development 17.1(i) to 17.1(iv) inclusive, prepare a bank deposit form for 17 May 19XX using the blank bank deposit form in your workbook.

Business Stationery Supplies Pty Ltd banks at the Wynnum branch. Its account number is 5594 9849494 888494.

RECONCILIATION OF DOCUMENTATION WITH MONEY CALCULATIONS

By comparing the information on the bank deposit form with a summary of receipts, merchant summary and cash register summary for the same day, a business can easily determine whether the correct amount has been prepared for deposit each day.

Reconciling cash received with amount listed on bank deposit form (and other money calculations)

The amount of money banked must agree with receipts written for the day, any cash received from cash sales, and the bankcard summary, if not included in the cash register summary. Cash sales are usually documented on a cash register summary, which shows the total amount received each day.

The following receipts, sales vouchers and cash register summary are available for 16 May 19XX.

- Look at each receipt and note whether the amount was received in cash or cheques.
- Cheque details are given below each receipt. Confirm that these details are recorded accurately on the bank deposit form on page 485 also.

The cash register summary for 16 May 19XX indicated that $2302.00 was received from cash sales, and this was made up as follows: $2290.00 in notes, $12.00 in coins. As well, four sales vouchers totalling $465.00 were recorded on the merchant summary.

Receipts issued for 16 May 19XX are as follows:

Business Stationery Supplies Pty Ltd
126 Wynnum Road
WYNNUM QLD

Official Receipt
No. 106

Date	Received from	Particulars	Discount	By Cash/Cheque (circled: Cheque)
16 May 19XX	C. Jones	Sales $200 00 GST $20 00		220 00

per: (x) A. Jackson (Signature)

Sold goods to C. Jones, $220.00, and received a CTB, Wynnum, cheque.

Business Stationery Supplies Pty Ltd
126 Wynnum Road
WYNNUM QLD

Official Receipt
No. 107

Date	Received from	Particulars	Discount	By Cash/Cheque (circled: Cash)
16 May 19XX	A. Hass	Sales $205 09 GST $20 51		225 60

per: (x) A. Jackson (Signature)

Sold goods for cash to A. Hass, $225.60.

Business Stationery Supplies Pty Ltd
126 Wynnum Road
WYNNUM QLD

Official Receipt
No. 108

Date	Received from	Particulars	Discount	By Cash/Cheque
16 May 19XX	T. Toms	April account	5 50	544 50

per: (x) A. Jackson .... (Signature)

T. Toms settled his April account by paying $544.50 using an NAB, Capalaba, cheque, and receiving a discount of $5.50.

Business Stationery Supplies Pty Ltd
126 Wynnum Road
WYNNUM QLD

Official Receipt
No. 109

Date	Received from	Particulars	Discount	By Cash/Cheque
16 May 19XX	S. Smith	Sales $200 00 GST $20 00		220 00

per: (x) A. Jackson .... (Signature)

Sold goods to S. Smith, and received $220 cash.

Business Stationery Supplies Pty Ltd
126 Wynnum Road
WYNNUM QLD

Official Receipt
No. 110

Date	Received from	Particulars	Discount	By Cash/Cheque
16 May 19XX	K. Karson	April account	6 00	594 00

per: (x) A. Jackson .... (Signature)

K. Karson paid her April account by paying $594, using a Bank of Queensland, Cleveland, cheque and receiving a discount of $6.00.

Business Stationery Supplies Pty Ltd
126 Wynnum Road
WYNNUM QLD

Official Receipt
No. 111

Date	Received from	Particulars	Discount	By Cash/Cheque
16 May 19XX	O. Mann	Sales $154 55 GST $15 45		170 00

per: (x) A. Jackson .... (Signature)

Sold goods to V. Mann for $170.00 and received NAB, Ormiston, cheque.

The merchant summary, cash register summary and receipts are summarised below. If you were checking the receipts for a day, you would draw up a rough summary like this, to simplify the process, although if only a few receipts are involved, it would be fairly easy to check the amounts directly from the bank deposit form. The table below helps to simplify the checking process, to ensure that the correct amounts are recorded on the bank deposit form, and also confirms that the summary table for notes and coins, the merchant summary, and the recording of cheques on the bank deposit form, are correct.

Source document	Cash	Credit sales vouchers	Cheques
Receipt no. 106			220.00
107	225.60		
108			544.50
109	220.00		
110			594.00
111			170.00
Cash register summary (CRS)	2302.00		
Merchant summary (MS)		465.00	
Totals	$2747.60	$465.00	$1528.50
To check the accuracy of each calculation:	refer to the table on page 483	refer to the merchant summary on page 482	refer to bank deposit form, point 10, page 485
To check the accuracy of the deposit, refer to the bank deposit form on page 485	points 1 and 2	point 3	points 4 and 6–10

Skill development 17.1(vi)

Reconcile cash received with amount listed on bank deposit form (and other money calculations)

The following receipts, merchant summary and cash register summary are available for 17 May 19XX.

- Look at each receipt and note whether the amount was received in cash or cheques.
- Summarise the merchant summary, cash register summary and receipts on the table provided in your workbook, refer back to the bank deposit form in Skill development 17.1(v), and tick each item to verify that the correct amounts are recorded, ready for banking.
- Cheque details are given below each receipt. Confirm that these details are recorded accurately on the bank deposit form also.

The cash register summary for 17 May 19XX indicated that $1286.25 was received from cash sales, and this was made up as follows: $1250.00 in notes, $36.25 in coins. As well, four sales vouchers totalling $242.00 were recorded on the merchant summary.

Receipts issued on 17 May are written as follows:

Business Stationery Supplies Pty Ltd
126 Wynnum Road
WYNNUM QLD

Official Receipt
No. 112

Date	Received from	Particulars	Discount	By (Cash)/Cheque
17 May 19XX	A. Marsden	Sales $86 36 GST $8 64		95 00

per: (x) A. Jackson .... (Signature)

Sold goods for cash to A. Marsden, $95.00.

Business Stationery Supplies Pty Ltd
126 Wynnum Road
WYNNUM QLD

Official Receipt
No. 113

Date	Received from	Particulars	Discount	By Cash/Cheque
17 May 19XX	Arc Products	Sales $45 45 GST $4 55		50 00

per: (x) A. Jackson (Signature)

Sold goods to Arc Products for $50.00 and received a CTB, Cleveland, cheque.

Business Stationery Supplies Pty Ltd
126 Wynnum Road
WYNNUM QLD

Official Receipt
No. 114

Date	Received from	Particulars	Discount	By Cash/Cheque
17 May 19XX	S. Brown	Sales $181 82 GST $18 18		200 00

per: (x) A. Jackson (Signature)

Sold goods to S. Brown and received $200.00 cash.

Business Stationery Supplies Pty Ltd
126 Wynnum Road
WYNNUM QLD

Official Receipt
No. 115

Date	Received from	Particulars	Discount	By Cash/Cheque
17 May 19XX	Baseline Promotions	Payment of a/c	2 10	100 90

per: (x) A. Jackson (Signature)

Baseline Promotions settled its April account by paying $100.90, using a Bank of Queensland, Belmont, cheque and receiving a discount of $2.10.

Business Stationery Supplies Pty Ltd
126 Wynnum Road
WYNNUM QLD

Official Receipt
No. 116

Date	Received from	Particulars	Discount	By Cash/Cheque
17 May 19XX	*Arc Products*	*Sales $500 32* *GST $50 03*		*550 35*

per: (x) *A. Jackson* (Signature)

Sold goods to Arc Products and received a $550.35 NAB, Capalaba, cheque.

Business Stationery Supplies Pty Ltd
126 Wynnum Road
WYNNUM QLD

Official Receipt
No. 117

Date	Received from	Particulars	Discount	By Cash/Cheque
17 May 19XX	*B. McCain*	*Sales $3777 95* *GST $377 80*		*4155 75*

per: (x) *A. Jackson* (Signature)

Sold goods to B. McCain for $4155.75 and received an NAB, Capalaba, cheque.

Reconciliation of cash receipts book with bank deposit form (optional)

The cash receipts book provides a useful extra check on the accuracy of money calculations. Because receipts are checked prior to entering in this book, errors should be limited to the transposing of incorrect entries into this record. As long as suitable checks are in place to ensure accuracy, the amount recorded in the cash receipts book on each day should be the same as the amount banked on that same date. (This is not specifically noted as a learning outcome for the chapter, although it is a very useful extra check, so ask your tutor/supervisor if he/she requires you to complete this section.)

Receipts are entered in a cash receipts book, which looks similar to the one presented opposite. Look at each receipt, and notice how it is entered. The amount shown as point 8 represents the amount to be banked on 16 May 19XX, and this agrees with point 5 on the bank deposit form (point 5 on page 485). Look at each receipt, and notice they are ticked to verify that the amount recorded in the cash receipts book is the same as that recorded on the receipts listed on pages 487–489 in this chapter.

Cash Receipts Book							
Date	Rec. no. (2)	(3) Particulars	(4) Discount expense	Accounts receivable control (5)	(6) Cash sales (excl. GST)	(7) GST recd.	(9) Cash at bank
(1) *May 16*	*106* ✓	*C. Jones*			*200.00* ✓	*20.00*	
	107 ✓	*A. Hass*			*205.09* ✓	*20.51*	
	108 ✓	*T. Toms*	*5.50* ✓	*544.50* ✓			
	109 ✓	*S. Smith*			*200.00* ✓	*20.00*	
	110 ✓	*K. Karson*	*6.00* ✓	*594.00* ✓			
	111 ✓	*O. Mann*			*154.55* ✓	*15.45*	
	CRS	*Sales*			*2092.73* ✓	*209.27*	
	MS	*Sales*			*422.73* ✓	*42.27*	*4741.10*

(8)

When verifying the cash receipts book:

1. Enter the date. Notice that the date in the cash receipts book is the same as that on the bank deposit form (see point 11 on page 485).
2. Check and tick the receipt number.
3. Ensure that the 'Particulars' are accurately recorded.
4. Check and tick the discount, if applicable.
5. Check and tick the 'Accounts receivable control'. Entries appearing in this column will be noted as being received in payment of an account; for example, 'April a/c' is listed in the 'Particulars' section of the receipt.
6. Check the 'Cash sales' amount. This will record 'sales' that have had a receipt, and will also record cash receipts from the cash register, verified by a cash register summary. GST is not included.
7. GST shown. Value recorded from invoices.
8. To check the accuracy of the amount recorded in the 'Cash at bank' column, add together all amounts from columns 5, 6 and 7, using entries relating to the date in question. *Do not* include column 4. This is referred to as the net method of recording (a more detailed explanation of which is outside the scope of this chapter).
9. The amount recorded for each date is listed in the 'Cash at bank' column, and this should agree with the amount being banked for that particular date (see the bank deposit form, page 485, point 5, in this chapter).

Skill development 17.1(vii)

Use the cash book below to verify the banking details recorded on the bank deposit form in your workbook for 17 May 19XX. Receipt numbers for verification are on pages 490–492.

Cash Receipts Book

Date	Rec. no.	Particulars	Discount expense	Accounts receivable control	Cash sales (excl. GST)	GST recd.	Cash at bank
May 17	112	A. Marsden			86.36	8.64	
	113	Arc Products			45.45	4.55	
	114	S. Brown			181.81	18.19	
	115	Baseline Promotions	2.10	100.90			
	116	Arc Products			500.30	50.03	
	117	B. McCain			3777.97	377.78	
	CRS	Sales			1169.32	116.93	
	MS	Sales			220.00	22.00	6680.25

Lodging deposits with banking institutions

Once all of the above checks have been performed, an employee should be confident that the amount being banked is the correct amount. This is very important, because later within the cycle the bank's own record will be used to check the accuracy of the organisation's cash receipts book. If these checks have been performed before banking, the checking process further down should be merely a formality.

When lodging cash at the bank, security procedures are very important. Go back now and revise the security procedures listed under *Occupational health and safety procedures* on pages 476–477 of this chapter.

Activity 17.3

Security procedures

1. On a separate sheet, list each security procedure that should be followed when lodging cash with a banking institution. In your own words, explain why this procedure is relevant in ensuring the safety of employees and the protection of cash.
2. Can you think of any other procedure that could be implemented? Explain fully.

Night safe

Banks provide a night safe for the use of customers. As previously mentioned, all cash should be banked intact daily; however, sometimes a large amount of cash is received after banking has occurred. To avoid the possibility of theft in this event the night safe should be used, as it is much safer than storing cash on the premises.

The bank deposit form and all other checks are prepared in the usual way, and the bank staff will process the funds the next day.

Some organisations prefer to use the night safe each day because this removes all cash from their premises each day, limiting the possibility of theft of cash.

Activity 17.4

Night safe

On a separate sheet, briefly list the advantages of using a night safe.

Skill development 17.2(i)

Business Stationery Supplies Pty Ltd requires you to complete the bank deposit forms in your workbook. You will need to sign the forms, because for this activity it is assumed that you will be depositing the daily takings.

Business Stationery Supplies Pty Ltd banks with the State Bank Limited, Capalaba Branch, and the account number is 49 1078 1088.

(i) Banking for 2 May 19XX
Cheque details

Drawer	Bank/Branch	Amount
Ace Stationery Services	Westpac, Toombul	$2000
Wainright Cleaners	NAB, Capalaba	$20
J. Brackson	CTB, Cleveland	$560
Bob Black (Aust) Limited	Westpac, Chelmer	$570

Bankcard sales amount to $4000.

Cash and notes received: 1 × $100, 20 × $50, 2 × $20, 19 × $10, 30 × $5, 24 × $2, 30 × $1, 60 × $0.50, 20 × $0.20, 10 × $0.10, 5 × $0.05.

Complete the table in your workbook to calculate your notes and coins, and use a calculator to add the totals. Check your answer before continuing.

(ii) Banking for 3 May 19XX
Cheque details

Drawer	Bank/Branch	Amount
Forde Pty Ltd	NAB, Toombul	$750
Swift Carriers	NAB, Capalaba	$50
H. Hogg Pty Ltd	CTB, Cleveland	$1000
J. Sampson & Co	Westpac, Capalaba	$700

Bankcard sales amount to $1000.

Cash and notes received: 3 × $100, 5 × $50, 20 × $20, 19 × $10, 30 × $5, 24 × $2, 30 × $1, 60 × $0.50, 20 × $0.20, 15 × $0.10, 7 × $0.05.

Complete the table in your workbook to calculate your notes and coins, and use a calculator to add the totals. Check your answer before continuing.

(iii) Banking for 4 May 19XX
Cheque details

Drawer	Bank/Branch	Amount
Blake Stationery	Westpac, Wynnum	$2000
J. Hampson	NAB, Capalaba	$250
Ace Stationery Services	Westpac, Toombul	$2000
F. Farmer & Co	Westpac, Wynnum	$700.85

Bankcard sales amount to $4500.

Cash and notes received: 2 × $100, 15 × $50, 2 × $20, 25 × $10, 60 × $5, 10 × $2, 20 × $1, 30 × $0.50, 10 × $0.20, 10 × $0.10, 2 × $0.05.

Complete the table in your workbook to calculate your notes and coins, and use a calculator to add the totals. Check your answer before continuing.

(iv) Banking for 5 May 19XX
Cheque details

Drawer	Bank/Branch	Amount
Forde Pty Ltd	NAB, Toombul	$850
J. Sampson & Co	Westpac, Capalaba	$700.50
Swift Carriers	NAB, Capalaba	$50

Bankcard sales amount to $700.

Cash and notes received: 20 × $100, 7 × $50, 10 × $20, 19 × $10, 30 × $5, 12 × $2, 10 × $1, 10 × $0.50, 20 × $0.20, 25 × $0.10, 2 × $0.05.

Complete the table in your workbook to calculate your notes and coins, and use a calculator to add the totals. Check your answer before continuing.

(v) Banking for 6 May 19XX
Cheque details

Drawer	Bank/Branch	Amount
Blake Stationery	Westpac, Wynnum	$2000
State Insurance	CTB, Fortitude Valley	$500

Bankcard sales amount to $3500.

Cash and notes received: 3 × $100, 20 × $50, 1 × $20, 12 × $10, 30 × $5, 15 × $2, 20 × $1, 30 × $0.50, 10 × $0.20, 20 × $0.10, 9 × $0.05.

Complete the table in your workbook to calculate your notes and coins, and use a calculator to add the totals. Check your answer before continuing.

Skill development 17.2(ii)

The following receipts, merchant summary and cash register summary are available for 2 May 19XX.

1. Look at each receipt and note whether the amount was received in cash or cheques.
2. Summarise the merchant summary, cash register summary and receipts on the table provided in your workbook, refer back to the bank deposit form in Skill development 17.2(i), and tick each item to verify that the correct amounts are recorded, ready for banking.
3. Cheque details are given below each receipt. Confirm that these details are recorded accurately on the bank deposit form also.
4. Summarise the merchant summary, cash register summary and receipts on the table provided in the workbook, and refer back to the bank deposit form (Skill development 17.2(i)) in your workbook, to verify that the correct amounts are recorded, ready for banking.

The cash register summary for 2 May 19XX indicated that $1093.25 was received from cash sales, and this was made up as follows: $980 in notes, $113.25 in coins. As well, five sales vouchers totalling $4000.00 were recorded on the merchant summary.

Receipts issued on 2 May are as follows:

Business Stationery Supplies Pty Ltd
126 Wynnum Road
WYNNUM QLD

Official Receipt
No. 45

Date	Received from	Particulars	Discount	By Cash/Cheque
2 May 19XX	K Karson	Sales $272 73 GST $27 27		300 00

per: (x) *J. Farthing*.................................. (Signature)

Sold goods to K. Karson and received $300.00 cash.

Business Stationery Supplies Pty Ltd
126 Wynnum Road
WYNNUM QLD

Official Receipt
No. 46

Date	Received from	Particulars	Discount	By Cash/Cheque
2 May 19XX	Wainright Cleaners	Sales $18.18 GST $1 82		20 00

per: (x) *J. Farthing*.................................. (Signature)

Sold goods to Wainright Cleaners for $20.00 and received an NAB, Capalaba, cheque.

Business Stationery Supplies Pty Ltd
126 Wynnum Road
WYNNUM QLD

Official Receipt
No. 47

Date	Received from	Particulars	Discount	By Cash/Cheque
2 May 19XX	S. Brown	Sales $181 82 GST $18 18		200 00

per: (x) *J. Farthing*.................................. (Signature)

Sold goods to S. Brown and received $200.00 cash.

Business Stationery Supplies Pty Ltd
126 Wynnum Road
WYNNUM QLD

Official Receipt
No. 48

Date	Received from	Particulars	Discount	By Cash/(Cheque)
2 May 19XX	Ace Stationery Supplies	Payment of April a/c	20 00	2000 00

per: (x) J. Farthing .... (Signature)

Ace Stationery Supplies settled their April account by paying $2000.00, using a Westpac, Toombul, cheque and receiving a discount of $20.00.

Business Stationery Supplies Pty Ltd
126 Wynnum Road
WYNNUM QLD

Official Receipt
No. 49

Date	Received from	Particulars	Discount	By Cash/(Cheque)
2 May 19XX	J. Brackson	Sales $509 09 GST $50 91		560 00

per: (x) J. Farthing .... (Signature)

Sold goods to J. Brackson and received a $560.00 CTB, Cleveland, cheque.

Business Stationery Supplies Pty Ltd
126 Wynnum Road
WYNNUM QLD

Official Receipt
No. 50

Date	Received from	Particulars	Discount	By Cash/(Cheque)
2 May 19XX	Bob Black (Aust) Ltd	Sales $518 18 GST $51 82		570 00

per: (x) J. Farthing .... (Signature)

Sold goods to Bob Black (Aust) Limited for $570 and received a Westpac, Chelmer, cheque.

Skill development 17.2(iii)

Use the cash book below to verify the banking details recorded on the bank deposit form (in your workbook) for 2 May 19XX (Skill development 17.2(i), part i).

Cash Receipts Book

Date	Rec. no.	Particulars	Discount expense	Accounts receivable control	Cash sales (excl. GST)	GST recd.	Cash at bank
May 2	45	K Karson			272.73	27.27	
	46	Wainwright Cleaners			18.18	1.82	
	47	S. Brown			181.82	18.18	
	48	Ace Stationery	20.00	2000.00			
	49	J. Brackson			509.09	50.91	
	50	Bob Black (Aust) Pty ltd			518.18	51.82	
	CRS	Sales			993.86	99.39	
	MS	Sales			3636.36	363.64	8743.25

Activity 17.5

Lodging deposits with banking institutions

1. Explain in detail how you would need to bundle notes and coins in preparation for banking, in order to satisfy the banking institution's guidelines.
2. What is the document that summarises credit card vouchers, in order to comply with the banking institution's guidelines?
3. List security procedures that should be followed when banking cash.

Summary

In this chapter you have learnt how to prepare and process documents for banking in accordance with enterprise policies and procedures, and in accordance with banking institution guidelines. You have also learnt how to check each document involved in the financial processes for validity, to ensure that mistakes do not occur. You should now be able to:

- check credit card vouchers and cheques for validity before processing;
- count and list cash, cheques and credit card vouchers on relevant pay-in banking forms for depositing funds and total according to banking institution guidelines;
- reconcile pay-in documentation with all money calculations;
- bundle cash, cheques and credit card vouchers according to banking institution guidelines; and
- lodge deposits with a banking institution.

Review questions

(These activities may be used for self-assessment, or for formal assessment.)

1. The following receipts are correct. Compare them with the cheques that follow, and advise your tutor/supervisor of any irregularity. A short note on a separate piece of paper will be sufficient.

Business Stationery Supplies Pty Ltd
126 Wynnum Road
WYNNUM QLD

Official Receipt
No. 136

Date	Received from	Particulars	Discount	By Cash/Cheque

per: (x) .... (Signature)

Business Stationery Supplies Pty Ltd
126 Wynnum Road
WYNNUM QLD

Official Receipt
No. 137

Date	Received from	Particulars	Discount	By Cash/Cheque
20 May 19XX	*R. Rodbury*	*Sales $54 45* *GST $5 45*		*59 90*

per: (x) *P. Marr* .... (Signature)

The NBA Bank
Gosford New South Wales

not negotiable
account payee only

20 May 19*XX*

Pay *Business Stationery Supplies Pty Ltd*

The sum of *Two Thousand Five Hundred*
Dollars and fifty cents

$ *2550.50*

G. Black

G. Black

4486 888638 191737

The Southpac Bank
Richmond New South Wales

15 June 19*XX*

Pay *Business Stationery Supplies Pty Ltd*

The sum of *Fifty-Nine Dollars*
and ninety cents

$ *59.90*

not negotiable
account payee only

R. Rodbury

4486 888638 191737

2. The following credit card voucher has been received. It has been checked against the list of lost and stolen cards and it is not included. Itemised prices have also been checked and are correct.

   The store limit for unauthorised transactions is $50. If any errors are found, write a note to your tutor/supervisor on a separate piece of paper.

SPECIMEN ONLY

4478 39078 1977 8888

1073 11/XX
R. ROBERTS

2088799
BUSINESS STATIONERY
SUPPLIES PTY LTD
WYNNUM QLD

557902

Authorisation No.	Day	Month	Year
	15	*06*	*XX*

Qty	Description	Dollars	Cents
5	*Boxes pens red*	*50*	*00*
4	*reams white paper*	*48*	*00*
Sales voucher	Total $A	*88*	*00*

☑

B/C
M/C ✓
VISA

Cardholder's Signature

(x) ....

3. (i) Prepare a bank deposit form for 20 July 19XX for Business Stationery Supplies Pty Ltd using the cheques, merchant summary and information on notes and coins. The bank account number is 07893444, and the branch is Wellington Point.

SPECIMEN ONLY

5088 1999 1688 9393

1068 10/XX — 9/X2
BUSINESS STATIONERY
SUPPLIES PTY LTD

2076849
BUSINESS STATIONERY
SUPPLIES PTY LTD
WYNNUM QLD

223543

List vouchers below or supply adding machine tape
335 00
14 00
5 00
Total $ 374 00

Important — Cross this box if credit value is greater than sales

Day	Month	Year
16	05	XX

Voucher type	No. of items	Dollars	Cents
Sales Vouchers	3	374	00
Less Credit Vouchers			
Merchant summary	Total SA	374	00

Merchant's Signature

(x) ....

This form and the vouchers herewith are legible and undamaged in any material respect

Cheque Details

Drawer	Bank/Branch	Amount
Watson Pty Ltd	NAB, Wellington Point	$700.50
Underwood Stationery	Westpac, Cleveland	$35.00
Martins Business Supplies	NAB, Murarrie	$95.00

Bankcard sales amounted to $374.00.

The following notes and coins have also been received. List and count them in your workbook prior to banking.

2 × $100, 7 × $50, 8 × $20, 10 × $10, 7 × $5, 18 × $2, 8 × 50c, 27 × 20c, 9 × 10c, 17 × 5c.

3. (ii) To reconcile pay-in documentation with money calculations, prepare a summary of receipts on the table provided in your workbook.

- Check the summary against the bank deposit form, and tick to verify each amount. If any errors are found, refer back to the merchant summary and the summary of notes and coins.
- The cash register summary for 20 July 19XX indicated that $707.15 was received from cash sales, and this was made up as follows: $700.00 in notes, $7.15 in coins. As well, sales vouchers totalling $374 were recorded on the merchant summary.
- Receipts issued for 20 July are listed below.

Business Stationery Supplies Pty Ltd
126 Wynnum Road
WYNNUM QLD

Official Receipt
No. 33

Date	Received from	Particulars	Discount	By Cash/Cheque
20 July 19XX	Watson Pty Ltd	Payment of account	10 50	700 50

per: (x) S. Morris .... (Signature)

Business Stationery Supplies Pty Ltd
126 Wynnum Road
WYNNUM QLD

Official Receipt
No. 34

Date	Received from	Particulars	Discount	By **Cash**/Cheque
20 July 19XX	*R. Ross*	*Sales $50 00* *GST $5 00*		*55 00*

per: (x) *S. Morris* .... (Signature)

Business Stationery Supplies Pty Ltd
126 Wynnum Road
WYNNUM QLD

Official Receipt
No. 35

Date	Received from	Particulars	Discount	By Cash/**Cheque**
20 July 19XX	*Underwood Stationery*	*Sales $31 82* *GST $3 18*		*35 00*

per: (x) *S. Morris* .... (Signature)

Business Stationery Supplies Pty Ltd
126 Wynnum Road
WYNNUM QLD

Official Receipt
No. 36

Date	Received from	Particulars	Discount	By Cash/Cheque

per: (x) .... (Signature)

Business Stationery Supplies Pty Ltd
126 Wynnum Road
WYNNUM QLD

Official Receipt
No. 37

Date	Received from	Particulars	Discount	By Cash/**Cheque**
20 July 19XX	*Martins Business Supplies*	*Sales $86 36* *GST $8 64*		*95 00*

per: (x) *S. Morris* .... (Signature)

Business Stationery Supplies Pty Ltd
126 Wynnum Road
WYNNUM QLD

Official Receipt
No. 38

Date	Received from	Particulars	Discount	By Cash/Cheque
20 July 19XX	R. Brady	Part payment of account		40 00

per: (x) S. Morris .... (Signature)

4. When preparing cash for banking, explain how you may need to bundle notes, coins, and prepare credit card vouchers in order to satisfy the banking institution's guidelines.
5. List four security procedures you should observe in order to avoid theft of cash and risk of personal injury.

Competency self-assessment and/or tutor/supervisor assessment

When you have finished studying the chapter, and have completed all of the Activities and Skill development exercises, complete this self-assessment and then hand it to your tutor/supervisor to complete.

Learning outcomes	**Self-assessment**		**Tutor/supervisor assessment**	
I have successfully completed the following learning outcomes	**Yes** (please tick)	**No**	**Yes** (please tick)	**No**
1. Prepare and process documents for banking				
1.1 check credit card vouchers and cheques for validity before processing	☐	☐	☐	☐
1.2 count and list cash, cheques and credit card vouchers on relevant pay-in banking forms for depositing funds and total according to banking institution's guidelines	☐	☐	☐	☐
1.3 reconcile pay-in documentation with all money calculations	☐	☐	☐	☐
1.4 bundle cash, cheques and credit card vouchers according to banking institution's guidelines	☐	☐	☐	☐
2. Lodge deposits with banking institution	☐	☐	☐	☐
Date:	Tutor/supervisor signature:			

INDEX